The Schools Council Low Attainers in Mathematics Project was based at the Centre for Science and Mathematics Education, Chelsea College, University of London from 1979 to 1982.

Project Team
Margaret Brown (honorary director)

General Report
Chris Stolz (deputy director)
Brenda Denvir (primary case study officer)

Literature Review
Linda Dickson
Olwen Gibson
Isobel Provan (secretary)

Children Learning Mathematics: A Teacher's Guide to Recent Research

Linda Dickson
Margaret Brown
Olwen Gibson

CASSELL

Cassell Educational Ltd
Artillery House
Artillery Row
London SW1P 1RT

British Library Cataloguing in Publication Data

Dickson, Linda
   Children learning mathematics,
   1. Mathematics—Study and teaching—
England
   I. Title   II. Brown, Margaret
   III. Gibson, Olwen
   510'. 7'1042   QA14.G7

ISBN 0 304 31449 8

Printed and bound in Great Britain by The Alden Press Ltd, Oxford.

First published in 1984 by Holt, Rinehart and Winston Ltd

Reprinted 1988, 1990

Last digit is print number: 9 8 7 6 5

Contents

Steering Committee

for the Schools Council Low Attainers in Mathematics Project.

J.W. Hersee (Chairman)	Executive Director, School Mathematics Project and Chairman of Schools Council Mathematics Committee
J. Bancroft	Head of Mathematics Department, Westfield School, Bedford (now retired)
R. Berrill	Lecturer in Education (Mathematics), School of Education, University of Newcastle-upon-Tyne
J. Brand	Warden, Hipper Teachers' Centre, Chesterfield, Derbyshire
M. Cahill	Deputy Headmaster, Frank F. Harrison School, Walsall
N. Cawley	Lecturer in Special Education, Avery Hill College, London
D.L. Davies	Headmaster, Christopher Whitehead Boys' Secondary School, Worcester
R. Edwards	Adviser for Remedial and Special Education, Knowsley
E.M. Hardwick	Adviser for Nursery and First Schools, Buckinghamshire
B.L. Hensman	Teacher responsible for Remedial Mathematics, Highfield Middle School, Keighley (now retired)
D.M. Jones	H.M. Inspector of Schools
G. Knights	Director of Mathematics and Tutorial Faculty, Redhill Comprehensive School, Arnold, Nottinghamshire (until 1980); Deputy Head, Broadland High School, Hoveton, Norfolk (from 1980)
E. Parkes	Headmaster, Gusford Primary School, Ipswich, Suffolk
M. Ward	National Foundation for Educational Research
K. Wimpenny	Deputy Head, The Denes High School, Lowestoft, Suffolk.

Foreword

What research has been undertaken into the learning of mathematics that throws light on the teaching of low attaining pupils? How can the findings of this research be made available to teachers?

These two questions led the Steering Committee of the Schools Council's Low Attainers in Mathematics project to propose that an 'annotated bibliography' of the relevant research should be compiled. However, it soon became clear that more could be done than had seemed possible when that proposal was drafted. The search of literature revealed more material than had been envisaged, but to make it accessible to the classroom teacher a summary of the findings, rather than simply a list of references with notes was required. This book is the result.

Although the work arose in connection with the Low Attainers in Mathematics project, the final text has a much wider appeal and relevance. Teachers who have read the draft have remarked on the interest and importance of what it contains for mathematics teaching to *all* pupils, not just to those with obvious learning difficulties, and I commend these pages to all who teach mathematics in primary and secondary schools.

It is also my pleasant duty to thank all those who have contributed to the work; the Steering Committee and their colleagues for valuable comment and suggestions; Dr. Olwen Gibson for her assistance with organisation and preliminary drafting over the first two years; Dr. Kath Hart who assisted with the preparation of the chapter on fractions; but especially Linda Dickson, who carried out so efficiently the bulk of the collation and writing, and Dr. Margaret Brown who produced and edited the final version of the text.

John Hersee
Chairman, Schools Council
Mathematics Committee

Introduction

The Low Attainers in Mathematics project was funded by the Schools Council and based at the Centre for Science and Mathematics Education, Chelsea College, University of London from 1979 to 1982. It had two aims; the first was to bring to the notice of teachers the current knowledge about learning processes in mathematics which has resulted in the writing of this book.

The second aim was to provide information to teachers and advisers about the ways in which primary, middle and secondary schools deal with pupils who have difficulty in mathematics. A report of this work has been published:

Low Attainers in Mathematics 5–16: Policies and Practices in Schools (Schools Council Working Paper 72) Denvir B, Stolz C, Brown M, Methuen Educational 1982.

THE AIMS AND SCOPE OF THIS BOOK

The intention here is to present a review of research in the learning and teaching of mathematics, especially that which is relevant to the teaching of low attainers in mathematics aged 5–16, for teachers and others in a readable form. The ultimate aim is to provide information and insight which will lead to improvements in the teaching of mathematics to such children, and indeed to all children.

Although the original intention was to restrict the research included only to that which specifically concerned low attainers, it was quickly realised that most research in mathematics education has implications for low attainers at some stage in their education. Conversely, there is little if any research which has implications only for the teaching of low attaining children and has no relevance in the case of more able children. Thus it is hoped that this book will prove useful to everyone responsible in some way for the mathematics education of children, regardless of the age or ability level concerned.

In particular, the book should assist the work of:
    classroom teachers;
    heads, heads of department and mathematics consultants;
    lecturers and advisers concerned with the organisation and provision of initial and
        inservice training;
    students and teachers studying for qualifications in mathematics education;
    researchers.
It is not intended that the book should be capable of being read at a single sitting, but
rather that it should be referred to as a resource over a period of time. In particular it
should be of use when decisions are being made concerning the teaching of a specific
topic (e.g. 'fractions' or 'area'), either by a teacher or teachers in a single school, or
within a local authority or curriculum development group.

The primary focus of the book is on the conclusions which can be drawn on the basis
of research evidence about the way children learn particular mathematical skills and
concepts. There is no attempt to describe more general theories of learning and teaching except where it is necessary to do so briefly in order to interpret empirical evidence.
Nor is there any emphasis on sociological aspects or on the teaching of general
mathematical processes.*

The book is divided into four sections:
    Spatial Thinking;
    Measurement;
    Number;
    Language: Words and Symbols.
The first three are concerned with specific aspects of the mathematics curriculum
while the fourth extends across all areas of mathematics. There is considerable overlap
between the sections; wherever possible this is indicated by cross-referencing.

The general criterion for inclusion of a mathematical topic was whether it would be
included in the mathematics curriculum of most low attaining children in British primary or secondary schools. (Here, as elsewhere in the 'Low Attainers in Mathematics'
project, 'low attainers' were generally thought of as constituting 'those pupils outside
special schools who fall, for whatever reason, into the bottom 20 per cent of mathematical attainment in their age group taken over the country as a whole'. However this
should not be taken to imply that this group is either constant or readily identifiable.)

The weighting given to different topics in the book depends mainly on the quantity,
and quality, of the related research rather than on the importance of the topic in the
mathematics curriculum, or in everyday life.

---

*For a more detailed account of all these areas see:

Bell, A.W., Bishop, A.J., Howson, A.G. et al (1983) A Review of Research in
Mathematical Education Prepared for the Committee of Inquiry into the Teaching of
Mathematics in Schools. NFER/Nelson.

SELECTION OF MATERIAL

Because of the extent of recent research in mathematics education, particularly in the United States, it was necessary to be selective about what research literature was to be included in the review. The project team attempted generally to focus on that research which they felt provided most insight and information to practising teachers. For this reason in some cases one or two studies are described and quoted in detail, including where possible the oral and written responses of particular children, in preference to providing a more comprehensive and balanced, but necessarily less full, review of a number of studies.

Some research has been omitted because the implications were unclear. This was sometimes because the literature available was insufficiently detailed, sometimes because the research design was considered to be unsatisfactory, or for other reasons. Almost certainly some useful studies will have been omitted unintentionally simply due to the difficulties of locating and updating information, given the number of journals, reports and theses involved.

Nevertheless the review contains references to approximately 300 sources, which are listed at the end of each relevant section – Spatial Thinking; Measurement; Number; Language – Words and Symbols.

Finally it should be emphasised that the selection of the literature, the interpretation of results, and the choice of frameworks for presenting the review must ultimately be subjective and thus be dependent on the particular viewpoints of the authors, based on their own experiences in teaching and research.

Nevertheless we hope that the material, and the way it is presented, may be found to be acceptable and of value to teachers and others involved in the improvement of mathematics education at all levels, both in Britain and in other countries.

REFERENCES

Care has been exercised to verify every reference and quotation used in this text, and also to trace ownership of copyright material. The Publishers will gladly receive information enabling them to rectify any reference or credit in subsequent editions.

SECTION 1: *Spatial Thinking*

SECTION 1: *Spatial Thinking*

1.1 INTRODUCTION: SPACE AND LANGUAGE –
THE TWO SIDES OF THE BRAIN

An example from Choat (1974) concerns Johnnie, a six year old in an Educational Priority Area school. He was sticking coloured shapes to produce a picture. The dialogue between him and the interviewer started with a conversation about what he had done so far.

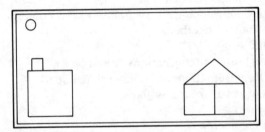

Fig. 1.1   (reproduced from Choat, 1974, *Mathematics Teaching*, **69**, 9.)

Then,

Interviewer:	Can you tell me which shape that is?
Johnnie:	A square.
Interviewer:	What is a square?

He stopped his sticking and looked hard at the squares which formed his house. He cut out a blue square, the same size as those which he had used, fitted it on one of them and rotated it.

Johnnie:	It's got four sides.
Interviewer:	Good. And what about the four sides?
Johnnie:	They're the same.
Interviewer:	The same what?

He gesticulated by spreading his hands apart and trying to maintain the same distance between them as he followed the direction of the rotation. He had grasped the configuration of the relationship but did not possess the language necessary to express his thoughts.

I placed the blue square before him. Running a finger along the edges from corner to corner I said, 'The sides are the same length.'

'The sides are the same length', he echoed...

I touched the picture of the house, and asked, 'How were you able to fit the two squares together?'

''Cos they fit ...'

'But what makes them fit?'

He pondered for some time, and eventually lifted his head. Perceptually he was aware of the relevance but rationalisation was beyond him.

I said, 'All of the sides are straight so this allows us to fit two straight sides together. By doing that the bottom edges carry on in a straight line as well.'

Tracing on the house accompanied the description. (Choat, 1974)

Choat's approach with Johnnie was first to emphasise visual and perceptual aspects with the help of finger tracing. In conjunction with this he encouraged the use of appropriate language. Later on Johnnie constructed a robot out of blocks, and when asked how this construction had been possible, he then made a reference to the straight edges of the blocks.

This article by Choat (1974) is particularly concerned with children from deprived backgrounds who enter school unable to express ideas because of language inadequacies. This restricts their learning of mathematics: a situation which seems to worsen as they progress through the educational system.

Wheatley and Wheatley cite the case of Paul, a low attainer, who received specifically designed instruction which was based on spatial activities. The effect was overwhelming. Not only did he enjoy a high degree of success but,

...he came alive intellectually and even began teaching others. The effect on him as an individual was profound. (Wheatley and Wheatley, 1979)

These two quotations illustrate the interplay between the two different aspects of spatial representation and language in the development and communication of mathematical ideas.

There is increasing physiological evidence that these two aspects of the learning of mathematics may be linked to the activities of different halves of the brain: spatial processing in general is performed by the right 'hemisphere', and language functions in the left.

Wheatley and Wheatley suggest that individual learners are not necessarily equally proficient in both types of activity, and, in particular, many children who appear to be low attainers, like Johnnie and Paul, find a spatial approach both more accessible and more congenial. Choat's quotation shows how such a spatial approach may lead on to the introduction of appropriate language in a meaningful way, while both authors stress the important role of spontaneous oral communication in the learning of mathematics.

The Left and Right Hemispheres of the Brain

Sharma (1979) gives a comprehensive account of much of the literature in this area. He shows that research indicates that for most right-handed people (it is more complicated for left-handers), the left and right hemispheres are mainly concerned with the following aspects of information processing:

The Left Hemisphere

The left hemisphere 'thinks' in words.

It processes information one bit at a time.

This is sequentially organised.

It processes detail working from the parts to the whole.

The left hemisphere is the centre for language communication in terms of reading and
    speaking and is concerned with the comprehension and organisation of language.

Visual material received by the left hemisphere is described in speech and writing.

Left hemisphere activity is concerned with processing information on the abstract level
    of language and words.

The Right Hemisphere

The right hemisphere 'thinks' in images.

It is concerned with spatial and visual aspects.

Information is processed in terms of its overall configuration.

Visual information is processed working from the whole to its parts.

The right hemisphere is the centre for intuition and creativity.

It can understand simple language and carry out some abstract thinking using
    symbols and mental operations associated with *simple* arithmetic but answers can
    only be expressed by pointing to visually presented options.

The right hemisphere memorises facts.

It records visually and communicates through actions and images.

The right hemisphere seems to be the centre for information to be perceived, com-
    prehended and remembered.

The information which the right hemisphere receives can be communicated through the
    left hemisphere by means of written and spoken language.

Sharma identifies two types of mathematical learning personalities.

> (A) ...one who has *left hemispheric* orientation. He is good in language and verbal expressions, is good in
>   solving those problems bit by bit... He is good in quantifying and in quantitative operations. He is spe-
>   cially good in operations which build up sequentially, such as: counting, addition, and multipli-
>   cation. This child when given a word problem, looks for a familiar algorithm to solve the problem.
> (B) ...one who has a *right hemispheric* orientation. He looks at the problem wholistically and explores
>   global approaches to solutions; this child is good in identifying patterns – both spatial and symbolic, is
>   more creative and faster in solving "real life" problems. This child when given a word problem seems to
>   play with the problem in a non-directed metaphoric way before he begins to solve it. (Sharma, 1979)

This classification is similar to one used by Pask (1976), in which learners are
categorised as 'serialists' (corresponding to left hemisphere preference) or 'holists'
(corresponding to right hemisphere preference). Pask designed parallel sets of learning
materials for a mathematical topic, one employing a 'serialist' approach and the other a
'holist' approach. Those students whose preferred learning style matched the style of
materials provided learned very much more effectively, than in the cases where a mis-
match occurred.

This supports the argument referred to previously in the cases of Johnnie and Paul,
that some children may be disadvantaged when the approach used to teach mathematics
in the classroom is out of step with their 'natural' style.

At the opposite extreme of ability Wheatley cites the example of Einstein:

> Einstein reported that his great discoveries came as flashes of images, not in words or symbols. It was often weeks before he could put the ideas into words and symbols... It should be noted that Einstein was unsuccessful in school: for example he was not good at the left-hemisphere tasks demanded in arithmetic.
>
> (Wheatley, 1977)

It should however be borne in mind that not only is much of this 'hemisphere theory' speculative, but also the 'holist/serialist' distinction may be misleading, in that although some learners may strongly favour one or other of these strategies, the majority may well operate in a more versatile way. For instance, when Krutetskii (1976) was studying gifted children, he found that although there was a small 'geometric' group who appeared to generally use spatial representations in solving problems, and a small 'analytic' group who rarely used spatial reasoning, the majority varied their approach depending on the context of the problem. Wheatley however feels that for all children spatial ('right hemisphere') development has been under-emphasised:

> Our curricula stress rule-oriented sequential activities so extensively that children expect to apply a rule immediately. Rule application is characteristic of left-hemisphere processing. The mulling over of a problem to determine an approach is more characteristic of right-hemisphere functioning.
>
> Our society emphasises and rewards left-hemisphere activities. This is particularly true of our schools. A premium is placed on being able to put ideas into words, to state them explicitly, and to operate with rules. ...such emphasis is detrimental to right hemisphere development. (Wheatley, 1977)

He suggests this imbalance as one reason for the relatively poor performances in problem-solving noted in national surveys of mathematics achievement (APU, 1980, 1981; NAEP, 1980; Carpenter et al., 1980). In support of this he quotes a study showing that, when left to their own devices, children pass through three stages in solving a problem:

(i) The all important first stage is a rather lengthy mulling over of the problem involving the perception of the overall picture of the problem – spatial organisation and visual representation, which is a right hemisphere activity.

(ii) Next is the application of a chosen method of solution – a left hemisphere activity.

(iii) Finally, a period of reflection and looking at the reasonableness of the solution – a right hemisphere activity.

Wheatley asserts that the less successful problem solver omits stages (i) and (iii).

Thus again it is demonstrated that the two aspects of mathematics represented, on the one hand, by language and symbols, and, on the other hand, by spatial representations, are entirely complementary in nature and should each receive a reasonable share of attention in any mathematics curriculum. Some children may, for various reasons, show a preference for one or the other. In these cases the best approach might be to introduce a topic in the style which is most congenial, but nevertheless to aim to use this, as Choat demonstrates in the case of Johnnie, to build up a capability in the less favoured medium as well.

1.2 HOW IMPORTANT IS SPATIAL THINKING?

> ... the current anxiety about children's acquisition of numerical skills tends to obscure the realisation that most people are more frequently faced with spatial problems than with numerical ones whether in their work as bricklayers or dress designers or draughtsmen or in their "every day" activities in parking cars or playing tennis or putting up shelves.

> ... if, as we believe, mathematics like literature offers a way of understanding and appreciating one's environment then a large part of this appreciation will come about through a spatial awareness and understanding simply because our physical environment is itself spatial.

> ... a spatial facility is an essential component of mathematical functioning.... Intuitive awareness of spatial properties seems to be at the heart of most mathematical thinking.

> These are quotes from the Teachers' Handbook for the Leapfrog television series (Hemmings et al., 1978) as cited by Delaney (1979).

Plunkett (1979) points out that the spatial matters involved in the teaching of mathematics are of two basic types: those to do with the real world and those concerning representations of the real world. Spatial representations involve pictures and diagrams.

Plunkett uses the term 'diagram' for a spatial means of representing non-spatial ideas, for example, the number-line to represent the sequence of numbers, Venn diagrams to represent sets, graphs to represent relationships, etc. In everyday life, he mentions clocks and price lists as being spatial means of representing the non-spatial ideas of time and cost.

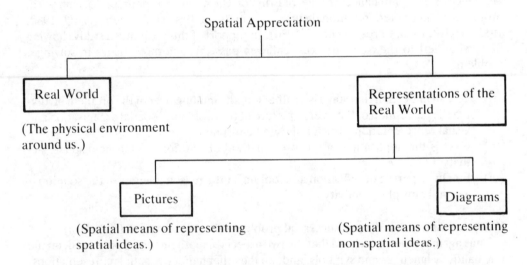

In contrast, he uses the term 'picture' for a spatial means of representing spatial ideas, for example, geometrical drawings, 3-dimensional models, maps, plans, pictures and drawings of physical objects etc. Inevitably these kinds of representation involves varying degrees of distortion of the real world. More will be said of the difficulties in this area in Section 1.7 page 37, with particular reference to 2-dimensional representations of 3-dimensional objects.

Since a spatial medium is so useful for communicating ideas, which are both spatial and non-spatial, much of mathematics teaching has its foundations in this medium and hence spatial difficulties can lead to problems in other areas. For as Lesh points out,

> Most of the models and diagrams teachers use to introduce arithmetic and number concepts presuppose an understanding of certain spatial/geometric concepts. Consequently misunderstandings about number concepts are often closely linked to misunderstandings about the models that are used to illustrate them.
> (Lesh, 1978)

In particular, difficulties arise when certain spatial representations are used to illustrate ideas which are not spatial or are of a spatial nature different from that of the respresentation. The over-riding spatial aspects of the representation may distract from the ideas being represented. This is particularly evident with say the problem of number conservation.

Fig. 1.2

Since the bottom row is more spread out a young child generally thinks there are more counters there than in the top row. The spatial aspect distracts from the idea of number. (There is a full discussion of this point in part 2.1 of the 'Number' section in this book.)

Kerslake gives instances of where the visual appearance of certain graphs distracts from the nature of the information they are portraying. Second, third and fourth year secondary pupils were shown the following:

> Which of the graphs below represent journeys?
> Describe what happens in each case.

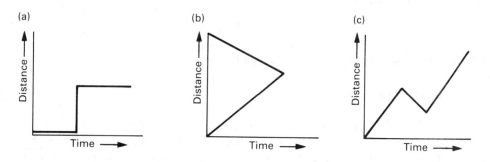

Fig. 1.3

Any answer indicating some understanding was accepted. A good answer might be, e.g.:

> Only (c) represents a journey. Graph (a) represents an object at rest, then changing position in zero time, then stopping again. Graph (b) has either two distances or none for any given time, so it could only represent two objects colliding and then disappearing. Graph (c) shows an object travelling steadily, then going backwards a little way, then forwards again.

Many saw (a) as:

'going along, up and along',
'climbing a vertical wall',
'going east, then north, then east' and so on.

(b) was often:

'going NE then NW'.

(c) was often interpreted as:

'going uphill, then downhill, then up again',
'climbing a mountain' etc.

The percentage of pupils giving this sort of answer was:

	Year 2		Year 3		Year 4	
	Boys	Girls	Boys	Girls	Boys	Girls
(a)	34	33	41	41	44	39
(b)	31	33	39	47	45	38
(c)	26	27	34	31	36	27

The percentage of correct answers was as follows, the rest giving no answers at all.

	Boys	Girls	Boys	Girls	Boys	Girls
(a)	12	5	19	16	31	20
(b)	14	3	23	8	27	25
(c)	23	7	34	26	35	39

(Kerslake, 1977)

Thus in this case a certain type of spatial interpretation has hindered the communication of a mathematical idea.

1.3   HOW DO SPATIAL CONCEPTS DEVELOP?

1.3.1   Introduction

The young child's first interactions with his environment, before the development of language, are based almost totally on spatial experiences, particularly through his senses of sight and touch. Later language develops and takes on meaning within the context of the physical environment.

Many psychologists, among them Piaget, Bruner (1967) and Dienes (1959), believe that the manipulation of 'concrete' objects forms the basis of human knowledge, and in particular of mathematics. Physical actions become internalised and generalised into concepts and relations, to which may be attached symbols, either words or mathematical symbols. For example, the arithmetical operation of addition is a generalisation of the physical act of putting together two sets of objects. Piaget would claim that children at his 'concrete operations' stage or below cannot deal in a meaningful way with symbols which are not firmly related to physical actions, either real or imagined. By the age of 16 the majority of children do not seem to have progressed beyond the 'concrete operations' stage (evidence for this is contained in Shayer and Wylam, 1978). Hence, they, and especially low attainers, are likely to be dependent on spatial concepts for their understanding in all areas of mathematics.

1.3.2 Piaget's Theory

Piaget, as a result of his numerous experiments (Piaget and Inhelder, 1956; Piaget, Inhelder and Szeminska, 1960), proposed a theory of the child's development of spatial concepts. He distinguishes between *perception*, which he defines as 'the knowledge of objects resulting from direct contact with them', and *representation* (or mental imagery) which 'involves the evocation of objects in their absence' (Piaget and Inhelder, 1956). The child's perceptual ability develops during the period up to about 2 years old (the 'sensori-motor' stage), while the power to reconstruct spatial images starts at around two years old and in most cases is perfected from around seven years upwards for the average child (the 'concrete operations' period). While tests of 'perception' might involve the ability to discriminate between different objects when presented visually, Piaget uses as his tests for 'representation' (mental imagery) the ability to identify shapes by touch, and the ability to reproduce shapes using matchsticks or drawings.

As an example of the time-lag involved between perception and representation, Lesh and Mierkiewicz (1978) quote the demonstration of Bower (1966) that babies of only 50-60 days could learn to discriminate rectangles from trapezia, whereas Piaget showed that only at around 5½ to 6 years could the average child differentiate when allowed only to touch the shapes which were hidden behind a screen, or manage to reproduce a trapezium successfully in a drawing.

Fig. 1.4

Within each of these periods of development Piaget further distinguishes a progressive differentiation of geometrical properties, starting with those properties he calls *topological* i.e. global properties which are independent of size or shape. He lists these as:

(i) nearness[1] ('proximity') e.g. drawing a man with the eyes close together, even if these are below the mouth
(ii) separation e.g. not overlapping head and body
(iii) order e.g. drawing the nose between the eyes and the mouth

---

[1] The sense in which Piaget uses 'nearness' does not in fact correspond to the mathematical definition of a 'topological property'.

(iv) enclosure e.g. drawing the eyes inside the head
 (v) continuity e.g. making the arms continuous with the body and not the head.

Fig. 1.5   Drawing of man by child aged 4 yrs 4 mths (from Eng, 1954).

   The second group of properties that Piaget says are distinguished by children are those he terms *projective*, which involve the ability to predict how an object will appear as viewed from different angles. For example young children may attempt to draw a face in profile and yet still put two eyes on it, or may not appreciate that a pencil will look like a circle to someone seeing it from one end. 'Straightness' is a projective property since straight lines appear straight from any angle of view.

Fig. 1.6   Mixed profile – about age 7 (from Eng, 1954).

   The third group of geometrical properties are *Euclidean* i.e. those relating to size, distance and direction, and hence leading to the measurement of lengths, angles, areas, and so on. Distinctions can be made between shapes, for instance, a trapezium and a rectangle, based on the angles and the length of the sides. (Projectively speaking the shapes are equivalent since a rectangular table-top appears as a trapezium if viewed from certain angles.) Children are now able to reproduce the exact position of a point on a page, or a geometrical figure, by deciding what lines and angles to measure.
   Although the above sequence is expressed in terms of representational abilities, Piaget claims that the same sequence is earlier repeated at the perceptual level.

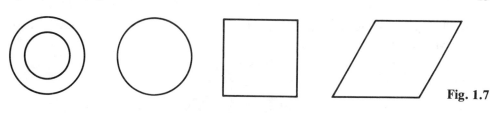

Fig. 1.7

Thus in a shape-fitting puzzle using the shapes in Figure 1.7, the ring would be expected to be differentiated first ('having a hole' relates to 'enclosure' which is a topological property), followed by the circle (the distinguishing feature between the quadrilaterals and the circle being 'straightness', a projective property) and finally the distinction would be made between the square and the rhombus ('angle' being an Euclidean property). Piaget in fact claimed that projective and Euclidean properties were discriminated during roughly the same period. Piaget gives details of numerous experiments to prove his point; some of the tasks were adapted for classroom use in the Nuffield Mathematics Project (1972) 'Checking Up II' Teacher's Guide.

1.3.3   Criticisms of Piaget

Piaget has obviously made a tremendous contribution to the study of the development of spatial thinking, both in providing much sound data based on carefully designed tasks, and in proposing interesting hypotheses. However his theory relating to this area, as outlined above, is probably more open to criticism than are other aspects of Piagetian theory. Three particular grounds for criticism are:

(a) As Lesh and Mierkiewicz (1978) note, the present tendency in psychology is to blur the distinction between perception and representation (mental imagery) which Piaget regarded as so significant. Perception is now seen as a complex organisational process differing only in degree from representation. To take an example, a two year old can learn to correctly name squares and triangles, but in order to do this he must have some mental representation of these figures against which to match his perception. This task would seem to be intermediate between Piaget's perception and representation tasks.

(b) Piaget's experiments have sometimes been shown to give very different results when the methodology is varied in apparently trivial ways. For instance in a study quoted in the next section it is demonstrated by Fuson and Murray (1978) that children can much more easily identify shapes by touch if the shapes are made smaller; thus the difficulty may not be in matching tactile to visual information so much as in using a systematic approach in exploring a larger shape.

Similarly, if one considers the following Piagetian task, which relates to projective ideas:

A child sits at a table on which is a model of three easily distinguishable mountains, one has snow on, another a house, and the third a cross. (See Fig. 1.8.) A doll sits opposite and the child has to select the picture – from a set of ten possible views – which portrays the doll's viewpoint of the mountain scene. (This task is adapted for classroom use in Nuffield's Checking Up II). Children up to the age of 8 or 9 years usually cannot do this and 6 and 7 year olds tend to select the picture portraying their own viewpoint.

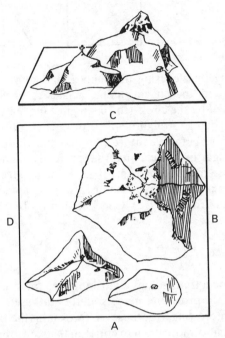

Fig. 1.8   (From Piaget and Inhelder, 1956)

Donaldson (1978) points out that one cannot draw from this the conclusion that Piaget does, namely that the child is egocentric and unable to see things from anybody else's viewpoint other than his own at any moment in time. She substantiates this by relating the following study:

> Two intersecting walls were placed on a table with a policeman doll positioned as shown in Fig. 1.9.

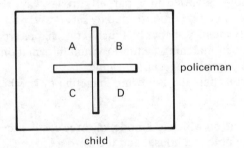

Fig. 1.9

Another doll representing a little boy was placed in section A and the child was asked if the policeman could see him. This was repeated for sections B, C and D. The policeman was moved to other positions as well. The child was gradually introduced to the situation making sure that he understood the task. Then the test proper began. This time two policeman dolls were used – one situated as shown above and the other at the end of the wall separating A and B, opposite the child. The child was told to hide the boy doll from both policemen. This was repeated three times with different positions for the policemen but leaving one section as the only hiding place.

Thirty children between the ages of 3½ and 5 years were given this task. Ninety per cent of their responses were correct. More complex arrangements of walls giving 5 or 6 sections were also used as well as a third policeman. Even so 60 per cent of the 3 year old responses were correct and 90 per cent of the 4 year olds.

Donaldson argues that although the Piagetian Mountains task differs in important respects from the doll situation, for example the child is not required to make left-right reversals, and he is only asked to determine what can be seen but not precisely how it will appear, nonetheless the child is able to co-ordinate differing points of view. In the doll tasks he is familiar with the motives and intentions involved in the situation: it has meaning for him as opposed to the rather cold, abstract nature of the mountains task. (Alternatively the factor causing difficulty may be the complexity and number of relationships with which the child has to deal in the 'Mountains' task.)

(c) As regards Piagetian theory, Piaget tried to explain the development of spatial ideas by using the logical structure of mathematics itself in hypothesising a topological – projective – Euclidean sequence. However as Weinzweig and Fuson (1978) note, Piaget does not use definitions of these properties which are mathematically acceptable. Nor does he test the full implications of his theory, which would predict that a child would distinguish a bead with a hole from a similar bead without a hole more easily than he would distinguish a bead with a hole from a tea cup, since the latter pair are topologically speaking more similar than the former pair! Darke (1982) also gives a comprehensive critique of the theory that topological ideas develop earliest, supporting this with considerable research evidence.

In light of the current position in research into the development of geometric concepts in children and particularly when considering such extensions of Piagetian research as just outlined, Coxford (1978) maintains it is more probable that some topological concepts develop early, while others, like topological equivalence, develop later after some Euclidean and projective ideas have been grasped, even though the more

Fig. 1.10 From SMP, Book B : 1971

difficult topological ideas may not be mathematically dependent on Euclidean or projective concepts. To demonstrate this, an example of topological equivalence is shown. The beetle is topologically equivalent to the Christmas tree and the donkey, since although the shapes are different the relationship of the various junctions, lines and regions remains fixed.

In an unpublished item on topological equivalence the APU (1980b) secondary survey found that less than 30 per cent of 15 year olds were able to select the figure which was topologically equivalent to a given figure. (The difficulty may have been due to the unfamiliarity of the terms and the concepts, although few children omitted the item.) However 73 per cent were able to successfully answer the following Euclidean item:

Which two of the following shapes could be fitted exactly on top of each other if you cut them out?

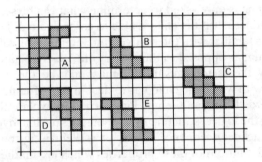

Fig. 1.11   (APU, 1980b)

Thus although Piaget's theory of spatial development is fruitful in terms of explaining some results, it cannot really be said to hold up in view of recent evidence. It seems likely that a developmental theory relating dimensions of psychological complexity to the child's increasing ability to process information will eventually provide a more successful explanation.

1.3.4   The Van Hiele Levels

Coxford outlines Van Hiele's theory of spatial development which is gaining in popularity and repute particularly with specific reference to school geometry curricula. It involves five levels of development:

Level I   Figures are distinguished in terms of their individual shapes as a whole and relationships are not seen between these shapes or their parts. For example a 6 year old can reproduce a square, rhombus, rectangle and parallelogram on a geoboard with rubber bands and memorise their names but he does not see the square as being a special sort of rhombus or the rhombus as a particular parallelogram. They are for him distinct and separate shapes.

Level II   Now there begins a development of an awareness of parts of the figures. These properties become realised through observations during such practical work as measuring, drawing, model making etc. For instance the child sees that a rectangle has four right angles, that the diagonals are of the same length, as are the opposite pairs of sides. Opposite pairs of sides of the general parallelogram are also recognised as being of the same length but the child still cannot see the rectangle as a particular parallelogram.

Level III   Relationships and definitions are beginning to be clarified but only with guidance. The square is seen as a special case of a rectangle which is a particular instance of a parallelogram. Logical connections are becoming established through a mixture of practical experimentation and reasoning.

Levels IV and V   are concerned with the development of deductive reasoning and theory construction culminating in complete abstraction devoid of concrete interpretation. These final stages are not discussed in detail since few low attainers are likely to reach them.

Wirszup (1976) maintains that if a child's introduction to geometry is with measurement and other concepts of Level II and III without a sound grounding in the visual geometry of Level I then he is doomed to failure.

Thus the Level I, in which activities should concentrate on individual figure recognition, production and naming, is of fundamental importance in providing a sound basis upon which to progress to work at Level II and possibly Level III.

In the following sections Van Hiele's levels are used as a framework for the discussion.

1.4   LEVEL I ACTIVITIES: SHAPE RECOGNITION

In a replication of a study reported by Piaget (Piaget and Inhelder, 1956) Fuson and Murray (1978) worked with 2 to 7 year old children studying their ability to recognise shapes by touch and their ability to construct and draw them. For the task where the children were to feel the shapes they used the following.

Fig. 1.12

Each was of a size which would fit comfortably into a child's hand, in contrast to the shapes used by Piaget, which were about 10cm across. The child felt and manipulated each shape behind a screen. From a visual display of all four shapes he pointed to the one which he thought he was holding. Most of the children were able to identify all the

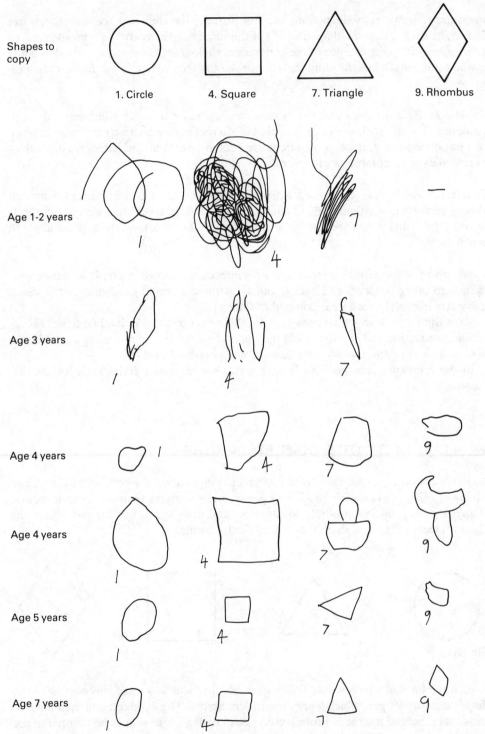

Fig. 1.13   Children's attempts to copy shapes (from Noelting, 1979)

shapes by the age of 3½ years. The circle was the easiest to identify followed by the square, then the triangle and lastly the 'diamond'. The authors suggest that the oblique sides of the triangle made it more difficult than the square. What they describe as the simultaneous co-ordination of right/left and top/bottom halves of the 'diamond' made this shape the most difficult to identify.

Each of these four shapes was presented to the child to draw using pencil and paper. In terms of accuracy, the order was again circle, square, triangle and most difficult of all, the 'diamond'. Even on a generous criterion of success, it was only the five year olds who succeeded with the triangle and 'diamond'. On stricter criteria, even at six and a half years, less than 20 per cent correctly reproduced the three straight-edged shapes.

Each child was also asked to construct the triangle and the square from sticks. He was given six sticks from which to choose for each construction. Although the constructions of figures were not so accurate as recognition by touch, they were more successful than the attempts by drawing. Thus most of the 3½ year olds managed a square, although it was not until 6½ that two-thirds succeeded with the 'diamond'. Fuson and Murray explain the greater success of construction over drawing as being because the latter does not allow for trial and error placements of the separate parts. The child has to be able to draw a straight line and co-ordinate subsequent lines with those he has already drawn. They are not as readily adjustable as the sticks. (This is a major advantage of geostrips or of a geoboard over the drawing of shapes.)

Noelting (1979) also performed a very thorough replication and extension of Piaget's work on the development of drawings of 2-dimensional geometric shapes. Shown in Fig. 1.13 are some of the children's attempts to copy the four shapes used by Fuson and Murray.

The naming of shapes follows a similar pattern of success as shown by the results of Ward's (1979) survey of 10 year olds. They were asked to give the names of each of the following shapes. The percentages reflect the answers judged correct.

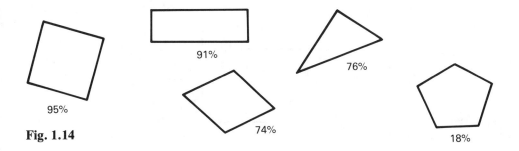

91%

76%

95%

Fig. 1.14

74%

18%

However the main drawback with a straightforward naming of shapes is that success is likely to be dependent mainly upon familiarity with the particular label rather than upon the ability to think spatially.

The first APU secondary report (APU, 1980b) found that 85 per cent of 15 year olds successfully named a square but only about 30 per cent could name the kite, rhombus and trapezium. Success at naming solids was shown by the first APU primary report

(APU, 1980a) on 11 year olds to range from 90 per cent for a cone to 65 per cent for a sphere.

Weinzweig (1978) points out that a child's first experiences in space are with solid 3-dimensional objects and that 2-dimensional figures are initially encountered as the surfaces of solid objects such as cubes, cones, cylinders, spheres, rectangular boxes, prisms and pyramids etc. Different sizes of each of these shapes should be at hand so that the concept of a particular shape is emphasised. For example, a short squat cylinder is often not recognised as being a cylinder at all (see Fig. 1.15).

Fig. 1.15

Weinzweig suggests that making prints with the flat surfaces of 3-dimensional solids helps to develop an awareness of different 2-dimensional shapes as well as some basic properties of solids. Solids which roll on a table top such as a sphere or cone on its side compared with those that do not roll e.g. cube, upright cone, help to develop a notion of flatness. However these are really verging on Level II activities.

Sorting activities according to shape can help the child to focus on the similarities and differences of solids and what aspects remain unchanged or invariant. Solids provide a good introduction to the notion of a point – i.e. the corner or vertex, and notions of straight and curved lines as seen on the edges.

Egsgard (1970) recommends using a collection of coloured polygons (flat, straight-sided shapes) for tracing around and sorting according to the number of sides. The child can then learn such names as triangle, quadrilateral, pentagon and hexagon etc. Names such as equilateral triangle, regular pentagon, parallelogram and rhombus come later when the child has begun to learn about the measurement of length and the idea of parallelism. Even at this initial stage however, it would be unfortunate if the child acquired the idea that, for instance, the shape in Fig. 1.16

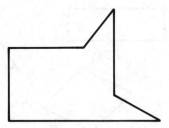

Fig. 1.16

was not just as valid a hexagon as the one in Fig. 1.17:

Fig. 1.17

1.5   LEVEL II ACTIVITIES: DEVELOPING AN AWARENESS OF THE PROPERTIES OF SHAPE

1.5.1   Fitting Shapes Together

Egsgard also emphasises the importance of working with 3-dimensional solids before concentrating on 2-dimensional surfaces. The activity of wall building – trying to fit solids together without any gaps – provides an important basis for developing the notion of a right angle for the 'square corner'. The idea of flatness is further developed and turning bricks round to fit in the wall provides the basis for future work on rotational symmetry (see later in transformations).

The patterns made by the building of walls with solids provide a useful source of reference for beginning work on covering flat surfaces with shapes fitting together without gaps (tessellations). There are many articles and suggestions concerning activities with tessellations in the Association of Teachers of Mathematics (ATM) Journal, *Mathematics Teaching*. Such activities also provide a grounding for later work on area.

Egsgard recommends that there should be a sufficient number of the same sized shapes available to cover a large surface – for example the cover of a book, a small table top. There should also be discussion on the various ways the shapes are used and a consideration of the merits of each. Such work with 2- and 3-dimensional shapes is essential groundwork for developing an understanding of the notions of area and volume and the measurement of these as well as providing a foundation for the study of geometric transformations.

Other useful activities for exploring the properties of shape come about through the dividing up of shapes as with the Chinese Tangram, rearranging them and fitting them together again to form a variety of shapes:

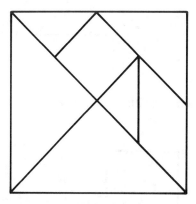

Fig. 1.18

Such activities are not merely a source of amusement but an integral part of learning mathematics. They can help develop notions of the right angle and parallelism. They create problem situations and play an important role in promoting spatial thinking.

Wheatley and Wheatley point out:

In addition to the obvious advantages (e.g. familiarisation with common geometric figures, motivation, developing problem solving ability, establishing a basis for area measurement) tangrams, if properly presented, can promote the development of spatial ability so important in many practical, as well as academic, pursuits. The mental comparison and manipulation of shapes provides practice in gestalt thinking and may help children develop an important dimension of thought. (Wheatley and Wheatley, 1979)

Ward found that 54 per cent of the 10 year olds he studied could satisfactorily divide up a quadrilateral into *just* four triangles but 23 per cent did this:

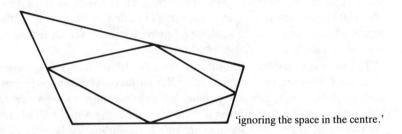

Fig. 1.19   'ignoring the space in the centre.'

Fielker (1973) suggests other activities involving the relationships between shapes. For instance discovering what shapes can be made from two congruent triangles (i.e. triangles of the same size and shape).

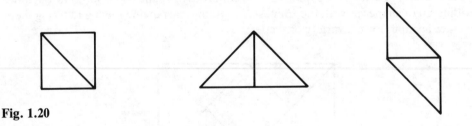

Fig. 1.20

Enlarging a shape by combining it with other congruent shapes. e.g. A larger L-shape from smaller L-shapes.

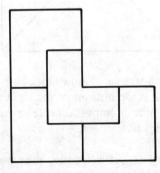

Fig. 1.21

Overlapping shapes e.g. squares

Two squares overlapping.
How many sides now?

Fig. 1.22

Considering possible relationships between two shapes, with appropriate discussion. e.g.

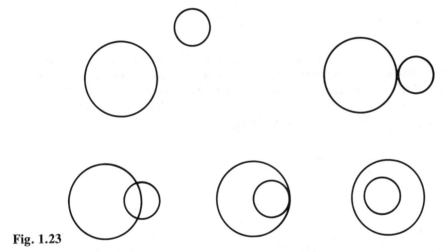

Fig. 1.23

1.5.2   Lines and Angles

Fielker (1973) also offers suggestions which he maintains help children to develop a notion of angle. His approach is based on the study of the relationships between two straight lines. He recommends that pupils draw a straight line on each of two pieces of tracing paper and place one on top of the other e.g.

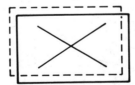

Fig. 1.24

By turning the top sheet they can experience the change in relationship between the lines, i.e. the angle. The special cases of perpendicularity and parallelism can be found and seen as being concerned with the angle between the lines.

Similarly the relations between two right angles can be explored e.g. forming a square

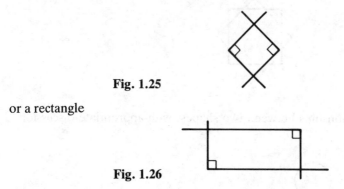

Fig. 1.25

or a rectangle

Fig. 1.26

He suggests other relations with which to experiment. For example combining sets of parallel lines on two sheets of tracing paper. These would help achieve an understanding of the Level III type mentioned earlier, for instance that a square is a special case of a rectangle which is a special case of a parallelogram. Although Fielker does not give the following diagrams – these are what may emerge from his idea:

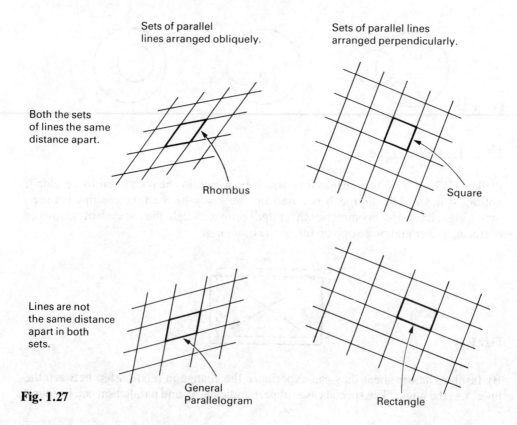

Fig. 1.27

The first APU primary report (APU, 1980a) shows that about half of the 11 year olds were able to distinguish between parallel and non-parallel sets of lines and 30 per cent were able to correctly select shapes with more than one pair of parallel sides from a selection of different quadrilaterals.

The first APU secondary report (APU, 1980b) shows that only 25 per cent of 15 year olds could satisfactorily define 'perpendicular'; 15 per cent gave definitions for 'parallel' instead. Forty per cent of 15 year olds gave the correct definition of a 'square' but almost a third of those not doing so mentioned only equal sides, omitting the necessity for parallelism of opposite sides or any reference to the angle sizes.

The national survey in the USA (NAEP, 1980) also reflected a relatively poor performance by secondary school children on items involving some understanding of the properties of figures. One example involved a diagram of a rectangle with the lengths of all the sides labelled. Thirteen and seventeen year olds had to select conditions which would guarantee the figure was a rectangle. Fourteen per cent of the 13 year olds and 20 per cent of the 17 year olds selected the correct choice, 'The angles are right angles'. Forty per cent and 34 per cent respectively selected the choice, 'The opposite sides are parallel.' A similar item was given as part of a practical test reported in the second APU primary survey (APU, 1981a). Here 20 per cent of the British 11 year olds referred to the right angle property, although a further 10 per cent mentioned the need for 'straight sides'. Each of the properties of equal opposite sides, and parallel opposite sides, were suggested by about 50 per cent of the 11 year olds interviewed. (More details on this item are given on page 33 in the next section).

Fielker (1981), in a series of articles, points out that there are ways of teaching properties of shapes other than having pupils copy down lists of shapes, each followed by the corresponding properties. For example, he suggests children should try to find quadrilaterals (or pentagons, hexagons etc.) with different types of symmetry. Or that they might look at how to fill in the cells in a cross-classification of quadrilaterals like that in Fig. 1.28:

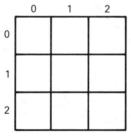

Pairs of parallel sides.

Pairs of equal sides.

Fig. 1.28

For instance a trapezium (non-isosceles) would have one pair of parallel sides and no pairs of equal sides, whereas a 'kite' would have two pairs of equal sides but no pairs of parallel sides. Alternatively, either of the labels on the table might be replaced by 'number of right angles'. Another possibility would be to discover how many quadrilaterals could be made with a set of four geostrips, and what happened if some of these four were equal. All these suggested activities lead on to what we have previously termed Level III work on classification.

A further activity suggested by Fielker is to generate different types of quadrilaterals from their diagonals. He supplied each pupil in a class of 11 year olds (who were not low attainers) with 2 (unequal) geostrips each, and asked them to make as many 'different' quadrilaterals as they could which had these strips as diagonals. Two responses are shown in Fig. 1.29a.

Karen

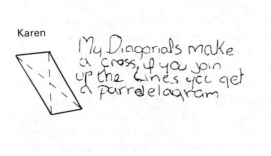

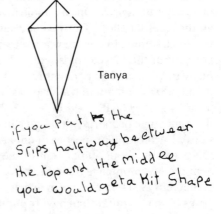

Tanya

My Diagonals make a cross, if you join up the Lines you get a parralelagram

if you put the Srips halfway beetween the top and the middle you would get a Kit shape

Fig. 1.29a

One girl even tried experimenting with a concave shape which had an external diagonal (Fig. 1.29b).

Susie

This is yet another quadrulatural. This a shape that is hard tell If you think a diagonal line should be outside the shape then it is right. If you think they should be out side the it is wrong

Fig. 1.29b

They then tried again with an equal pair of geostrips as diagonals (see Fig. 1.29c).

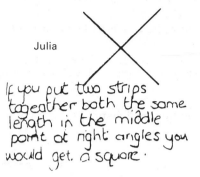

Julia

If you put two strips together both the same length in the middle point at right angles you would get a square.

Fig. 1.29c

(from Fielker, 1981, *Mathematics Teaching*, **97**, 36-37.)

1.6   HOW DO CHILDREN FORM MISCONCEPTIONS ABOUT SPACE?

Many of the misconceptions children develop about space seem to be primarily due to inadequate teaching where children have focused on the wrong criteria and hence developed limited or false concepts. The following examples illustrate this.

Fielker reports the case of a class of 11 year olds who were presented with:

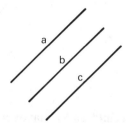

Fig. 1.30

The class told Fielker that:

'a is parallel to b and b is parallel to c.'
'Then a is parallel to c' he said. 'No', they replied, 'Because b is in the way.'

(Fielker, 1973)

Greenes points out that often when a geometric form changes position the child believes its character (shape, size etc) has changed as well. She quotes the case of Robbie aged 9 years. He was presented with three triangles.

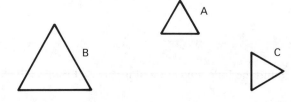

Fig. 1.31

Robbie selected B as being most like A. He said A and B were triangles. When the interviewer pointed to C and asked, 'Is this a triangle?' He replied, 'No'. 'Because it fell over.'

Greenes claims that Robbie had a learned misconception.

Robbie had only seen triangles in standard position, i.e. with their baselines horizontal. He had for recognition purposes, used the horizontal baseline as one of the identifying characteristics.

(Greenes, 1979)

Kerslake (1979) agrees that the usual manner for presenting geometric figures is the following.

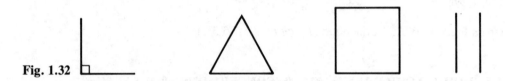

Fig. 1.32

It is difficult for children to generalise these concepts when they are rarely faced with any non-standard illustrations such as

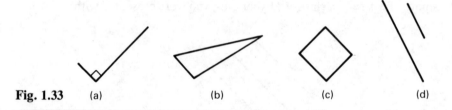

Fig. 1.33   (a)   (b)   (c)   (d)

In studies involving 5 to 11 year old children Kerslake reports the following findings.

Age	% recognising (c) in Fig. 1.33 as a square
5 years	54
6	56
7	80

Age	% recognising (b) in Fig. 1.33 as a triangle
5 years	38
6	47
7	24
8	65
9	50
10	67

With 10 year olds she found the following percentages recognising each figure as a right angle:

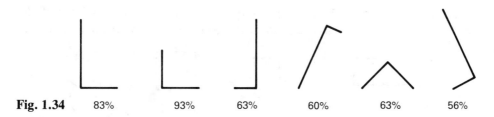

Fig. 1.34   83%   93%   63%   60%   63%   56%

The following percentages of 10 year olds recognised each as a set of parallel lines:

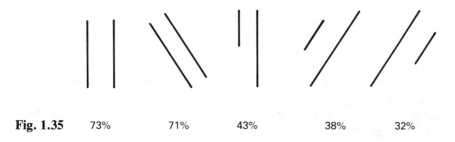

Fig. 1.35   73%   71%   43%   38%   32%

It seems equality of length had become a criterion for parallelism.

Fisher (1978) studied students' bias in favour of 'upright' figures by experimenting with teaching schemes which contained only 'upright' figures, only 'tilted' figures, or a mixture. She discovered that even those taught using only 'tilted' figures found 'upright' figures easier, and that in general the preference for 'upright' figures was stronger when right angles were present. This suggests that the effect may be due to natural orientation tendencies rather than instruction, although obviously the effects of any previous instruction could not be excluded.

Kerslake also points out the difficulties involved with rectangles and squares (bearing in mind a square is a specific case of a rectangle). She says some teachers and books like to distinguish between rectangles which are not square (these they call oblongs) and rectangles which are squares.

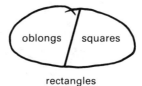

Fig. 1.36   rectangles

Others prefer a classical approach in defining a square as a special rectangle.

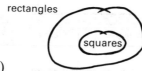

Fig. 1.37 (From Kerslake, 1979)

One hundred and fifty-five children aged 5 to 10 years were asked to point to all the rectangles on the card. The percentage identifying each as being a rectangle is shown:

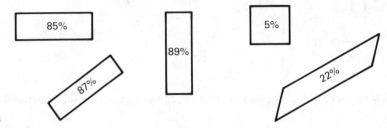

Fig. 1.38

Of the 5 per cent identifying the square as a rectangle some also thought the non-right-angled parallelogram was a rectangle.

One of the practical items given to 11 year olds in the second APU primary survey was similar to Fig. 1.39.

The reasons given for rejecting (d) included the insistence that rectangles 'lie down', or, are 'flat, long and not very wide', again demonstrating the importance of orientation. The diagram (f) was rejected because it was 'too thin' – or because 'a rectangle is about twice the size of a square'. As in Kerslake's study, the square was almost universally excluded from the class of rectangles; one tester reported 'No child included a square in the selection of rectangles, and a few were even outraged that one should contemplate such an act'.

Kent gives an example of a boy who had a total misconception of the idea of an angle, which was not realised until he was having difficulties working with derivatives of trigonometric functions. Kent relates:

> Mentioning the word "angle", I could see a strange look on Richard's face, especially when I demonstrated an angle of over 360° by doing a very poor imitation of a skater's spin.
> 'Where's the lines?' asked Richard.
> 'What do you mean?'
> 'For the angle'
> 'No I'm not with you', I said.
>
> Richard went on to explain to me. An angle is the distance between two lines. These angles are not the same.

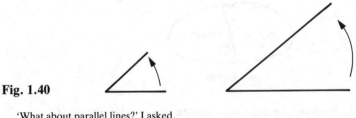

Fig. 1.40

> 'What about parallel lines?' I asked.
> It is, of course,

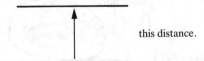

this distance.

Fig. 1.41   (Kent, 1978)

Shape		Percentage choosing shape as rectangle	Percentage choosing shape as non-rectangle
a	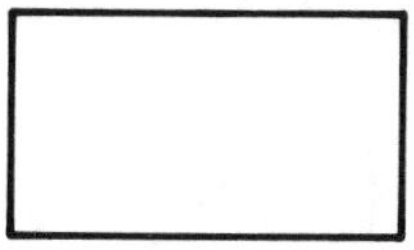 All	91	6
	Boys	93	4
	Girls	89	8
b	 All	91	6
	Boys	94	3
	Girls	87	10
c	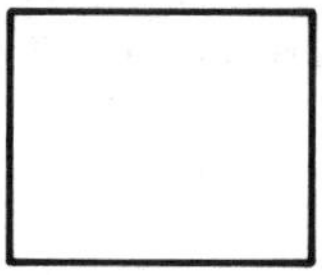 All	60	36
	Boys	64	32
	Girls	56	41
d	 All	78	18
	Boys	81	16
	Girls	75	21
e	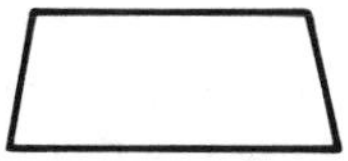 All	15	82
	Boys	15	81
	Girls	15	82
f	 All	84	13
	Boys	85	12
	Girls	82	14
g	All	8	88
	Boys	8	88
	Girls	9	88
h	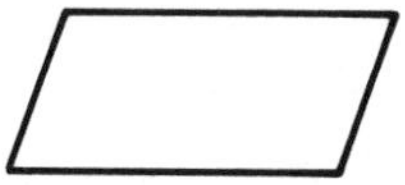 All	25	71
	Boys	26	71
	Girls	25	72
i	 All	82	14
	Boys	86	11
	Girls	78	18

*402 pupils took this topic.   **Fig. 1.39**   (APU, 1981a)

Charles (1980) cites instances where the nature of examples given to illustrate a concept have resulted in the formation of misconcepts. He relates that a class of 10 year olds were being taught 'polygons'. They were told that a polygon was a simple closed shape made up of line segments. They were shown standard examples and then some non-examples. These non-examples were

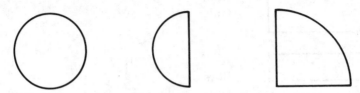

Fig. 1.42

The main characteristic of these non-examples was the presence of curves rather than just straight lines. Consequently most children concentrated on this point and went on to say that the following figure was a polygon.

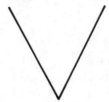

Fig. 1.43

Charles recommends the following guidelines in selecting examples and non-examples.
1) Identify the relevant and most frequently occurring irrelevant characteristics of the concept under consideration e.g. for the polygon there are three relevant characteristics:
     i) A closed shape (i.e. it has no 'gaps').
     ii) A simple closed shape (i.e. it does not cross itself).
     iii) Made up of line segments i.e. straight lines.
All these must be present.
Irrelevant characteristics are:
     i) Regularity or irregularity of the figure.
     ii) The number of line segments (providing there are at least three to make it closed).
2) Select examples so that the most frequently occurring irrelevant characteristics are varied.
    e.g:

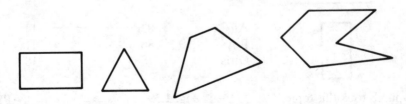

Fig. 1.44

3) Select a variety of non-examples so that the relevant characteristics are varied.

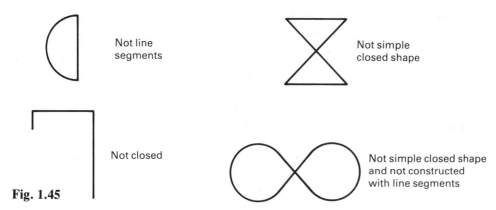

Not line segments

Not simple closed shape

Not closed

Not simple closed shape and not constructed with line segments

Fig. 1.45

4) Draw the pupils attention to the relevant or irrelevant characteristics by questioning and explaining. For example
'Do the line segments have to be the same length for the figure to be a polygon?' or 'How can both of these be polygons when one has 4 sides and the other 6?' or 'Why isn't this a polygon?'

Fig. 1.46

'But this one is made up of line segments. Why isn't it a polygon?'

Fig. 1.47

Zykova gives a wealth of examples of misconceptions which Russian 12 year olds have formed due to the nature of their geometry teaching where it has been confined to the presentation of standard examples. He states that a significant number of pupils 'do not conceptualise beyond the limits of the figures shown, or go only slightly beyond them, within the limits of a symmetric and ordered disposition of the figures'. For example, the interviewer asks a 12 year old pupil to draw a circle and a diameter. The pupil draws two diameters as follows.

Fig. 1.48

Interviewer	'How many diameters can be drawn in a circle?
Pupil	'Two'
Interviewer	'And can you draw any more?'
Pupil	'No more'

Interviewer shows a diameter in a new position as in:

Fig. 1.49

Interviewer	'What's this?' – indicating the diameter
Pupil	'A diameter'.
Interviewer	'So how many diameters can be drawn?'
Pupil	'Two more' drawing
	'Four diameters in all.'

Fig. 1.50

Interviewer demonstrates a diameter in a new position

Fig. 1.51

Pupil	'As many as wanted. I was wrong; I thought only two.'
Interviewer	'Why did you think that?'
Pupil	'I was confused because that's how they showed them to us.'

(Zykova, 1969)

Zykova also conducted a study across several schools and concluded in line with Charles that:

> It was experimentally confirmed that variation of the form and position of geometric figures alone, without the organizational strength of the teacher's explanations, does not foster correct mastery of concepts. Only when the teacher's explanations play the leading role in instruction do variations in geometric illustrations help the pupils to abstract essential features and to master the true geometric relationships. Under these conditions, the mastery of geometric concepts is based on a large accumulation of visual images which mediate the use of concepts in problem solving. (Zykova, 1969)

For instance in a small study of 12 year olds Zykova found about a third of the children, when presented with:

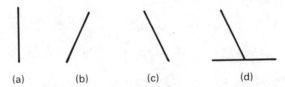

Fig. 1.52   (a)   (b)   (c)   (d)

identified (a) as a perpendicular. The following kinds of explanation were given by the children:

> 'Because the perpendicular is dropped vertically.'
> 'Because it's in a vertical position.'
> 'Because it's dropped straight.'
>
> (Zykova, 1969)

However in one school

> when the concept of perpendicular was being studied ... the teacher made special efforts to emphasise the relationships between the two lines and stressed the importance of this feature, explaining that the perpendicular cannot exist alone, and that a line becomes a perpendicular only when it has a certain relationship with another line (Zykova, 1969).

Drawings of this relationship were well varied. This teacher's pupils 'never called an isolated line a perpendicular'.

This provides evidence that a systematic attempt to avoid the misconceptions that a teacher knows may arise, can be very effective.

1.7   2-DIMENSIONAL REPRESENTATIONS OF 3-DIMENSIONAL SPACE

As mentioned briefly at the beginning of this section any 2-dimensional representation of 3-dimensional objects must always involve the distortion of some of the properties of the shape/space. Fuson states,

> Projections are a common experience in the life of a child. Discovering the invariance of a 3-dimensional object as it is seen from different viewpoints is a major accomplishment. Building up images of three dimensional objects when one can never see them from more than one viewpoint ... at a time is a long and difficult task.

However as Lappan and Winter say,

> In spite of the fact that we live in a 3-dimensional world, most of the mathematical experiences that we give our children are 2-dimensional. We use 2-dimensional books, containing 2-dimensional pictures of 3-dimensional objects, to present mathematics to our children. Surely this use of 'pictures' of objects introduces (for the child) another difficulty in the process of understanding. Yet it is necessary that children learn to cope with 2-dimensional representations of their world. In our modern world information will continue to be disseminated through books and pictures, possibly through moving pictures as on television, but still 2-dimensional representations of the real world. (Lappan and Winter, 1979)

One kind of relationship between these different dimensions is the 'net' of a solid. A net is the shape which may be drawn on card and cut out and folded to give a particular object; in a certain sense it is a 2-dimensional form of the solid.

Piaget (Piaget and Inhelder, 1956) showed

Fig. 1.53

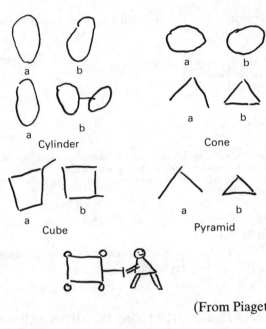

Fig. 1.54

(a : side view
 b : net)

(From Piaget and Inhelder, 1956)

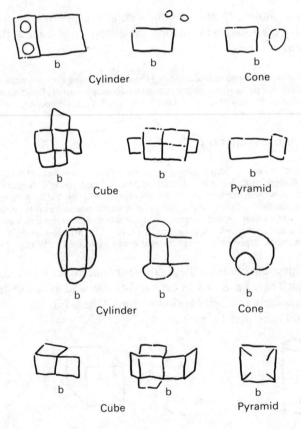

Fig. 1.55

a cylinder, cone, cube and pyramid (tetrahedron) to a number of children between 4 and 13 years. After having demonstrated, by unfolding a fifth shape, what was meant by a net, he asked the children to draw first the 3-dimensional shape as they saw it, and, second, the shape that they thought it would make when unfolded flat.

At around 5 to 8 years, the children did not distinguish between their view of the object and its net. The same children tended to draw a picture of a horse and cart showing all four wheels, again confusing a net with a view (see Fig. 1.54).

By 7 to 9 years, reasonable attempts were made at drawing nets by coordinating different viewpoints (see Fig. 1.55)

From the age of about 8½ years, the children could draw the nets for the cylinder, cone and cube, but they were only successful in the case of the pyramid at 11½ years or more. However Piaget acknowledged that children who had had experience working with nets at school might be as much as three years ahead of other children lacking this experience.

This experiment was one of those Piaget undertook tracing the understanding of projective properties of shapes.

Ward found that 62 per cent of his sample of English 10 year olds could produce one of the nets for a box with a lid having been shown a net for a box without a lid.

Fig. 1.56

The first APU Primary Survey (1980a) reports that

> ... about a quarter of the pupils were able to select from a set of various plane shapes made from six congruent squares the three which were nets of cubes.

One of the practical tests in the APU Primary Survey involved looking at pupils' performances in translating pictures of 3-dimensional models into the actual models using wooden bricks. They report that the vast majority of 11 year olds did most of the simpler tasks quite easily such as making models from these pictures:

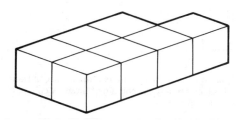 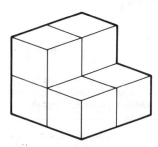

Fig. 1.57

However about 10 per cent showed great difficulty with these tasks. One arrangement which was a particular source of difficulty was:

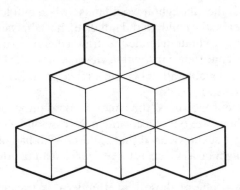

Fig. 1.58

The problems with this involved the alignment of bricks and the appreciation of the depth of the model. Many pupils found it difficult to place the bricks at an angle to themselves and would place the front row face to face as

Fig. 1.59

Some would continue building on these three bricks with no regard for the depth of the model. Some built with a mixture of 'square' and 'turned' layers so that the bottom layer may have been

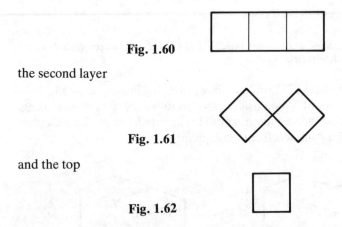

Fig. 1.60

the second layer

Fig. 1.61

and the top

Fig. 1.62

Many successful pupils were very methodical, building the whole bottom layer first before the second and the top, or forming the front row, then the second row back and finally the column of three. A common fault however was to leave out the two back supports for the top brick.

Lappan and Winter report some work they carried out with 8 to 11 year olds which has some connection with the work of Piaget and Donaldson quoted earlier on page 17. It involved the construction of a building with bricks which was then to be represented on squared paper (the side of each square was the same length as the edge of a brick) by,

i)   a base plan (i.e. those bricks touching the ground)
ii)   a front view
iii) a side view.

It was then the task of another child to reconstruct the 3-dimensional building from the 2-dimensional views.

The authors identified four steps involved in developing an understanding of such a representational scheme and they describe the following activities for each step.

Step 1   This is aimed at seeing how a plan and its building are related to each other. The activities for this initial step involved children trying to match each of five buildings with five sets of views. There were only two different base plans involved and each building was constructed on an oriented sheet of paper, with peg people viewing it from the front and from the left side.... the elevations were defined as 'what the peg people would see if they looked straight at the building with one eye'. The buildings were arranged, so that they could be viewed by the children at eye level from all sides.

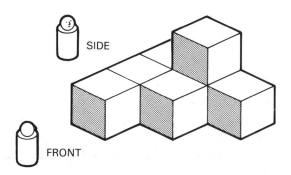

Fig. 1.63

The most frequent kind of mistake involved the left-right orientation of the side elevations. So, for instance, from two sets of views/plans where base and front view are the same the side view

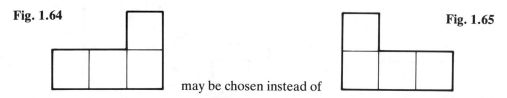

Fig. 1.64

may be chosen instead of

Fig. 1.65

for the side elevation of the model shown above. This sort of error was overcome by encouraging the children to be more attentive and to take more care with this particular aspect.

Step 2   Drawing the plan/views for an existing building. As with Step 1 activities, buildings were set out on squared paper with peg people and the front and side views labelled. Children were now to draw the plans. The authors mention that,

> One of the buildings in the set involved two balanced blocks (i.e. blocks forming a bridge). The question of how to indicate this on the view was asked by at least half of the children.
>
> 'How do you think?' invariably brought a correct response. Moving from associating a brick with a square on the grid paper to associating a brick with a square region equal to a square on the paper is another step towards abstractness. (Lappan and Winter, 1979, *Mathematics Teaching*, **87**).

Again a common cause of error was orientation. For example, one boy aged 9 years drew:

Fig. 1.66

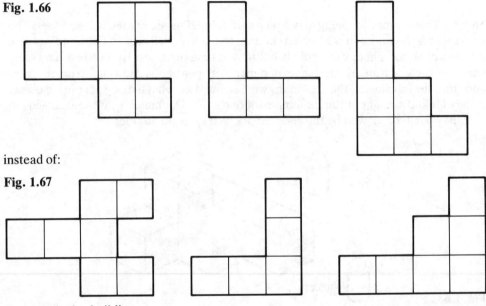

instead of:

Fig. 1.67

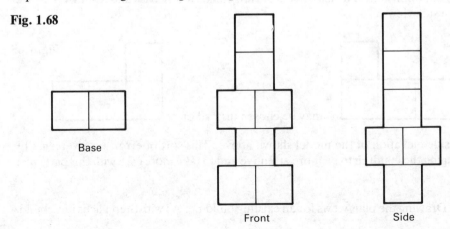

for a particular building.

Step 3   Constructing a building from a given set of plans/views. One set of plans was:

Fig. 1.68

and many children following their initial attempts thought it was impossible to build. However

> As they focussed on each step of the plan (here's a second layer; does it match the front? does it match the side?) they were able to figure out the perpendicular staggering of the bottom three layers and were absolutely delighted with their success. The importance of each of the three parts of the plan – base, front and side – is seen in this example. (Lappan and Winter, 1979)

Another example used by Lappan and Winter is the following

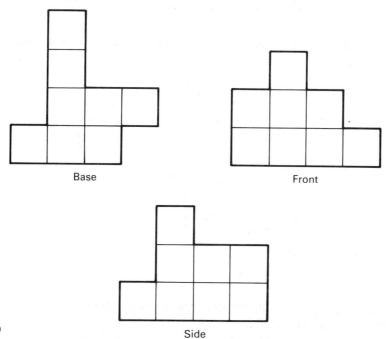

Fig. 1.69

Apparently there are ten different buildings which can be constructed from these plans using from 12 to 15 bricks. A whole field of exploration is set up and children can experiment to find whether or not specifying the number of bricks makes a building unique.

Step 4 Constructing, representing and reconstructing. The final stage involved children designing their own buildings and drawing the representational plans and elevations and then evaluating these by reconstructing each other's buildings from the plans. Initially a limit of 12 bricks per design was enforced. The children tried to design as complex buildings as possible experimenting with bricks being displaced by fractional quantities other than a half, such as a third, and using as few bricks in the base as possible.

Bishop (1977) points out that 2-dimensional representations of 3-dimensional objects are often a matter of convention. He notes that while working in Papua/New Guinea, he found university students who were unable to interpret the following figure (from J.B. Deregowski, 1974):

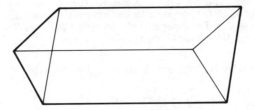

Fig. 1.70

as a triangular prism, or who were happy to represent the windows on a house, of which the outline was given, as in Fig. 1.71.

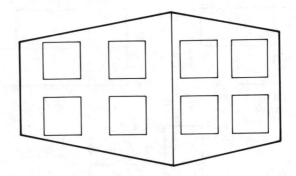

Fig. 1.71

(from Bishop, 1977, Mathematics Teaching, **81**, 32.)

However such students were able to perform feats of spatial memory far in advance of those normally met in our culture.

   This suggests that such representation does not develop "innately", but has to be learned by example.

1.8   THE DEVELOPMENT OF REFERENCE SYSTEMS

Movement in space involves the utilisation of reference points as a means of locating position and direction. Research indicates that an important factor in the development of spatial appreciation is the ability to use some sort of reference system. A paper by Dietz and Barnett (1978) is concerned with just this point. They state that:

   Piaget and Inhelder regard the conceptualisation of a frame of reference as fundamental to an individual's ability to deal with orientation, location and movement of objects; and hence the very culminating point of the entire psychological development of Euclidean space (Dietz and Barnett, 1978).

The essence of a reference system is the relating of the moving parts to some stationary, unchanging aspect of space, e.g. a horizontal surface, the axes of a graph, the notion of a northerly direction. These are reference points which provide a framework in which to study the motion of, say, building bricks, a triangle or a ship.

According to Piaget and Inhelder the development of reference systems is founded upon the natural ability to use what they describe as the natural reference frame, namely that concerned with horizontal and vertical. (We shall see evidence for this in studies quoted in the section on Transformation Geometry.

An important factor in the satisfactory use of reference systems is an awareness of direction. Greenes maintains that normally spatial relationships are initially explored along the vertical axis – looking up and down. Up/down, high/low, above/below etc. have very distinct meanings, e.g. what you see when you look up at the ceiling is different and separate from what you see when you look down at the floor. Next horizontally orientated relationships develop. These are not quite so clear cut. Although while facing in a particular direction what you see is to the front and that which you do not see is to the back if you turn, what was once in front is now behind and similarly what was once to the left is now to the right. Horizontal orientation is later in developing than vertical orientation because the relative ease of bodily movement in the horizontal plane confounds orientation. According to Greenes the differentiation of left from right is later to develop than front from back. This, she says, is because it requires an

understanding of the existence of the midline of the body. Left-right distinctions are most difficult because the left-right domains are simultaneously within the visual field (Greenes, 1979).

Piaget and Inhelder studied children's ability to utilise horizontal and vertical frames of reference in various ways. One of these was the following which is concerned with the horizontal axis of reference.

A child was shown a jar about a quarter full of coloured water.

Fig. 1.72

An empty jar of the same shape and size was tilted in front of the child and he was to indicate where he thought the water line would be in various positions.

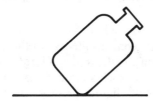

Fig. 1.73

To study the use of the vertical reference axis a cork with a matchstick stuck through it was floated on the water so that the matchstick was perpendicular to the surface of the water, and was termed the 'mast of a ship'.

Fig. 1.74

The child was asked to draw the position of the mast of the ship for various orientations of the jar. (Another variation involved a toy fish suspended by thin string from the top of an unfilled jar.)

Piaget and Inhelder maintain that it is not until a child is roughtly 5 to 7 years old that he begins to compare the water level with the position of another object. Initially he uses the jar as a reference point rather than some immobile aspect of the situation. The next stage of development is when he realises the need for some system of reference other than the jar and water and he begins to relate the water level to the surface of the table or floor. However, according to Piaget, it is not until around 9 to 12 years of age that the child is fully able to coordinate all the angles and parallels throughout the whole spatial set up which is under consideration.

The sequence of development observed by Piaget is shown in Fig. 1.75. (Alongside the picture of the jars are shown corresponding stages in development of children's drawings of 'a hill with a house and trees on it,' which again illustrates the tendency to use the side of the hill as a 'local' frame of reference.)

Shayer (Shayer, Kuchemann and Wylam, 1976) replicated these tests to assess the understanding of the idea of a frame of reference, using a representative sample of British children aged 9 to 13, and found that the ages at which children attained the various stages were in general much later than those given by Piaget, although there was a considerable spread in each case.

It also is of interest to note that Dietz and Barnett found that on another replication of the Piagetian jar task with student primary school teachers in the United States, about 45 per cent of the water levels drawn deviated sufficiently from the horizontal to be considered incorrect. On a series of related tasks this lack of 'ability to make comparisons in orientations between real objects and to use natural reference systems' occurred in the majority of student teachers. The authors see this as having very serious implications since 'Obviously, if the teacher cannot understand the concepts which need to be taught … he will be even less able to design instruction to teach these concepts' (Dietz and Barnett, 1978).

In the study by Schultz (1978) there is also an account of the child's use of reference points. Schultz's main comments concern the size of the object which was being transformed in relation to its distance from the frame of reference employed by the child. She suggests that the pupils were using such stationary external points of reference as the walls of the room. The smaller the figure the more distant it is from these reference

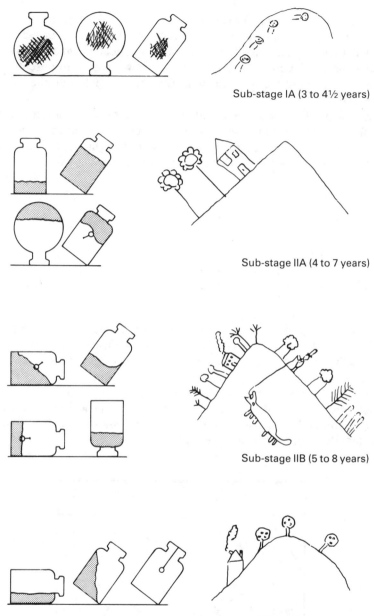

Sub-stage IA (3 to 4½ years)

Sub-stage IIA (4 to 7 years)

Sub-stage IIB (5 to 8 years)

Fig. 1.75

Intermediate level IIB-IIIA (6½ to 8 years)

(Piaget and Inhelder, 1956)

points and hence it becomes more difficult to orientate in the spatial field. She feels that an important line of research would be to investigate just what size is too small for children to cope with adequately in relation to the reference system they are using. (This study is also referred to in part 10.2 on page 60.)

Piaget, Inhelder and Szeminska (1960) claim that along with the use of horizontal and vertical axes of reference the ability to use coordinates also develops. They presented children with two congruent sheets of paper. On one sheet was marked a dot. The child was asked to mark a dot on the second sheet, which was semi-transparent, so that if this sheet were to be placed directly on top of the first sheet with the dot, then the dots would coincide exactly.

Performance on this task showed that at the earliest level of development the child relied totally on a visual estimate which then led later to an estimate by means of a crude usage of rulers and sticks. The next stage is where the child appreciates the need to measure but still only operates with one measurement such as from a nearby corner to the dot. e.g.

Fig. 1.76

Then develops an awareness for the need for two measurements. The procedure involves a lot of trial and error, the child often using one measurement and estimating the second. Finally (around 9 years of age according to Piaget) both measurements are coordinated using the sides of the paper as axes of reference.

Kerslake (1977) gave secondary school age children a task involving an unfamiliar coordinate system not based on the usual horizontal and vertical axes of reference. They were asked to find the coordinates of B in the system shown in Fig. 1.77 in which they were told that the coordinates of A were (4, 6).

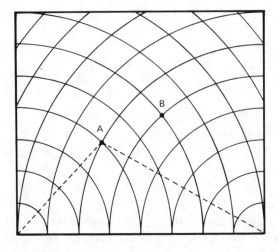

Fig. 1.77   (from Kerslake, 1977)

Kerslake reports,

The percentages of pupils who could do this were.

Year 2		Year 3		Year 4	
Boys	Girls	Boys	Girls	Boys	Girls
34%	22.5%	47.5%	35%	57%	37%

There was quite an improvemment with age and it is also noteworthy that the girls were less successful than the boys in each case.

In the case of a more routine question involving the plotting of points expressed in the form of conventional coordinates, around 90 per cent of children were successful in each year group.

1.9 CONSERVATION

From the time he is born the child begins to organise his world according to its unchanging aspects. He comes to recognise people, objects and situations through familiarity based on his five senses. His perceptions of the spatial world around him rely particularly on sight, touch, and to a lesser extent, on hearing. The development of his spatial awareness depends upon an appreciation of what remains unchanged, i.e. what is conserved or invariant despite alterations in time and space. If he walks to and from school by the same route is it the same distance both ways? It is not until the child can appreciate the unchanging aspects of such situations that any deep understanding of space and geometry, particularly of a Euclidean nature, can occur.

Musick (1978) studied 142, 3½ to 9 year olds in various situations concerned with distance. She had a doll which walked from A to B and then back again from B to A. The children also walked the doll back and forth and each child walked the distance by himself from A to B and back again. Another aspect of the study was for the child to walk one way by himself, collect a heavy basket and return. A further activity was to walk one way and run or jump the other direction. Finally each child jumped the distance in one direction and ran in the other. All these activities involved travelling between the points A and B which were on opposite sides of a room.

Musick gives examples of children's comments when they judged the same distances to be different.

'It's always farther to go some place than to come back.'

'It's more far for the dolly to go from me to you than to come back to me 'cause she was farther away from me – see?'

'It went farther when I ran 'cause it's faster than jumping.'

Also there were instances of correct judgments but for inappropriate reasons.

'It's the same both ways 'cause there's a window on both sides of the room.'

(It is of course possible that the child making this last comment had not fully understood the nature of the judgment he was supposed to make.)

Responses which were judged by Musick to be correct and adequately justified included:

> 'See it doesn't matter what you do, the room stays the same size. It's all the same space.'
>
> 'I could just walk back and forth and count my steps ... it would be the same both ways.'
> (The beginnings of measurement are reflected here.)

Musick found that it was not until around 6½ years of age that the children could understand that the distance travelled in one direction was the same as the return journey. There was no particular preference as to which direction was longer – going or coming back – incorrect judgments for walking tasks were more or less evenly split.

She points out that Piaget sees the child as initially regarding time, space and movement as a whole, being unable to clearly differentiate among them. Where tasks involved some effort such as carrying the heavy basket, the distance travelled was judged to be longer than it was for walking.

> 'When I carried the basket it made that path longer.' 'Do you mean it stretched it out.'
> 'Yes it pulls it out longer.'
>
> 'If it's hard to carry something, that way is the longest.'

Similar judgments were made when jumping or running in one direction was compared with walking in the other direction. Yet on the tasks comparing running with jumping, although the jumping involved more effort, the direction which was covered by running was often judged to be longer: this was because it was faster. Children's explanations of their various judgments included the following.

> 'Runs are always more far than jumpers, 'cause jumpers is slower and they take a longer time.'
>
> 'Jumping is real slow, so you've gone farther to get there than when you run.'

There were many children who could appreciate that the distances travelled from A to B and from B to A were the same when they or the doll walked it but who did not conserve the equality of these distances on the run/jump tasks.

> 'It's the same distance when the doll walks it and when I walk it, but it's not the same distance to run it or jump it.'
> 'Why is that?'
> 'Well when a person runs, it's just longer that way.'
> 'But you just told me walls don't move, so how come you're saying it's farther when you run?'
> 'It just makes it kind of stretchy. ... Oh I don't really know why. It's just farther.'

Musick claims that, on the basis of her findings, some caution should be taken in the utilisation of such gross motor tasks (i.e. using the whole body) when trying to help very young children learn about space since

> too many extraneous factors ... distract the child and impede his ability to grasp the concept and its underlying structure.

However she does emphasise the value of such gross motor tasks with children who have some understanding of the symmetry of distance. Those children in her study who displayed some such awareness appeared to derive a 'bonus' from such activity. There was obvious enjoyment in performing the actions and

in the intellectual exercise involved in separating the extraneous from the relevant variables in the gross motor acts.

For example a child of 7½ years told the experimenter,

'Even if you go fast, or slow, it doesn't change the space. The space always stays the same.' or 'I could carry 16 baskets and take a real long time, and be *so* tired, but that can't make the walls move back and forth. Now, if I'd walked a big crooked, curly way, I'd have done more distance, but I just went *straight* back and forth both times in this same piece of space' (Musick, 1978).

The classic conservation of length test, Piaget (1969), involves two sticks of the same length which are presented as follows

Fig. 1.78

One stick is then displaced

Fig. 1.79

and the child is posed questions aimed at a comparison of the relative lengths of the sticks – whether they are still the same length or whether one is shorter/longer than the other. Most studies of this type have led to the conclusion that children display a classical conservation of length around 6 to 8 years of age on average, i.e. they recognise that despite the displacement the lengths of the sticks remain the same. Before this stage, the length of the stick is not fully distinguished from the position of the end-points.

Detailed descriptions of conservation of length, area and volume occur in the Measurement section in this book.

1.10 TRANSFORMATION GEOMETRY

1.10.1 Introduction

In recent years much of school geometry has been concerned with the movement of geometrical figures from one position to another. Some movements involve a change of size or shape as well. The study of the transformation of shapes has increasingly taken over from the more formal approach to geometry involving theorems and proofs and the deductive method. This trend has been based on the belief that geometric transformations help provide a unified picture of mathematics as a whole by virtue of their links with vectors and matrix algebra. However Küchemann points out that this unification is proving to be

> ... to many children as inaccessible as the (Euclidean) deductive geometry.

It seems that the main value for most children lies in the study of certain transformations for their own sake which ...

> can be based on easily performed actions (such as folding and turning), which can be used both to generate discoveries about the transformations and to check children's predictions and inferences
>
> (Küchemann, 1980).

Thomas also points out that an American report on school geometry decided that

> the main value of motion geometry is in achieving the objective of an informal, intuitive appreciation of geometry (Thomas, 1978)

Also it helps highlight certain more traditional aspects of Euclidean geometry namely, congruence and similarity.

Below is a brief outline of the basic transformations involved in school geometry.

REFLECTION   (Folding, mirror images etc.)

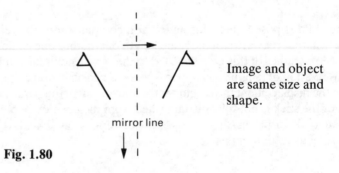

Image and object are same size and shape.

mirror line

Fig. 1.80

ROTATION   (Turns)

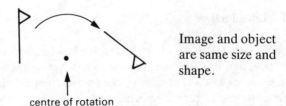

Image and object are same size and shape.

Fig. 1.81   centre of rotation

TRANSLATION (Slides)

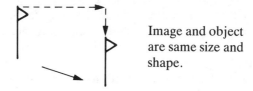

Image and object
are same size and
shape.

Fig. 1.82

Can be described in terms of 'horizontal' and 'vertical' displacement, or in terms of distance moved and direction.

ENLARGEMENT (Making larger/smaller) – Similarity

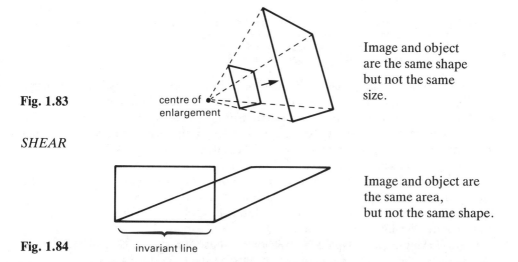

Image and object
are the same shape
but not the same
size.

Fig. 1.83 centre of
enlargement

SHEAR

Image and object are
the same area,
but not the same shape.

Fig. 1.84 invariant line

While reflections, rotations and translations often occur throughout primary and secondary curricula, enlargements and shears are usually met only at secondary level.

1.10.2 Translations, Reflections and Rotations

Thomas (1978) gave the classic length conservation test (described on page 51) to 30 children, 10 each of the ages 6, 9 and 12 years. She found the following numbers of children to be non conservers:

*Non-conservers of length in
the classic sense*

	6 Year Olds	*9 Year Olds*	*12 Year Olds*
Number Out of 10	8	7	0

She then ran a series of tasks which involved the transformation of a triangle. The transformations used were rotations, reflections and translations. The children had to compare the length of a specified side of the triangle before and after each particular motion, saying whether it was 'shorter than', 'longer than' or the 'same length as' before. (It was ascertained that the children were fully familiar with this vocabulary before the tests began.)

Thomas found that the crucial factor for the recognition of the invariance of length of the specified side of the triangle under transformation was the ability to conserve length in the classical Piagetian sense. This finding is also borne out in a study by Kidder – see page 58.

Most of Thomas' pupils considered length to remain invariant for rotations and reflections, but for translations the non-conservers saw the length of the sides of a geometric figure as having changed.

Similarly, for tasks involving a before-after comparison of only one figure, most (pupils) said that length(s) would stay the same; however, when there was a congruent copy close by with which the (pupil) could make a visual comparison, non conservers were significantly more apt to believe that a transformation changed the length (of sides) of a figure (Thomas, 1978).

Another series of tasks used by Thomas were aimed at discovering whether children understood that a particular point on the side of a triangle would remain in the same position on that same side following some transformation of that triangle. She used two squares of transparent plastic and on each was drawn a right-angled triangle – the triangles being congruent (same size and shape). There were two pennies at hand.

Thomas introduced the task to each child with the triangles superimposed. She then transformed the top copy. She positioned one of the pennies to represent a point on the side of the bottom copy triangle which remained stationary. The child was then to place the second penny on the transformed triangle on 'the spot to which the point would have moved under the given transformation'

The following table, adapted from Thomas, illustrates the percentages of correct placements of the second penny for each of the specified transformations.

Percentage of pupils answering each item correctly

Item	10 pupils from each age group		
	6 year olds	9 year olds	12 year olds
¼ turn clockwise	40%	50%	100%
¼ turn counter clockwise	60%	60%	100%
½ turn clockwise	50%	60%	100%
½ turn counter clockwise	40%	60%	100%
Horizontal Slide (translation)	80%	80%	90%
Vertical Slide (translation)	90%	60%	100%
Reflection in vertical axis 	80%	70%	100%

Fig. 1.85

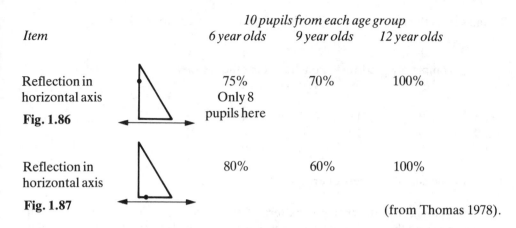

Item	10 pupils from each age group		
	6 year olds	9 year olds	12 year olds
Reflection in horizontal axis **Fig. 1.86**	75% Only 8 pupils here	70%	100%
Reflection in horizontal axis **Fig. 1.87**	80%	60%	100%

(from Thomas 1978).

The errors made by the children were classified in the following way. They are illustrated by means of the reflection transformation.

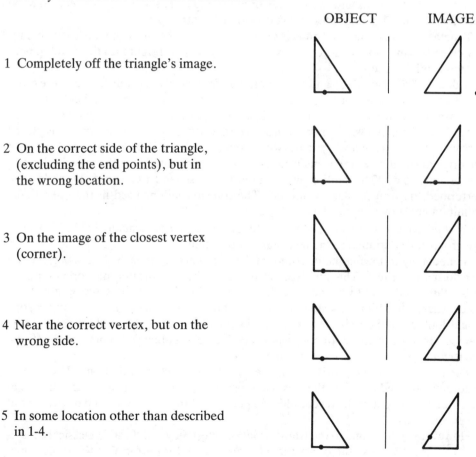

OBJECT   IMAGE

1 Completely off the triangle's image.

2 On the correct side of the triangle, (excluding the end points), but in the wrong location.

3 On the image of the closest vertex (corner).

4 Near the correct vertex, but on the wrong side.

5 In some location other than described in 1-4.

Figs. 1.88-92 (from Thomas 1978).

Thomas gives the following classification of the errors made for each age group.

	1 completely off	2 correct side	3 on the vertex	4 near the vertex	5 other
Age 6	0	18	0	4	8
Age 9	0	10	10	3	11
Age 12	0	0	0	0	1

These are the actual numbers of errors.

Thomas emphasises the changing pattern of errors with age. The youngest children most often placed the penny on the correct side of the triangle but in the wrong position on that side – apparently focusing on the sides rather than the vertices. The 9 year olds made the same number of errors of types 2 and 3 showing that they were becoming more aware of both the sides and the vertices of the triangle. The 12 year olds seemed to be successful in coordinating sides and vertices in their attempts.

Thomas points out that her study is only a beginning in pointing the way to further research with many more children in the quest for a deeper insight into the development of their spatial concepts.

Kidder (1978) studied 8, 9 and 10 year olds (20 from each age group). They were given the classic conservation of length test. They were then given operational definitions of basic transformations, translations, reflections and rotations. These involved a stick about 10 cm long and wire arrows indicating the various motions. The stick was placed in front of the child together with the wire arrow. An identical stick was placed on top of the original and the desired transformation was demonstrated with the top stick leaving the original fixed. The motion was demonstrated several times and the child then performed the transformation himself. The transformations used in the operational definition are shown in Fig. 1.93.

Only children who were successful with these transformations made up the 20 in each age group and continued to the next stage of the study.

Immediately following the operational definition activities each child was given the transformational test. This involved an object stick, an indicated motion (similar to those illustrated) and five other sticks, only one of which was the same length as the object stick. The child was asked to use one of the sticks to show how the object stick would look after the indicated motion. Each child was told he could measure if he wanted – this was to ensure that he was aware that he was allowed to compare the sticks. Each child was encouraged to explain his actions.

The results showed (surprisingly in view of Piagetian claims) that only 31 of the 60 children conserved length in the classical sense during the initial stage of the investigation: i.e. 40 per cent of the 8 year olds, 55 per cent of the 9 year olds and 60 per cent of the 10 year olds.

However when it came to the transformational test only 7 of these 31 classical conservers consistently chose the correct length image stick for performing the transformations. (They were judged to be consistent if they chose the correct length image stick for

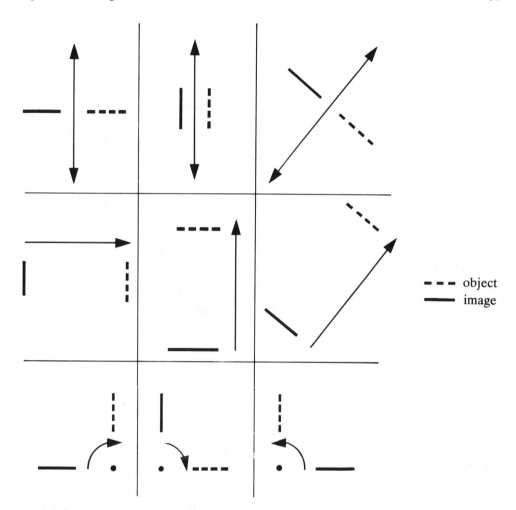

Fig. 1.93   Reproduced from
                Kidder 1978

9 or more of the 12 motions they were given.) Nineteen of the 31 classical conservers consistently failed to use the correct length image stick. (Consistent failure was judged to be the choosing of the correct stick 4 or less times out of the 12.) Kidder (1978) sees these results as:

indicating that classical conservation of length is not sufficient to ensure length conservation on more complex mental operations. (Kidder, 1978)

Of the 29 children who did not conserve length in the classical sense only 6 consistently chose the correct length image stick in the transformation test. It appeared that this was as a direct result of the instructional period from which they continued to follow the instructions to the letter i.e. putting an image stick directly on top of the object stick before performing the motion and by so doing they kept trying sticks until one fitted exactly on top of the object. These 6 children were given the classical conservation test a second time following the transformation test and again failed to conserve length in the classical sense.

However 18 of the 29 children failing the original test of classical conservation consistently failed to choose the correct length image stick in the transformation test. (The remaining 5 had partial success.) Kidder concludes that conservation in the classical sense is probably prerequisite for conservation of length in more complex situations. With each of the 60 children performing 12 transformation tasks this made a total of 720 performances. Of these length was *not* conserved in 427 instances. Of these 427 failures 290 did involve a correct positioning of the image stick. Kidder points out that even though the image stick is placed in the correct position, if its length has not been preserved then it is not a Euclidean transformation (where in this situation length is the essential invariant).

In a similar study Kidder investigated 9, 11 and 13 year olds and their performance on transformational tasks, this time using a triangle.

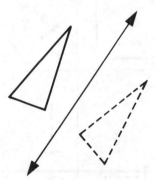

Fig. 1.94   Kidder (1978)

Each child was given seven sticks from which to select three for the image, the other four being of inappropriate lengths. Sixty-seven per cent of all the errors made were to do with failure to conserve length. Of the 72 pupils involved in this investigation only 4 performed all 12 transformational tasks with no conservation of length errors.

Kidder suggests that maybe it is not until a person is at a Piagetian level of formal operational thought that he is able to separate out all the various factors involved in a task of this nature where he has to construct an image. Before this he can only operate by attending to one aspect at a time while forgetting or ignoring others, for example, he is concerned with performing the motion in terms of finding the position of a 'like' image, ignoring the necessity for it to be a 'congruent' image.

Perham (1978) studies translations, reflections and rotations with 6 year old children. The first two were investigated in terms of horizontal, vertical and oblique orientation of movement. Rotations of 45°, 90° and 180° were studied. Also she investigated the ability of children to construct for themselves the image of an object under a certain transformation as compared with their performance in choosing a correct image from a selection of already drawn images – i.e. multiple choice format. Her study also involved a comparison of performance on these tasks before and after specific instruction with one group and with no specific instruction in another group. Her results are tabulated below.

Translations

Direction of movement	Horizontal/Vertical	Oblique
Performance before specific instruction.	Pupils showed a good understanding at both levels of representation, i.e. multiple choice and constructing image themselves.	Difficulty was experienced.
Comparison of perform-ance after specific instruc-tion by those receiving it and those not.	As above.	Continued difficulties. No difference between groups.

Reflections

Orientation of Reflection Line	Horizontal	Vertical	Oblique	Horizontal or Vertical Internal to the Figure
Before instruction	The 6 year olds showed no understanding of reflection at either level of representation of the image i.e. neither when having to select from a choice of images, nor when the image had to be constructed by the child himself.			
Following instruction	There were significant gains in performance of the group receiv-ing instruction as compared with those who had not – at both levels of representation.	No significant difference between the groups at either level of represen-tation.		Group receiving instruction did significantly better at both levels of repre-sentation.

Rotations

Before instruction there was no evidence of understanding for any of the turns. Instruc-tion was somewhat effective for all turns but only where images were presented in a mul-tiple choice format. Unfortunately Perham does not give details of the objects used in the study save that they were simple cardboard figures.

She sees this study as bearing out Piaget's conclusions that children learn transformations in the order: translations, reflections and then rotations. Also the ability to select the correct image from a number of alternatives precedes the ability to construct an image from scratch by oneself. She points out that the results suggest that rather than different types of transformation being the all important factor in teaching transformation geometry perhaps a more pertinent consideration is the orientation of the transformation, i.e. whether it is horizontal, vertical or oblique, particularly since the latter repeatedly caused difficulties even after specific instruction.

Perham advocates the inclusion of work on horizontal and vertical translations and reflections at both the levels of representation, i.e. multiple choice and construction, for 6 year olds. Work on rotations in a multiple choice medium also seems appropriate.

Schultz (1978) looked at children's performance on transformational tasks with particular attention to such factors as the nature and size of the object as well as the complexity of the nature of the transformation. The Figures below which have been reproduced from Schultz's article illustrate these points.

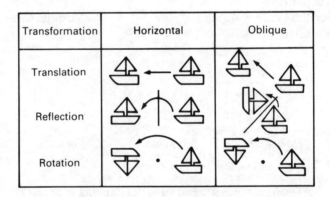

Fig. 1.95   (From Schultz, 1978)

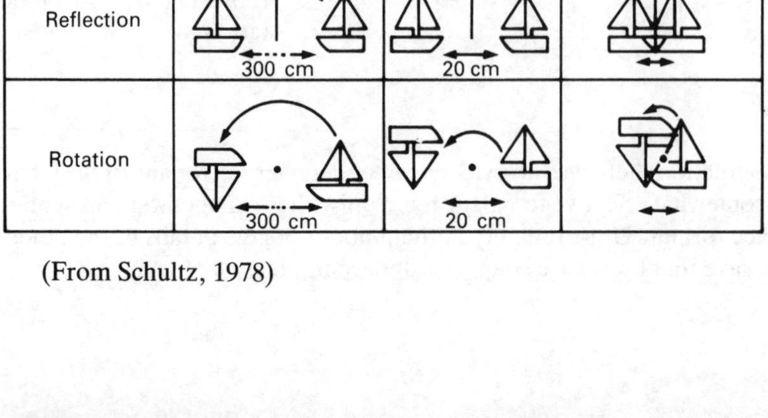

Fig. 1.96   (From Schultz, 1978)

Length was equal to width in both configurations.

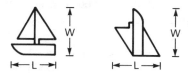

Fig. 1.97   (From Schultz, 1978)

The three distinct wooden shapes were arranged to form 'meaningful' and 'non-meaningful' configurations i.e. a yacht and an abstract shape. Two sizes were employed, large and small where overall dimensions were 80 cm and 8 cm square respectively. The children (who were individually interviewed) were:

40	6 year olds	80	7 year olds
70	8 year olds	10	10 year olds

They were shown a particular configuration mounted on a square sheet of 'plexiglass'. On top was another sheet which the interviewer moved according to which transformation was being studied. For example, Fig. 1.98 represents an oblique translation.

Child to place transformed
object on Sheet 2.

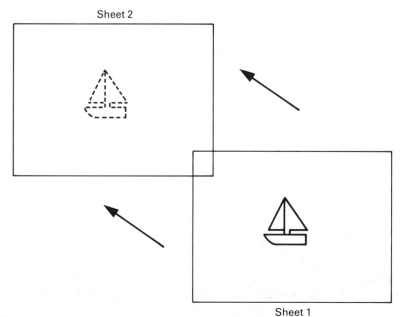

Fig. 1.98

The child was then to imagine the transformed position of the object and represent this with a set of matching shapes on the transformed top sheet to show how the interviewer's object would look if it had moved in exactly the same way as the sheet of 'plexiglass'. The main results were:

1. Tasks involving translations were performed much more successfully than reflection and rotation tasks and reflections were more easily done than rotations.
2. Short translations were easier than long or overlapping ones.
3. Horizontal translations were far easier than oblique ones. With the oblique translations it was often the case that a configuration appeared to be orientated in the direction of the displacement, e.g.

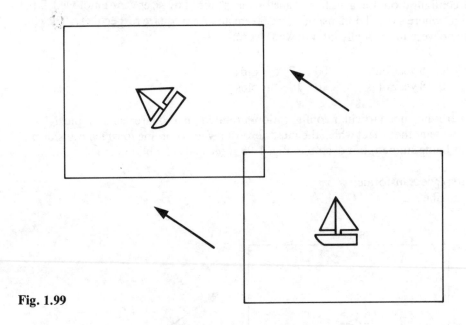

Fig. 1.99

4. Images which necessitated an overlap onto the original proved difficult. Often a child would move the 'plexiglass' clear of the original or would represent the image superimposed exactly on the original, part on and part off the transformed sheet.
5. The 6 year old children frequently changed the 'non-meaningful' configuration into a 'meaningful' one.
6. Schultz reports an obvious enthusiasm amongst the children towards the larger 'meaningful' configuration. Generally the larger configurations were easier to translate than the smaller ones.
7. The most significant error with oblique reflections was to do with a fixation for either vertical or horizontal displacements rather than coordinating the two. (Further discussion on oblique reflection occurs later in the context of Kuchemann's work.)
8. In general overlapping reflections were harder than long or short reflections.
9. Schultz found that the most striking errors made with reflections or rotations were to do with spatial orientation. Children tended to turn an image so that it faced the

direction of the reflection or turn. This kind of error was made slightly more often with the non-meaningful configuration where a rotation was involved.

The first APU Primary Survey also shows that 11 year olds were more successful in reflecting with the mirror line in a vertical position than in an oblique orientation. About 80 per cent were successful with the following item:

B1   Draw the reflection of the shape in the mirror.

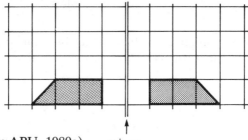

Fig. 1.100   (From APU, 1980a)   mirror

whereas only 14 per cent succeeded with the following:

B4   Draw the reflection of the ⌐ shape in the mirror.

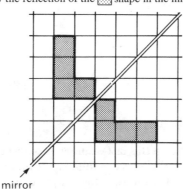

Fig. 1.101   mirror

The APU Secondary Survey (APU, 1980b) had

items which required pupils to name the type of transformation a given shape had undergone. Around 30% (of 15 year olds) correctly named a rotation, a reflection and an enlargement and some 10% were able to name a translation and a shear.

Further details of these items were not published.

Küchemann (1980) in his study of children's difficulties with reflections looked at the effect of the slope of the line of reflection. In addition he varied the complexity of the object under reflection: a dot • or a flag ▷ . He writes about his study with 449, third-year secondary pupils (about 14 years old).

He found that a common error was to ignore the slope of the mirror line and reflect horizontally or vertically. This horizontal/vertical fixation was more apparent with more complex figures such as a line or flag, rather than with a single point, especially when the slope of the object was itself horizontal or vertical in the first place.

e.g.

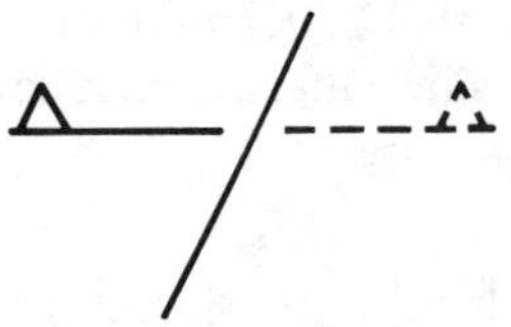

Of the 37 per cent who made errors on this item the majority reflected horizontally so that the image was parallel to the object.

Fig. 1.102

When it was necessary to coordinate the slope of the object and the slope of the mirror line, there was more difficulty than in cases where only one slope had to be considered as with:

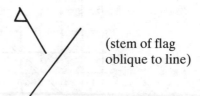

Fig. 1.103

where only 7 per cent made blatant errors compared with 27 per cent who were clearly wrong with:

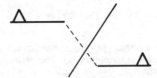

(stem of flag oblique to line)

Fig. 1.104

To correctly reflect the flag in this last example it is necessary to preserve the size of the angle between the two slopes. This was found to be quite difficult and often the slope of the mirror line was ignored or the two slopes were considered independently as in:

Fig. 1.105

Küchemann identifies two approaches to these reflection problems. The first involves a sequence of steps where for example in the following instance

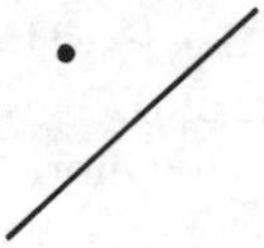

Fig. 1.106

it is necessary to control the direction in which the point is moved and then the distance. Performance on this task is improved by the presence of a grid.

However such a sequence of steps is inadequate for coordinating two slopes. Where there is a more complex figure such as a flag another approach is necessary. Both end points of the stem have to be located before the image can be drawn. The object has to be broken down into separate parts and these borne in mind while the next step is performed.

Küchemann (1980) also investigated rotations of a flag involving quarter turns. He found that major difficulties arose when the centre of rotation was not on the object. Horizontal and vertical starting positions were again easier than oblique orientations.

In contrast to the case with reflection a grid seemed to act as a distractor for all but the most able children. The following are some of the items he used together with the percentage of 14 year olds who could successfully indicate the image for each under an anti-clockwise quarter turn.

Fig. 1.107

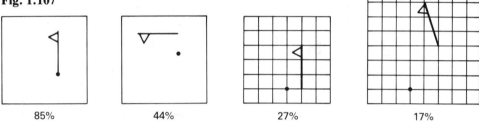

| 85% | 44% | 27% | 17% |

Küchemann's results indicate that some children at 14 years of age find difficulty in envisaging the result of even the simplest transformation. It may be the case that work of this type would have greater application in real life than the more traditional content of school geometry.

1.10.3 Transformations and Figure Symmetry

Thomas (1978) studied children's understanding of the effects of reflections and rotations upon the orientation of a plane figure using a sample of ten children from each of the four age groups of around 9 years, 12 years, 15 years and 17 years. The tasks involved the transformation of a cardboard square with a letter of the alphabet printed on it. Each child was to imagine what it would look like after a specified motion which was demonstrated by the investigator using a blank square.

Rotations were half turns clockwise and anti-clockwise with the lower right-hand corner as the centre of rotation.

Fig. 1.108

Reflections were performed with either the right-hand vertical or the top horizontal as the reflection line.

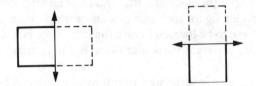

Fig. 1.109

Each pupil was given a choice of four images from which to choose the correct one for each transformation. If he thought none of the four was correct he could draw his own. The letters of the alphabet were chosen so that Thomas could study any special effects arising from the symmetries of the letters themselves. She chose the following letters:

A   T   B   C

Vertical axis of symmetry   Horizontal axis of symmetry

N   S   F   J

Fig. 1.110   Rotation symmetry   No symmetry

Her main findings were:

1. Rotating a figure which already has rotational symmetry – **S** and **N** – was very difficult for pupils of all ages to visualise. Often the incorrect images **ƨ** and **И** were given.
2. Horizontal reflection with the non-symmetric **J** was difficult.
3. The youngest group (9 year olds) scored much lower on all transformations than the other three groups.
4. There were no striking differences between direction of turning on the rotation tasks nor between horizontal and vertical on the reflection tasks.

On page 67 a table shows the number of pupils answering each item correctly.

With regard to symmetry of figures the first APU Primary Survey showed that approximately 65 per cent of 11 year olds could successfully draw in the line of symmetry for:

Fig. 1.111

**Alphabet tasks – numbers of pupils
answering each item correctly (from Thomas, 1978)**

	Item	Grade 3 9 years (10 pupils)	Grade 6 12 years (10 pupils)	Grade 9 15 years (10 pupils)	Grade 11 17 years (10 pupils)	Total (40 pupils)
Reflect Vertically	A	4	9	10	9	32
	B	7	10	8	8	33
	F	6	9	7	10	32
	N	6	10	7	10	33
½ turn clockwise	C	5	7	10	9	31
	F	2	8	9	10	29
	S	1	0	2	1	4
	T	3	10	9	9	31
Reflect Horizontally	T	7	10	9	8	34
	C	2	7	8	6	23
	J	0	6	3	5	14
	S	7	9	7	8	31
½ turn anticlockwise	B	6	9	9	9	33
	J	2	9	6	8	25
	N	1	2	2	3	8
	A	6	10	9	9	34

50 per cent succeeded with:

Fig. 1.112

but in a case where more than one line of symmetry was involved only 19 per cent were successful:

Fig. 1.113

The first APU Secondary Report (1980b) on 15 year olds contrasts primary and secondary performance on items where all lines of symmetry were to be drawn for:

Figure	% correct at 11 years	% correct at 15 years
E	55	65
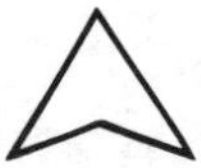 △	65	80
L	50	65

Fig. 1.114

Holcomb (1980) recommends the geoboard as a very useful piece of apparatus for developing an understanding of the properties of figures. It is useful for constructing open and closed figures and for investigating symmetry

e.g. open figures

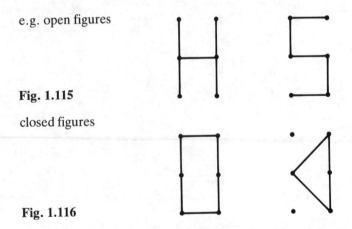

Fig. 1.115

closed figures

Fig. 1.116

He also maintains that copying designs onto dot paper gives useful experiences for counting and transposition.

Holcomb also advocates the use of geoboards for investigating turning and matching, e.g.

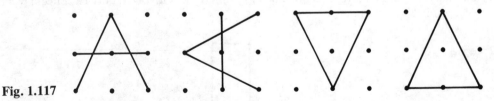

Fig. 1.117

where two boards are used for each figure.

1.10.4 Further Suggestions for Investigating Transformations

Another effective way of investigating rotational symmetry as well as other transformations such as reflections and translations is through tessellations of regular and non-regular shapes. An article by Oliver (1979) provides many ideas drawn from an historical context. Below is a copy she made of a tile pattern from a wall of the Alhambra Palace which illustrates translations, reflections and rotations.

Fig. 1.118   (Reproduced from Oliver, 1979)

She outlines ways in which interesting patterns can be built up from basic geometric shapes – such as a square, and then tessellated according to various transformations. For example

> Start with a square (Fig. 12) with sides labelled A, B, C and D. Translate the square up, down, left and right, keeping track of each side. Similarly translate each new square always keeping track of the sides. As this process is continued (*ad infinitum*) the lattice is completed. Notice that sides A and C and sides B and D always fall adjacent. This means that any modification to side A will effect side C and any modification to side B will effect side D.
>
> In Figure 13a, sides A and C are modified in exactly the same way, and in Figure 13b sides B and D are modified in the same way. Now using tracing paper (or by cutting a piece of card to the desired shape) this new modified version of the 'square' may be copied onto each square in the completed lattice of Figure 12.
>
> The final touches are added to produce the tessellating starfish (Fig. 14).

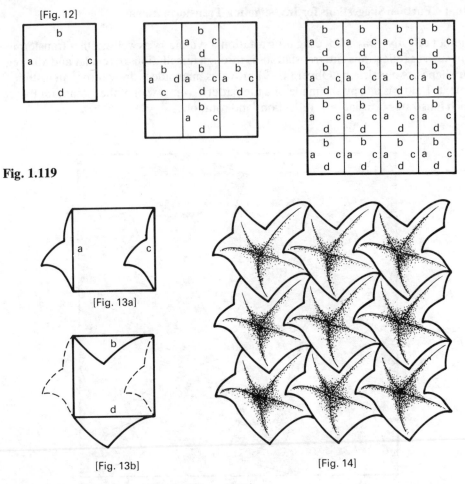

Fig. 1.119

Fig. 1.120   (Reproduced from Oliver 1979)

This is only one of many articles on tessellations; as has previously been noted the journal *Mathematics Teaching* is a particularly rich source of ideas for such activities.

Eba (1979) gives examples of 3-dimensional tessellations (i.e. filling space with repeating patterns of congruent solids), using 3-D pentominoes such as those in Fig. 1.121,

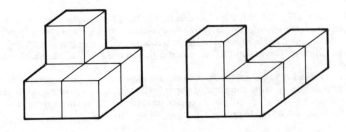

Fig. 1.121

for interlocking in various ways to fill space. Such activities provide good groundwork for developing notions of volume and capacity and the measurement of these.

The first of the above references relates to the application of mathematics in the field of art and design, in particular the design of wallpaper and fabrics. Three-dimensional tessellations are also of interest to designers and architects, especially in the fields of modular furniture, unit housing and packaging. Such tessellations also occur in chemistry, in investigations of the arrangement of molecules and crystal structure. More general transformations e.g. enlargement, are important in the making of maps and plans, and therefore relate to work in geography (for example, see *Understanding Maps*, Schools Council Geography Committee 1979), as well as to engineering and building.

Topics such as these can provide a rich source of ideas for teaching material, with the added benefit of demonstrating to pupils the importance of being able to think spatially, in both work and leisure activities.

References for Section 1

Assessment of Performance Unit (APU) – see Department of Education and Science.

Bishop, A. (1977) Is a Picture Worth a Thousand Words? *Mathematics Teaching*, **81**, 32–5.

Bower, T.G.R. (1966) The Visual World of Infants. *Scientific American*, **215(6)**, 80–92.

Bruner, J.S. (1967) *Towards a Theory of Instruction*. Cambridge, Mass: Belknap Press.

Carpenter, T.P. et al. (1980a) Results and Implications of the second NAEP Mathematics Assessments: Elementary School. *Arithmetic Teacher*, **27(8)**, 44–47.

Carpenter, T.P. et al. (1980b) Results of the second NAEP Mathematics Assessment: Secondary School, *Mathematics Teacher*, **73(5)**, 329–338.

Charles, R.I. (1980) Some Guidelines for Teaching Geometry Concepts. *Arithmetic Teacher*, **27(8)**, 18–20.

Choat, E. (1974) Johnnie is Disadvantaged; Johnnie is Backward. What Hope for Johnnie? *Mathematics Teaching*, **69**, 9–13.

Coxford, A. (1978) Research Directions in Geometry. In *Recent Research Concerning the Development of Spatial and Geometric Concepts*. (Ed) Lesh, R. The Ohio State University: ERIC Clearinghouse for Science Mathematics and Environmental Education.

Darke, I. (1982) A Review of Research Related to the Topological Primary Thesis. *Educational Studies in Mathematics*, **13(2)**, 119–142.

Delaney, K.C. (1979) A Place for Space. *Mathematics Teaching*, **86**, Primary Supplement, xvii.

Department of Education and Science, APU – Assessment of Performance Unit (1980a) *Mathematical Development, Primary Survey Report No. 1*. HMSO.

Department of Education and Science, APU – Assessment of Performance Unit (1980b) *Mathematical Development, Secondary Survey Report No. 1*. HMSO.

Department of Education and Science, APU – Assessment of Performance Unit (1981a) *Mathematical Development, Primary Survey Report No. 2*. HMSO.

Department of Education and Science, APU – Assessment of Performance Unit (1981b) *Mathematical Development, Secondary Survey Report No. 2*. HMSO.

Deregowski, J.B. (1974) Teaching African Children Pictorial Depth Perception: In Search of a Method. *Perception*, **3**, 309.

Dienes, Z.P. (1959) The Growth of Mathematical Concepts in Children through Experience. *Educational Research*, **2**, 9–28.

Dietz, C. and Barnett, J. (1978) Understanding of Frames of Reference by Preservice Teacher Education Students. In *Recent Research Concerning the Development of Spatial and Geometric Concepts* (Ed) Lesh, R. The Ohio State University: ERIC Clearinghouse for Science Mathematics and Environmental Education.

Donaldson, M. (1978) *Children's Minds*. Fontana.

Eba, P.N. (1979) Space Filling with Solid Polyominoes. *Mathematics in School*, **8(2)**, 2–5.

Egsgard, J.C. (1970) Some Ideas in Geometry that can be Taught from K-6. *Educational Studies in Mathematics*, 478–495.

Eng, H. (1954) *The Psychology of Children's Drawings*. London: Routledge and Kegan Paul.

Fielker, D.S. (1973) A Structural Approach to Primary School Geometry. *Mathematics Teaching*, **63**, 12–16.

Fielker, D.S. (1981) Removing the Shackles of Euclid 1. Classification; 2. Interpretations; 3. Context. *Mathematics Teaching*, **95**, 16–20; **96**, 24–28; **97**, 321–327.

Fisher, N. (1978) Visual Influences of Figure Orientation on Concept Formation in Geometry. In *Recent Research Concerning the Development of Spatial and Geometric Concepts* (Ed) Lesh, R. The Ohio State University: ERIC Clearinghouse for Science Mathematics and Environmental Education.

Fuson, K. (1978) An Analysis of Research Needs in Projective Affine and Similarity Geometries Including an Evaluation of Piaget's Results in These Areas. In *Recent Research Concerning the Development of Spatial and Geometric Concepts* (Ed) Lesh, R. The Ohio State University: ERIC Clearinghouse for Science Mathematics and Environmental Education.

Fuson, K. and Murray, C. (1978) The Haptic-Visual Perception Construction and Drawing of Geometric Shapes by Children Aged Two to Five: A Piagetian Extension. In *Recent Research Concerning the Development of Spatial and Geometric Concepts* (Ed) Lesh, R. The Ohio State University: ERIC Clearinghouse for Science Mathematics and Environmental Education.

Greenes, C.E. (1979) The Learning Disabled Child in Mathematics. *Focus-On Learning Problems in Mathematics (Framingham, Massachusetts)*, **1(1)**.

Hart, K.M. (Ed, 1981) *Children's Understanding of Mathematics 11–16*. John Murray.

Hemmings, R; Last, D; Rodgers, L; Sturgess, D; Tahta, D. (1978) *Leapfrog-Teachers' Handbook*. Central Independent Television plc.

Holcomb, J. (1980) Using Geoboards in the Primary Grades. *Arithmetic Teacher*, **27(8)**, 22–25.

Kent, D. (1978) The Dynamic of Put. *Mathematics Teaching*, **82**, 32–36.

Kerslake, D. (1977) The Understanding of Graphs. *Mathematics in School*, **6(2)**, 22–35.

Kerslake, D. (1979) Visual Mathematics. *Mathematics in School*, **8(2)**, 34–35.

Kidder, F.R. (1978) Conservation of Length: A Function of the Mental Operation Involved. In *Recent Research Concerning the Development of Spatial and Geometric Concepts*. (Ed) Lesh, R. The Ohio State University: ERIC Clearinghouse for Science Mathematics and Environmental Education.

Kilpatrick, J. and Wirszup, I. (Ed, 1969) *Soviet Studies in the Psychology of Learning and Teaching Mathematics. Vol. I*. Stanford California: School Mathematics Study Group.

Krutetskii, V.A. (1976) *The Psychology of Mathematical Abilities in Schoolchildren*. University of Chicago Press.

Küchemann, D. (1980) Children's Difficulties with Single Reflections and Rotations. *Mathematics in School*, **9 (2)**, 12–13.

Lappan, G. and Winter, M.J. (1979) Buildings and Plans. *Mathematics Teaching*, **87**, 16–19.

Lesh, R. (Ed, 1978) *Recent Research Concerning the Development of Spatial and Geometric Concepts*. The Ohio State University: ERIC Clearinghouse for Science Mathematics and Environmental Education.

Lesh, R. and Mierkiewicz, D. (1978) Perception Imagery and Conception in Geometry. In *Recent Research Concerning the Development of Spatial and Geometric Concepts* (Ed) Lesh, R. The Ohio State University: ERIC Clearinghouse for Science Mathematics and Environmental Education.

Mathematics Teaching, Journal of the Association of Teachers of Mathematics (ATM) Kings Chambers, Queen Street, Derby.

Musick, J. (1978) The Role of Motor Activity in Young Children's Understanding of Spatial Concepts. In *Re-

cent Research Concerning the Development of Spatial and Geometric Concepts (Ed) Lesh, R. The Ohio State University: ERIC Clearinghouse for Science Mathematics and Environmental Education.

National Assessment of Educational Progress (1980) *Mathematics Technical Report: Summary Volume.* Denver Colorado: National Assessment of Educational Progress.

Noelting, G. (1979) *Hierarchy and Process in the Construction of the Geometrical Figure in the Child and Adolescent.* University of Warwick: Proceedings of the Third International Conference for the Psychology of Mathematics Education.

Nuffield Mathematics Teaching Project (1972) *Checking Up II.* John Murray and W. & R. Chambers for the Nuffield Foundation.

Oliver, J. (1979) Symmetry and Tessellations. *Mathematics in School,* **8 (1)**, 2–5.

Pask, G. (1976) Styles and Strategies of Learning. *British Journal of Educational Psychology* **46**, 128–148.

Perham, F. (1978) An Investigation into the Effect of Instruction on the Acquisition of Transformation Geometry Concepts in First Grade Children and Subsequent Transfer to General Spatial Ability. In *Recent Research Concerning the Development of Spatial and Geometric Concepts* (Ed) Lesh, R. The Ohio State University: ERIC Clearinghouse for Science Mathematics and Environmental Education.

Piaget, J; Inhelder, B; Szeminska, A. (1960) *The Child's Conception of Geometry.* London: Routledge and Kegan Paul.

Piaget, J. & Inhelder, B. (1956). *La conception de l'espace chez l'énfant.* Paris: Presses Universitaires de France.

Plunkett, S.P. (1979) Diagrams. *Mathematical Education for Teaching,* **3 (4)**, 3–15.

Schools Council Geography Committee (1979) *Understanding Maps: A Guide to Initial Learning – Occasional Bulletin.* London: Schools Council.

School Mathematics Project (1971) *Book B.* Cambridge University Press.

Schultz, K. (1978) Variables Influencing the Difficulty of Rigid Transformations During the Transition Between the Concrete and Formal Operational Stages of Cognitive Development. In *Recent Research Concerning the Development of Spatial and Geometric Concepts* (Ed) Lesh, R. The Ohio State University: ERIC Clearinghouse for Science Mathematics and Environmental Education.

Sharma, M.C. (1979) Children at Risk for Disabilities in Mathematics. *Focus – On Learning Problems in Mathematics.* (Framingham, Massachusetts.) Volume **1, (2)**, 63–64.

Shayer, M; Küchemann, D; Wylam, H. (1976) The Distribution of Piagetian Stages of Thinking in British Middle and Secondary School Children. *British Journal of Educational Psychology,* **46**, 164–173.

Shayer, M. & Wylam, H. (1978) The Distribution of Piagetian Stages of Thinking in British Middle and Secondary School Children II: 14 to 16 year-olds and Sex Differentials. *British Journal of Educational Psychology,* **18**, 62–70.

Thomas, D. (1978) Students' Understanding of Selected Transformation Geometry Concepts. In *Recent Research Concerning the Development of Spatial and Geometric Concepts* (Ed) Lesh, R. The Ohio State University: ERIC Clearinghouse for Science Mathematics and Environmental Education.

Ward, M. (1979) *Mathematics and the 10 year-old – Schools Council Working Paper 61.* Evans/Methuen for the Schools Council.

Weinzweig, A.I. (1978) Mathematical Foundations for the Development of Spatial Concepts in Children. In *Recent Research Concerning the Development of Spatial and Geometric Concepts* (Ed) Lesh, R. The Ohio State University: ERIC Clearinghouse for Science Mathematics and Environmental Education. College of Education.

Wheatley, C.L. and Wheatley, G.H. (1979) Developing Spatial Ability. *Mathematics in School.* **8**, **(1)**, 10–11.

Wheatley, G.H. (1977) The Right Hemisphere's Role in Problem Solving. *Arithmetic Teacher,* **25**, **(2)**, 37–38. Material reprinted from the *Arithmetic Teacher*, copyright © 1977 by the National Council of Teachers of Mathematics, Virginia, USA. Used by permission.

Wirszup, I. (1976) *Breakthroughs in the Psychology of Learning and Teaching Geometry.* In *Space and Geometry: Papers from a Research Workshop* (Ed) Martin, J.L. Columbus, Ohio: ERIC/SMEAC.

Zykova, V.I. (1969) *The Psychology of Sixth-Grade Pupils' Mastery of Geometric Concepts.* In *Soviet Studies in the Psychology of Learning and Teaching Mathematics, Vol. I, 149–187* (Ed Kilpatrick, J. & Wirszup, I. Stanford, California: School Mathematics Study Group.

SECTION 2: *Measurement*

SECTION 2: *Measurement*

2.1 A GENERAL INTRODUCTION TO MEASUREMENT

2.1.1 Introduction

This chapter is concerned with children's learning of measurement concepts and the difficulties they encounter with these ideas. It includes the measurement of physical space in terms of length, area, volume and angle and also the measurement of less easily perceived phenomena such as weight, time and money. These latter measurement systems differ because,

> For example, in coping with mass or weight, the child either (a) reads a scale or (b) observes whether a balance beam is balanced. In (b) not only is the perceptual evidence removed physically from the object being weighed, but the child must incorporate evidence of what he does to the other pan into his perceptual schemata. Consider the case of temperature; the attribute of the length of the mercury column and the position of its end point on a number scale bear little (no) resemblance to the attribute of the object under consideration. (Osborne, 1976)

The whole of this first section is intended as an introduction to measurement concepts in general but many of the examples used to illustrate various points refer to perceptual measurement such as that of length and area because their direct visibility makes them more accessible and hence they have been more widely researched. Also they tend to dominate school mathematics curricula in the field of measurement. Osborne states,

> If you accept the task of examining the research literature concerning the learning of measure concepts, you realize little is known except for the preliminary encounters of children with the more perceptually simple measure systems of numerousness, length and area.... The more complex measure systems, such as volume, mass, rate, energy and work, have not been explored. (Osborne, 1976)

In Section 1 ('Spatial Thinking') emphasis was placed on the importance of becoming familiar with shapes and their properties. It is through experiences with shapes that questions concerning size arise. Once attention is drawn to comparing size, in whatever way, then one is embarking upon the realms of measurement.

One of the problems concerning a child's grasp of measurement is revealed by Brookes (1970) as being the fact that children in our culture are brought up in a world of measurement through the use of sophisticated measuring instruments. They have missed out on the historical development of measurement which means that they do not appreciate the need for measurement in the first place and how this emerged from a 'socially agreed notion of equality' when comparing size, worth, value etc. in trading situations. He says that children are not aware of the subtleties of the notion of the replication of unity, i.e. the repetition of a single unit of measure from which man arrived at number and counting. And from this grew the need for standard units of measure. Children's experience of measurement usually begins with, and is often limited to, number with little, if any, opportunity to explore the earlier principles upon which measurement is founded. Brookes suggests the following to illustrate this:

> Infants are often introduced to measuring by being encouraged to give exact numerical answers when estimations between numbers would be a different activity. We see the consequences in many examples,
>
> 'Go and find how many pints there are in a gallon.'
> Six year old goes off to the tap with pint and gallon 'measures'. He comes back.
>   'Thirteen'
>   'Go and try again'
>
> On his next return—
>   'Twelve'
> ... 'Do it more carefully.'
>
> Eventually he may show the boy there are eight'. (Brookes, 1970)

This boy has counted the number of 'pourings' he has made to fill the gallon measure with no regard as to how full the pint measure was each time. The teacher because of his own 'exact knowledge' was insensitive to the boy's lack of understanding of the meaning of a full measure. Brookes points out that another child may not even appreciate that he can repeat the pourings with only one pint measure and he may request more measures for the activity.

Similarly, Osborne states,

> ... much of the learning about measure is of an incidental nature in today's schools. Measure concepts are encountered in settings where the goal is teaching and learning about number. It is assumed that the measure is intuitive and sufficiently possessed and understood by the learner to serve as an intuitive embodiment for explaining numerical operations. This assumption should be questioned. Further, the nature of how children learn and use measure in this outside transfer setting needs careful attention.
>
> (Osborne, 1976)

We return to this number/measurement relationship later on in this section.

2.1.2   The Approximating Nature of Measurement

Measurement, then, has become too bounded by precision through the exactitude of number for children to grasp any deep insight into the measurement process. School mathematics seems all too often to be restricted to measurement where precision is paramount. Carpenter and Osborne claim that most instructional material appears,

> ... to build a perceptual regularity and nicety (problems come out right) that is missing from real world measurement. (Carpenter and Osborne, 1976).

There is thus a conflict between the real world and the educationally designed world;

and according to Carpenter and Osborne, there is no evidence as to how the child copes with this and how it affects his understanding and learning of measure systems.

As a result such ideas as upper and lower limits of measurement, and estimations within these boundaries, as are necessary for realistic, practical measurement situations, are often lost or their importance minimised. The perceptual reality of measurement always involves approximation and error and this becomes evident when, for example, a length to be measured does not conveniently align itself with the marks on a ruler, or where units of area do not cover a surface in a neat and precise manner, or a scale pan does not exactly balance against a given mass. Unfortunately there is a marked lack of research into the development of measurement concepts concerned with this sort of reality.

An example from Wheeler (1975) serves to illustrate this dichotomy between the educationally designed and the realistic worlds of measurement. He uses the context of the measurement of area. Take for example the task of finding the area of this rectangle:

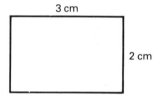

Fig. 2.1

In the educationally designed world what we are really saying (although it is probably never explicitly stated) is that as a theoretical model we consider a rectangle which we say is precisely 3 cm long and 2 cm wide and in which the angles are exactly 90 degrees. Under these ideal conditions what exactly is the area of this figure?

For a more realistic approach to the measurement of area Wheeler uses an irregular shape which does not lend itself to the tools of precise calculation available to the primary and secondary school pupil.

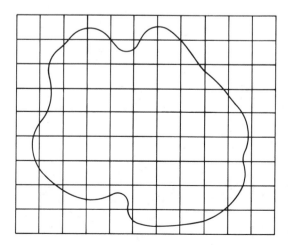

Fig. 2.2

He points out that the whole squares within the perimeter are really countable. Now what to do with the part squares? Wheeler says,

> There are customary rules-of-thumb–say counting one for each part-square which is bigger than a half square and ignoring the others, or counting each part-square inside the perimeter as if it were a half-square. (Do these two rules give the same answer?) We notice that both of these stop the process of measuring instantly and make it impossible to consider improving the answers. Perhaps we should do more with the awareness that we are engaged in approximating (as in all measurement) and let the question of how to make the approximation 'better' come to the surface. (Wheeler, 1975)

Wheeler then suggests returning to the point of having counted the number of complete squares within the figure and regarding this as a 'lower bound' to the area in question i.e. a measure which is definitely less than the one we are supposedly after.

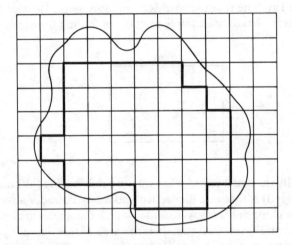

Fig. 2.3

Now we can continue in this vein and have an 'upper bound' i.e. counting not only all the whole squares which are totally within the perimeter but also those which are partly within. This will give us an area which is definitely more than we want.

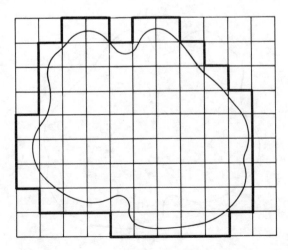

Fig. 2.4

It can now be said with certainty that the area of the figure lies somewhere between the lower and upper bounds. Given that these two 'bounds' will be quite far apart it can be seen that there is a need to find some way of reducing the gap between them. Reducing the measuring unit is one way of doing this, i.e. using a finer grid. Wheeler states that,

> Although there is a practical limit to the extent we can subdivide, a limit reached pretty quickly, the action of doing it once can generate in us the awareness that it could be repeated at the virtual level (i.e. in the imagination) as many times as we want ...'

He thus develops some notion of the approximating nature of measurement.

2.1.3 What is Fundamental to the Measurement Process?

The work of Piaget has made a significant contribution to our understanding of the development of measurement concepts in the child and again has provided an invaluable foundation for debate, controversy and need for further research in this area. Piaget identifies two fundamental operations upon which the measurement process is dependent. These are conservation and transitivity.

2.1.3.1 Conservation

The notion of conservation was introduced in the previous section on 'Spatial Thinking' (page 49) and concerns the invariance of certain crucial aspects of a situation. For instance the length of a corridor remains the same whichever direction one walks along it or runs or jumps along it. The tea in a packet will still be the same weight whether it is in the packet or all poured into a caddy. Six eggs in an egg carton are still six eggs if they are all taken out and arranged in the door of a refrigerator. The appreciation of the invariant aspects of a situation is basic to the development of the measurement process. Further discussion concerning conservation appears throughout this section within the context of each measure system discussed.

2.1.3.2 Transitivity

The notion of transitivity is best illustrated by an example. Suppose a child is shown a tower built from bricks and is asked to construct another tower of the same height some distance away on a base which is at a different level. He has at his disposal some bricks and a stick which is at least as long as the height of the tower. If he shows an understanding of the use of the stick as a measuring instrument by using it to measure the height of the original tower and then building his tower to this mark on the stick then he is developing the notion of transitivity. Symbolically this is often expressed in the following way. If the height of the original tower is A and the length marked on the measuring stick is B and the height of the tower he constructs is C then if he performs the process correctly we have the situation where is he displaying his grasp of the fact that if A=B and B=C then A=C i.e. his tower is the same height as the original by virtue of his use of the intermediary B (the measurement on the stick) as a comparison.

Whatever the measuring situation, the meaningful use of an instrument of measure rests on this notion of transitivity. Further discussion of this appears below in the outline of Piaget's developmental theory of measurement concepts.

However, it should also be noted that transitivity is also an exact 'idealised' concept, and does not take into account the approximate nature of measurement. For instance, a carpenter wishing to cut ten equal pieces of wood each 1 metre long could theoretically measure each new piece against the one he has just cut . However, in practice this tends to a cumulative error, so that the tenth piece is quite likely to be too long, or short, by a significant margin. In practice, the carpenter goes back to the standard measure each time to avoid this danger.

2.1.4   What are the Stages of Development in the Child's Grasp of Measurement? –The Piagetian Position

Most of the research concerning the development of measurement concepts emanates from Piagetian studies and relates mainly to the measurement of spatial entities such as length (Piaget, Inhelder and Szeminska, 1960).

2.1.4.1   Initial Stages

Initially the young child often up to the first year or two of infant school displays no grasp of conservation. His judgments are based primarily on a single perceptual feature. Consequently in the classic conservation of length test the child judges these lines to be unequal because their end points are not aligned.

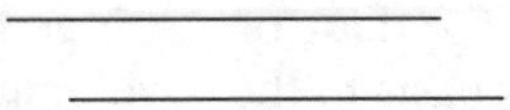

_____

Fig. 2.5   _____

Area and volume judgments are usually based on the longest linear dimension ('it's bigger because it's longer').

Typical of these earliest stages is the child's reliance on visual estimates. For example, with the activity involving the building of a tower of the same height as a given one the child will move them closer together to improve the visual comparison, but he only concentrates on the tops of the towers ignoring the different levels of the bases. He is unable to apply any measuring instruments with any meaning. If given a unit of measure he will either overlap the units, leave gaps or only cover part of the dimension to be measured. He displays no understanding of the reiteration of a unit or of subdivision into equally sized sections. He lacks conservation and transitivity of the moving middle term.

2.1.4.2   The Stage Where Conservation and Transitivity Begin to Emerge

The child shows he is beginning to develop some idea of conservation and transitivity when he uses a crude middle term as a measuring instrument, such as arm span or some reference point on the body such as the height of his shoulder. This often happens roughly round 6–7 years of age. The child at this stage is also starting to appreciate by trial and error experimentation, that if it takes more units to cover A than it does to cover B then A is greater. He still cannot realise the necessity of the units of measure all being the same size nor can he coordinate measures in all dimensions.

2.1.4.3 The Stage Characterised by the Onset of Operational Conservation and Transitivity

Once the child can measure the height of a tower and construct one of a similar height using a measuring stick which is at least the same length as the height of the tower and he can mark the height on this with his finger, or a pencil line or a sticker etc., and use this to guide his construction of the second tower, then he has reached Piaget's third stage of development. However, at this stage, reached typically at around 7 or 8 years of age, the child cannot use a measuring instrument which is shorter than the tower.

He is now beginning to appreciate 2-dimensional measurement in the sense of area enclosed by a boundary and can conserve amounts of matter, for example a quantity of liquid. If a certain amount of liquid is poured from a tall, thin vessel into a wider, squat one he is beginning to grasp that the amount of liquid is the same despite the initial visual impressions being to the contrary. He can begin to coordinate the two dimensions e.g. although one container is wider, at the same time the water level is lower, and this may compensate for the extra width.

2.1.4.4 The Stage Where the Idea of Smaller Units of Measure is Grasped

It is not until roughly around 8 to 10 years of age that the average child can appreciate measurement in terms of covering whatever is to be measured, with smaller units of measure. Up to this stage, development of the measurement process has been characterised by a trial and error approach. Now the child is able to proceed by means of a more calculated approach. Piaget claims that so far the development of linear, area and capacity concepts have occurred concurrently. However the measurement of volume in terms of the amount of space occupied by a particular object lags behind. This is because it is not possible to cover or fill the occupied space with units of measure because the object e.g. a potato, is solid, impenetrable and 'in the way'; the surface alone is visible. More is said later in this section about the distinction between capacity and volume.

2.1.4.5 The Final Stage in the Development of Measurement Concepts

It is not until the child has reached the onset of this stage of Piaget's formal operational thought that he can be said to be fully operational in his grasp of measurement concepts. Piaget has found this to be around the age of 11 or 12 years but many studies in Britain and the USA, e.g. Shayer et al. (1976, 1978) would put this at a much later age for most children and above the school leaving age, if at all, for many.

The child who has reached this stage of development is able to measure area and volume by calculation based on linear dimensions. According to Piaget, the real hallmark of this final stage is the awareness of space as a continuum containing an infinite and continuous set of points. He claims that it is not until concepts of infinity and continuity develop that measurement concepts become fully operational. This assertion of Piaget's is currently being questioned by many researchers but much more research is required.

2.1.5 Summary and Appraisal of the Piagetian Position

Piaget claims that length and area concepts are the first to develop and these occur simultaneously at about 6 to 7 years. Next, the child begins to grasp the idea of the conservation of mass (substance) and weight at approximately 7 to 8 years and 9 to 10 years of age respectively. Conservation of volume does not develop until 11 to 12 years of age. Although this hierarchy of development has been substantiated by many studies, Carpenter (1976) does question the concurrent development of length and area concepts.

Rothwell Hughes (1979) claims that his results from a survey of area, weight and volume concepts in over 1,000 primary school children in England and Wales show that there is as much variation between different aspects of conservation of each quality, as there is between the three quantities.

There is also much evidence of age variation. For instance Fogelman (1970) summarises studies in several countries replicating Piaget's conservation of weight experiment. These in general support 9 to 10 years as the average age of attainment, but generally suggest that around 25 per cent of all 11 year olds, and, less certainly, 15 per cent of 12 year olds, are still without conservation of weight. Rothwell Hughes (1979) also pointed out that 'age gives little guidance to the ability of the children to conceptualize' and found that, of those aged 9 years 11 months with IQ less than 89, about 40 per cent could not conserve weight, and up to just over 50 per cent, depending on the context, could not conserve area.

The NAEP national survey in the United States (Carpenter et al., 1980) found that most 13 and 17 year olds were familiar with basic measurement concepts and skills involving one dimension i.e. simple length measurement. However many had not grasped basic area and volume concepts. There was a tendency to use formulae in a rote fashion and even with the concepts and skills they had learned there was much difficulty in applying them to even relatively simple problem situations.

Carpenter and Osborne identify one aspect in the development of measurement concepts which is critical yet little researched. This is the transition from the filling of space with the reiteration of a unit to the approach of multiplying linear dimensions as in area and volume typical of Piaget's final stage of development. Can the child understand the relationship between, say, counting the number of unit squares required to fill a given area

Fig. 2.6

and the area formula, Area = Length x Width?

2 units

Fig. 2.7

4 units

Or as Dienes points out,

> Children repeat the phrase 'length times width equals area', but they cannot see how inches are suddenly turned into square inches as though by magic, by a mere multiplication. Suddenly, the lengths have become an area, whereas just now they were just lengths. If one allows them to build a rectangle out of, say, 24 little squares, they will soon see that, for example, there are 6 rows of 4 or 4 rows of 6. It does not need many such exercises before they realise that the number of rows by the number in each row gives the total number of squares. ... What the ... child cannot do is to identify the number of rows with the width, and the number in each row with the length, say. (Dienes, 1959)

Carpenter and Osborne (1976) also point out that the Piagetian research is limited in its usefulness for offering some model of conceptual development which will generalise across all measure systems. It does not throw light on how understanding of the measurement of length, say, transfers to developing an understanding of the concepts involved in the measurement of volume or time. Although conservation and transitivity are common elements of all operational measurement, how the attainment of these concepts in one system of measurement affects their development in another is not known.

Also there is some difference of opinion in relation to the sequence of events in the development of measurement concepts. Does the child have to be able to conserve quantity and display a grasp of the notion of transitivity before he can be taught how to measure the size of this quantity? Carpenter states,

> It is generally accepted that appropriate training can accelerate the development of specific measurement concepts. Furthermore training in measurement seems to accelerate rather than depend upon the development of concepts of conservation and transitivity. (Carpenter, 1976)

2.1.6 The Unit of Measure

It has been mentioned several times that measurement involves the repetition of a unit of measure i.e. some notion of subdivision in terms of some unit of measure being repeated throughout the extent of whatever is under consideration whether it be area, time etc. and this repetition must be such that the interval to be measured is covered or filled by the unit so that there are no gaps and no overlapping. Also one of the distinguishing features of the measurement process is that different units may be used to measure the same quantity. So another crucial aspect in the development of measure concepts is the realisation of the relationship between the size of the unit and the number needed to measure a given quantity, i.e. the smaller the unit the more that will be required. Secondary children were given the following item as part of the Concepts in Secondary Mathematics and Science Project–CSMS (see Hart 1978/81). It shows that many children could not appreciate the need to know the relative sizes of the units being used.

John measures how long paths A and B are, using a walking stick.
Then he measures how long C and D are, using a metal rod.
The answers are: Path A : 13 walking sticks
                    Path B : 14¼ walking sticks
                    Path C : 15 rods
                    Path D : 12½ rods
Draw a ring round the answer you think is true in each question.
(a)   Path B is longer than Path A: True/False/Cannot Tell.
(b)   Path C is longer than Path B: True/False/Cannot Tell.
(c)   Path D is longer than Path C: True/False/Cannot Tell.

51% of 1st year secondary pupils ⎫
34% of 2nd years   ⎬ judged (b) to be true.
27% of 3rd years   ⎭

e.g., Paul, aged nearly 13 years,

Paul:	'Yes it's got more.'
Interviewer:	'What's got more?'
Paul:	C's got 15 and B's got 14½.'
Interviewer:	C's got 15 what?'
Paul:	'Rods.'
Interviewer:	'What about B?'
Paul:	'It's got 14½. Sticks.'
Interviewer:	'So?'
Paul:	'C's longest.'

(Hart, 1978)

Carpenter (1976) studied the effect of using different units of measure for measuring a fixed quantity of liquid. He was interested in children's performance in comparing equal quantities of liquid where they were told the quantity of each in terms of different units of measure (e.g. 5 decilitres or 500 millilitres); despite this apparent numerical discrepancy in quantity they were confronted with what could clearly be seen to be equal amounts (see the following figure.)

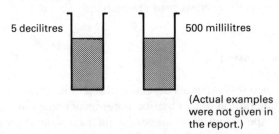

5 decilitres   500 millilitres

(Actual examples
were not given in
Fig. 2.8   the report.)

Carpenter contrasted the performance of 6 and 7 year olds in this situation with that in the Piagetian classical conservation test where equal quantities look different,

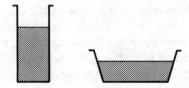

Fig. 2.9

It was found that in the first situation children tended to judge the quantity measured with the smaller unit as being more. For example children might judge 500 mls to be more than 5 dls.

Carpenter found that most children were just as likely to be distracted by the numerical cues, as in the first situation, as by the perceptual ones, as in the standard Piagetian test.

However, in a further part of the study, almost all Carpenter's 6 to 7 year olds,

> could correctly identify the relation between quantities in different shaped containers that were subsequently measured with a single unit. These results seem to imply that (6–7 year old) children do naturally attend to the results of measurement operations, and measurement operations used appropriately may facilitate conservation judgments. (Carpenter, 1976)

So, for example if these children were told that A contained 5 dl and B contained 5 dl then they are more likely to judge the quantities equal than if they just relied on the visual appearance without any measures being given.

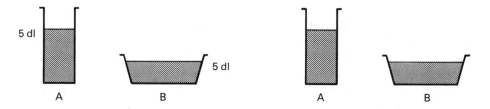

Fig. 2.10

Further studies with 6 and 7 year olds also indicate that although children of this age are unable to apply a compensatory relationship to a situation, i.e. if smaller units are used then more are required; nonetheless they do show evidence of realising that such a relationship exists. Carpenter found that,

> A significant number of (children) who failed the above measurement problem in which equal quantities were measured with equal units could successfully predict that a quanity measured with the larger unit would measure fewer units than an equal quantity measured with the smaller unit. (Carpenter, 1976)

2.1.7   Children's Understanding of Standard Units of Measure

2.1.7.1   *Familiarity with the Size of Standard Units*

For measurement to be meaningful and communicable it is necessary for children to become familiar with the standard units of measure. Not only must they know which units relate to which system of measure (e.g. that litres are a measure of volume or capacity) but also the relationship between different sized units within any one system of measure (e.g. 1 litre is equivalent to 1000 cubic centimetres or 1000 millilitres).

NAEP (See Carpenter et al., 1980) found that American pupils performed much better with their more familiar imperial units of measure. (USA does not employ metric measures to the same extent as Britain.) At the ages of 13 and 17 years there were about 35 percentage points difference between performance with customary units and performance with metric units of measure. Only just over a third of the 13 year olds and about half of the 17 year olds were able to make even fairly reasonable estimates of weight and length in metric units.

Furthermore for a child to employ a unit of measure in an appropriate way he must have some idea of and some feeling for what it actually represents. This enables him to estimate, or to judge whether a particular measure is reasonable. For example something roughly the same mass as a packet of granulated sugar is about 1 kg; ten minutes is about the period of time it takes him to walk to school etc.

The first APU survey reports (1980a,b) indicated that a quarter of 11 year olds did not have a meaningful grasp of the 'metre' by virtue of their selection of 30m or 200m as the likely height of a man from a list comprising 2m, 20m, 200m, 2000m. Seventy per cent of these 11 year olds correctly chose 2m as compared with about 80 per cent of the 15 year olds.

Another APU item required 11 and 15 year olds to choose the most appropriate height for a table from the list 10 millimetres, 1 centimetre, 10 centimetres and 1 metre. About twenty per cent of the 11 year olds and ten per cent of the 15 year olds did not select 1 metre.

On a similar item Ward (1979) found about forty-four per cent of 10 year olds unable to make the best guess for the height of a table from 1 cm, 10 cm, 1 m, 10 m.

The APU (1980b) found that fifty-five per cent of 15 year olds were able to choose an appropriate weight for an elephant from a list ranging from 2 tonnes to 2000 tonnes, and that eighty per cent of 15 year olds selected a reasonable temperature for a warm room from a range of 2°C to 2000°C.

2.1.7.2   Relationships between Standard Units

The following table adapted from the first APU results (1980a,b) indicates pupils' performance in relating similar units of measure.

APU Primary – 11 year olds – (1978 Survey)
50% indicated that 1Kg was equivalent to 1000 grams
30% indicated that 100 grams was 1/10 of one Kg.
40% indicated that 8 oz was equivalent to ½ lb.
75% indicated that 30 seconds was equivalent to ½ minute.
75% indicated that 15 minutes was equivalent to ¼ hour.
50% indicated that 1000 ml in a litre.

APU Secondary – 15 years olds – (1978 Survey)
77% knew 1m = 100 cm.
50% knew 1mm = 1/10 cm.

2.1.8 The Relationship between Measurement and Number Concepts

2.1.8.1 *The Distinction between Counting and Measuring*

We saw at the beginning of this section (page 79) that Brookes (1970) takes an historical stand in regarding the basis of measurement as the reiteration of a unit of measure, leading to a situation where number concepts and counting have to be developed. Therefore an operational concept of measurement cannot exist apart from the related number concepts. However, the associated number concepts in the measurement process involve more than just counting in the ordinary sense. Leaving aside the difficulties arising when different units are used to measure the same quantity, the units themselves may be indistinguishable from each other in the measurement process. Ordinarily, counting is concerned with what are termed 'discrete variables' i.e. it applies to a situation where each individual unit to be counted is a separate and distinguishable entity, it is the assigning of a number to a set (e.g. how many people are in the hall? – count heads; how many quarters make a whole? – count four; how many bricks will fit in this box? – count the bricks).

However, supposing we want to divide a real cake into four quarters. How do we measure a quarter of the cake? We could call upon a visual approximation of what looks like a quarter in terms of volume or surface area if the cake appears to be symmetrical and of uniform depth. But each unit of measure of volume or surface area is not individually distinguishable in the same way that it is in a counting situation. Therefore such measurement can only approach precision and the more refined and sophisticated the measuring instrument the nearer we can get to exactitude. We may decide that weighing the cake is a more accurate way of dividing it into quarters but again the more sensitive the weighing machine, the more accurate our measurement. 'How much?' is in this case a very different question from 'How many?'

In any situation where measurement can be refined by smaller and smaller units of measure then measurement is concerned with what is called a 'continuous variable'. This further explains the point at the beginning of the chapter where it was noted that the provision of educationally designed problems typical of school mathematics detracts from the true nature of the measurement process. Dividing a rectangle exactly into square units imposes a 'discrete' type of measurement on the situation – simply count the squares. In reality a certain number of squares may just cover the area with a bit left to cover or may just overlap being a bit too much. One can envisage a situation, such as that given by Wheeler (1975) where refining the measuring unit to smaller and smaller squares leads to a closer approximation of the area and shows measurement to be 'continuous' in nature.

A consequence of this fact that measurement deals with continuous quantities is that the unit of measurement has to be repeatedly divided into smaller units e.g. metres are subdivided into centimetres, which are further subdivided into millimetres, and so on. In practice, in classroom measuring situations a unit is reached which cannot realistically be further divided without very sophisticated instruments; in the case of length this is the millimetre, in mass the gram, and in capacity the millilitre.

The alternative to working in terms of such small units is to use the corresponding larger units but to introduce fractions or, more realistically, decimal fractions to deal with the 'extra bits'. This however requires an understanding of the nature and symbolism of fractions/decimal fractions, and therefore introduces an extra complexity. For instance Brown (1981b) showed that interpolation to one decimal place was not possible for many low attainers, e.g.

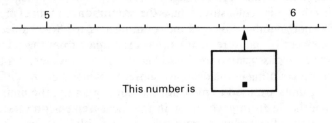

Fig. 2.11

Age	12	13	14	15
Facility	62	74	83	85 per cent

and the situation was rather worse if the number of subdivisions given on the scale between two whole numbers was five, rather than ten, or if more decimal places were involved, e.g.

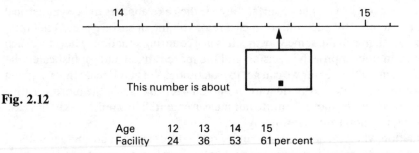

Fig. 2.12

Age	12	13	14	15
Facility	24	36	53	61 per cent

Although the most common application of decimals is to measurement, Brown showed that if children were asked to 'write a story to match the sum $6.4 + 2.3 = 8.7$' only thirty per cent of 12 year olds rising to forty-one per cent of 15 year olds could think of an appropriate context, even after they had been given examples of such contexts in previous questions. She also found several children who seemed to have little conception that decimals could be used for measurement or indeed anything other than routine computations e.g. Raymond (12) gave the following replies:

Interviewer: (after long pause) Might you use decimals in a job?
Raymond: No, I don't think so.
Interviewer: Or at home? (long pause) What about in another lessons in school?
Raymond: Not really ... you only do them in maths lessons.

The actual process of measurement involves having some 'feel' for the situation in terms of its size. The process requires decisions to be made as to what degree of accuracy is called for and hence how small a unit of measure and how refined a measuring tool is needed. Often, these aspects of the measurement process i.e. judgments of estimation, approximation etc. are never realised at the classroom level because of the preoccupation with the number/counting part of measurement. There are two schools of thought concerning whether or not number concepts should play the dominant role in the development of measurement concepts and these viewpoints are now outlined in terms of instructional techniques for the teaching and learning of measurement. They are reported by Carpenter (1976) and relate to (i) the Piagetian school and (ii) studies from Russia.

2.1.8.2 The Piagetian Position

The Piagetian School (Inhelder, Sinclair and Bovet, 1974) based their training procedures concerning measurement on their evidence that the conservation of number (e.g. 5 counters are 5 counters however they are arranged spatially) is acquired by the child some couple of years before he can conserve length. They 'hypothesized that elementary concepts of linear measurement could be facilitated by exercises in which numerical operations could be used to evaluate length'. Therefore their position is one of using the child's grasp of number concepts upon which to develop measurement concepts.

Their programme of instruction was devised to induce conflict in the child. They see conflict as being a major factor in the development of fundamental operations and they induce it by first presenting the child with the most difficult of their tasks. Each child was studied in a one-to-one clinical setting typical of the Piagetian approach.

All the 6 year olds who participated in the following programme could conserve number, but none could conserve length before their individual training sessions. They were shown the following figures, each made from 5 matchsticks, in the sequence:

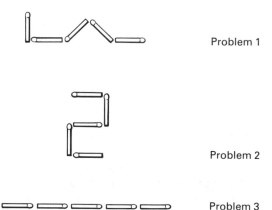

Problem 1

Problem 2

Fig. 2.13   Problem 3

Their task was, for each figure, to construct a straight line using a different set of matchsticks so that the length of their line was the same as that of the total length of the line in the given figure. The difficulty was that the matches given to the child for his construction were shorter than those used in the figures – in the ratio 5:7. All three problems remained in the child's view. The child, after completing the three constructions was asked to explain what he had done and was encouraged to reconsider earlier solutions.

It was found that,

> The first task generally induced an incorrect response based strictly on length. The second task generally induced an incorrect response based strictly on number. The basis for the two responses were in direct conflict. It was not until the third problem that sufficient information was provided to resolve this conflict.

Following this training session children were again tested for their ability to conserve length. Thirty-five per cent of them had made no progress. The responses of these children indicated that they only centred their attention on one dimension at a time, either length in terms of alignment of end points e.g.

Fig. 2.14

or number of matchsticks. They saw no contradiction in using different criteria.

The 37 per cent who made some limited progress by the post-test seemed, during training, to be aware of a contradiction in their method but were unable to resolve it. Some offered compromise solutions such as breaking a match to ensure the same number but with the same alignment of end points. Others ignored the instructions and constructed a path which was not straight.

The remaining 28 per cent of the children showed conservation of length in the post-test. They reached it during the training sessions where they recognised that more matches were needed when they were shorter and were able to use their results from problem 3 to correct their attempts at problems 1 and 2.

The Piagetian school advocates that it is the child's active efforts to discover compensatory and coordinating actions and not just the visual results of the tasks which lead to a grasp of the concepts.

2.1.8.3   The Russian Approach

The Russian investigation into the development of measurement concepts in children (Galperin and Georgiev, 1969) differs from that of the Piagetian school in two main respects. Firstly their training was carried out in the classroom setting and secondly they approached the teaching of measurement concepts not from a grounding in basic number concepts but from measurement concepts per se. This approach mirrors the historical development of measurement.

The Russians argue that young children who are taught by the traditional emphasis on number concepts do not recognise that a unit of measure

> … may not be directly identifiable as an entity and that the unit itself may consist of parts. They are indifferent to the size and fullness of a unit of measure and have more faith in direct visual comparison of quantities than in measurement by a given unit. (Carpenter, 1976)

This has been substantiated by American studies with 6 year olds.

The Russian training programme involved 50 kindergarten children aged 6 to 7 years, whose entire mathematics curriculum was based on the concept of a unit of measure. It comprised 68 lessons divided into 3 groups.

1. The first series of lessons introduced basic measuring skills and spatial concepts. They concentrated on comparing various quantities by a systematic application of measuring units thereby trying to overcome the misleading habit of simple visual comparisons. The programme involved the identification of appropriate measuring units according to the situation. A variety of units was employed such as matchsticks, spoons, cups, etc. Not only were these things used as single units of measure but were arranged to be units of several parts e.g. a unit of three matchsticks, and also fractions were used as a unit e.g. half a cupful. All these activities were performed at this stage without assigning numbers to the quantities e.g. fill this bottle with spoonfuls of water.
2. During the second part of the programme the concept of number was introduced. How long is this line in matchsticks? How many half cupfuls of sand are needed to fill this container? etc.
3. The third part of the training sessions was aimed at getting the children to discover the inverse relationship between the size of the unit used and the number of units required: so if a line were measured in terms of matchsticks then there would be more matchsticks than if it were measured in terms of drinking straws. The larger the unit of measure the fewer required.

It was found that these children made significant progress in their development of measurement concepts as compared to other groups not having this training programme. In fact the average success rate on 15 measurement tasks was 98 per cent as compared with 41 per cent in the control group.

An unexpected bonus was that children began to appreciate the nature of multiplication and division, for instance in the task in which a 'unit measure' was defined as 3 beads, and children were asked to find how many 'units' were in 8 beads.

Carpenter considers this approach to teaching measurement concepts as worthy of very serious attention particularly with respect to the much longer duration of the programme as compared with the Piagetian approach.

This general introduction to the child's development of measurement concepts has served to highlight crucial aspects involved in the process i.e. estimation and approximation, conservation and transitivity, and the repetition of the unit of measure. These themes recur throughout the remainder of this section where individual measure systems are discussed in relation to the child's learning difficulties in the areas of Length and Area, Angular Measurement and Geometrical Constructions, Weight and Volume, and Time and Money.

2.2 LENGTH AND AREA

2.2.1 Length

2.2.1.1 *Conservation of Length*

When a line alters position in space children sometimes judge its length to have changed as well. The child who makes this judgment is considered unable to appreciate conservation of length. Reference to this was made in Section 1 (pages 51 to 58), where examples concern the Piagetian classical conservation of length test within the context of transformational geometry. For instance when children were required to construct the image of, say, a triangle following a certain transformation, for example, a reflection, they often neglected to ensure that the sides of the image were the same length as the original.

Another facet to the question of conservation of length is the situation where children judge lines to be equal in length when in fact they are not. Piaget has found this to be particularly evident when the end points of the lines under comparison are in alignment. This is illustrated by results of a study carried out by the Concepts in Secondary Mathematics and Science Project (CSMS). Hart (1981) gives instances from the work of this project which reflect the lack of understanding experienced by many children on this aspect of measurement.

One item was the following:

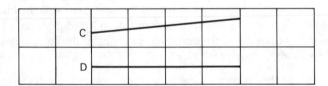

Fig. 2.15

		12 yrs	13 yrs	14 yrs
(i)	Line C is longer ...	42%	45%	52%
(ii)	Line D is longer ...	0%	1%	0%
(iii)	C and D are the same length ...	48%	48%	45%
(iv)	You cannot tell ...	4%	2%	1%

Another item comparing a segment and curve led to a higher success rate as shown.

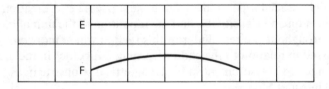

Fig. 2.16

		12 yrs	13 yrs	14 yrs
(i)	Line E is longer ...	1%	0%	0%
(ii)	Line F is longer ...	72%	78%	82%
(iii)	E and F are the same length ...	20%	16%	13%
(iv)	You cannot tell ...	6%	5%	3%

A further CSMS item was the following:
    The following 8-sided figure A is drawn on *centimetre square* paper

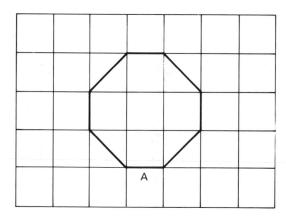

Fig. 2.17

Draw a ring round the correct answer:

The distance all round the edge of A is:

8 cm   more than 8 cm   less than 8 cm   you cannot tell.

Responses were:	*12 yrs*	*13 yrs*	*14 yrs*
8 cm	43.2%	43.2%	41.6%
more than 8 cm	38.5	36.7	46.9
less than 8 cm	13.6	14.9	9.1

Hart (1978) points out that there are quite serious implications to these results. One, in particular, is that Pythagoras' Theorem can have very little meaning for many children to whom it is introduced, if they start with the preconception that the hypotenuse is equal to one of the other sides, as in the following figure.

Fig. 2.18

2.2.1.2 *Using a Scale for Measuring Length*

The measurement of length usually involves the comparison of the length to be measured with some standard scale such as a ruler or calibrated thermometer tube where the length of the mercury column is measured. In this latter case the length to be measured is already in position against the scale but for practical measuring situations involving say a ruler, then the line segment has first to be appropriately positioned before a reading is taken.

The first APU Primary Report (1980a) asked 11 year olds:

What is the reading on each of these thermometers?

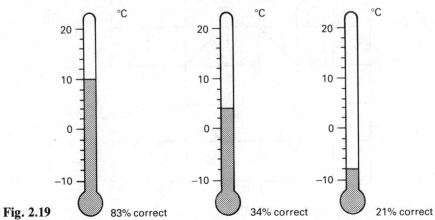

Fig. 2.19   83% correct   34% correct   21% correct

> It can be seen that over 80% of pupils were able to read a temperature correctly if its value was marked on the scale but the facility fell to around 30% if the temperature to be read was at an unmarked point or between the scale points which were graduated at 2° intervals. A negative temperature produced a further decline in the success rate to about 20%, a further 27% reading up the scale from −10°, ignoring the negative sign, and giving the temperature as 11° or 12°.

Eighty-five per cent of the 11 year olds in the APU practical test successfully measured the length of a 13 cm line to within 5 mm. However five per cent aligned the end of the line with the '1' on the ruler thus obtaining an answer one centimetre longer than it should have been. Other children aligned the edge of the ruler with the end of the line regardless of the position of the zero mark on the scale.

Hart (1980), as part of the CSMS Measurement Study, showed the following line to secondary pupils.

The marks on the line show centimetres. How long is the line?

Fig. 2.20

The answers given were:

	12 yrs	13 yrs	14 yrs
9 cm	79%	85%	90%
10 cm	18%	13%	10%

Those giving the answer of 10 cm seemingly counted the marks rather than the repeated unit of length.

On another CSMS item, pupils were required to give the length of the line drawn above the scale

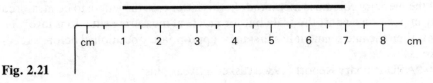

Fig. 2.21

The responses were:

	12 yrs	13 yrs	14 yrs
6 cm	49.1%	64.9%	76.1%
7 cm	46.2%	30.6%	22.8%

The answer 7 could have occurred because the right hand end of the segment was aligned with the 7 on the scale or because the number of marks, rather than gaps, were counted from the 1 to the 7 inclusive.

Another aspect concerning the use of a linear scale of measure is the child's ability to construct a scale. This often occurs within the context of graphical work where axes of reference are required for defining points in space by means of coordinates.

Ward (1979) asked British ten year olds to show the following information on a graph. They were provided with a 6 x 6 square grid.

These were the temperatures at 9 am. for five days during a week:

Monday	12°C
Tuesday	10°C
Wednesday	5°C
Thursday	11°C
Friday	11°C

Most children used the vertical axis for temperature and the horizontal for the day with each day placed in the appropriate column space thus giving a bar chart. The vertical scale was constructed in various ways, which included:

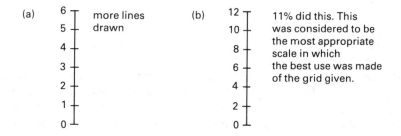

Fig. 2.22

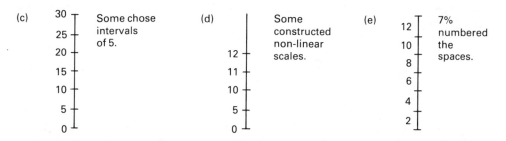

These last two examples show a basic lack of understanding of measurement in terms of the display of the reiteration of a unit of measure on a linear scale.

Once they had chosen their scales 28 per cent were able to mark the temperatures correctly.

2.2.1.3 *Practical Measurement of Length and Estimation*

Ward also asked British 10 year olds to describe what they would do to find the height of the tallest person in the class. There turned out to be a certain ambiguity in the way the task was phrased. Thirteen per cent of the children took it to mean that the tallest person had been identified and the task was to measure him. Others interpreted it as it was intended i.e. that the tallest person was not already known – he had to be found by measurement. Twenty-two per cent suggested measuring all the class. Six per cent suggested measuring 'a selection of the tallest selected by sight'. The most common method suggested by eleven per cent was standing up against a wall or post.

> Some children with a feeling for accuracy mentioned that shoes should be taken off before the measurement.... The name of a measuring instrument such as a ruler, metre stick, string or tape measure, was mentioned by twenty-one per cent of the children. (Ward, 1979)

The APU (1980a) conducted practical tests with 11 year olds where interviewers investigated children's ability to estimate and measure the length of straight and curved lines. About 75 per cent of the pupils could estimate the length of a 13 cm line to within 3 cm. However some answers reflected serious difficulties experienced by some children in their ability to estimate meaningfully. For example some of the estimates for the 13 cm line includes: 25 inches, 16 millimetres and 60 centimetres. Sometimes pupils did not seem to appreciate the need for giving any units at all. For example:

```
Pupil          '17'
Interviewer    '17 what?'
Pupil          '17.3'
```

After measuring straight lines with a ruler these pupils were then asked to estimate the length of curved lines and gained 80 per cent success. One child commented: 'I remembered the distance on the ruler and I put it against these (lines) in my mind'? This illustrates how familiarity and experience with the units of measure facilitate the ability to estimate and develop an overall feeling for the measurement situation.

In another practical APU test, 11 year olds were asked how they could measure curved lines

Fig. 2.23   (a)   (b)

Most suggested string but they report that one boy '... hardly believed a curved line could be measured. Another simply put a ruler straight from one end to the other.' The APU reports the following strategies adopted by pupils for measuring with string.

> When following the curves with string most pupils were careful and accurate; three quarters were successful, allowing ± ½ cm. Answers were given in both the forms '8 cm and 5 mm' and '8.4 cm, say'. Some pupils tended to make straight line sections round the curve and were generally rather clumsy. Some marked the string where the line ended while others just held the place with their fingers. Some pupils matched the string along the curve and held the end, but could not think to transfer it to the ruler until prompted.

It appears from these strategies that some 11 year olds had not grasped the idea of measuring length in terms of having to use some means of covering the curved line, and were confusing it with distance between the end points. Others seemed not to have fully grasped the crucial aspect of transitivity in terms of transferring the string to the ruler.

The National Survey in the USA (NAEP–Carpenter et al., 1980) found that only 53 per cent of 13 year olds could measure the length of a line to the nearest ¼ inch whereas 81 per cent of the 17 year olds could do so.

2.2.1.4   Measuring Perimeter

The NAEP survey also found that approximately one-third of 13 and 17 year olds were unable to measure the perimeter of a triangle. This corresponds quite closely to the results of the first APU Secondary Survey (1980b) in which 15 year olds were asked to estimate the perimeter of a rectangle as part of one of the practical tests. Sixty per cent gave an estimate of between 18 cm and 28 cm for the perimeter – which was about 22.6 cm. Ten per cent gave estimates below 15 cm and a further ten per cent gave estimates above 30 cm. Ten per cent did not give units with their estimates. When they actually measured the perimeter just over fifty per cent were correct to the nearest millimetre with a further twenty per cent just one millimetre out.

Pupils were also asked how they could check their estimates for the circumference of a circle. Most initially suggested using the formula but often seemed confused as to whether it was $C = \pi r^2$ or $C = 2\pi r$ or $C = 2\pi r^2$. Following on from this about half the 15 year olds suggested using string, about 10 per cent a protractor and about 10 per cent suggested compasses. If they managed to obtain two different answers they were asked to consider which was the more accurate. Sixty per cent judged the formula method to be better for such reasons as it's 'difficult to fit the string on the circumference'.

2.2.1.5   Children's Performance on Non-Practical Perimeter Problems

Many of the problems which children meet concerning perimeter are of a neatly designed standard textbook variety. As discussed earlier these problems do not offer the opportunity for the child to exercise any real insight into the measurement process and may often mislead in assessing the child's understanding of the situation. This is particularly evident when the task involves finding the perimeter of a figure where all measurements are indicated as compared with relatively simpler figures where certain measurements are not given but have to be deduced from others through the geometry of the figure.

For instance Ward (1979) found that 64 per cent of his 10 year olds gave the perimeter of this figure

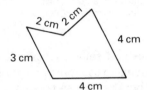

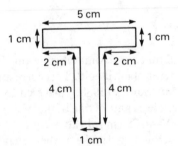

Fig. 2.24

In the first APU Primary Survey (1980a) there was a higher success rate for finding the perimeter of

Fig. 2.25

than for finding the perimeters of squares and rectangles where only minimum information was given. With reference to finding the perimeter of the T-shape figure, APU reports that:

> It is notable, that in contrast to the rectangle examples, all the lengths are given along the perimeter, and the higher success rate may indicate that in this kind of example some pupils need all the lengths to be stated explicitly if they are to get them right.

In the first APU Secondary Survey (1980b) when 15 year olds were 'given a diagram of a rectangle with sides of 24 cm and 11 cm marked on it, just under 60% of the pupils selected the correct value of the perimeter from alternatives including the size of the area, which was selected by 25%'. When asked for the perimeter of a more complex shape – a semi-circle attached to a square – only 10 per cent were correct. A similar low level of success was found for an item requiring the length of the arc for a 60° sector of a circle.

The NAEP Elementary Report (Carpenter et al., 1980), posed an interesting situation to 9 and 13 year olds which involved their performance on three perimeter items which although mathematically identical were presented in three different ways. Two of the presentations involved a picture of the rectangle:

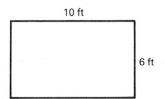

Fig. 2.26

the only difference between them, being that one involved the word 'perimeter', the other asked for 'the distance all the way round'. The third presentation was as a 'word problem' in a familiar context but with no diagram. (The complete questions and results are given in Section 4 ('Language – Words and Symbols' on page 344.)

The percentage of children giving the correct answer ranged from 8 per cent to 40 per cent for nine year olds, and 31 per cent to 69 per cent for 13 year olds, depending on the format. It is interesting that a higher percentage of 13 than 9 year olds confused perimeter and area for whatever reasons: not least that they probably have a greater facility for performing the required multiplication than do the 9 year olds which opens up their options for error.

More is said later in this section (page 112) in connection with the difficulties and confusions children experience with perimeter and area.

2.2.2   Area

2.2.2.1   Conservation of Area

A study by Hutton (1978) illustrates what is involved in the notion of the conservation of area and some of the difficulties children experience with it. If children are unable to grasp this idea then they inevitably have problems in understanding the derivations of such formulae as 'the area of a triangle is equal to half the base times the perpendicular height' and will thus often resort to a meaningless rote application of such formulae. Even this is jeopardised by an insufficient grasp of the geometry of the figures in question. This becomes particularly apparent when children perform badly on the so called standard area problems of the mathematics curriculum about which more will be said later on in the section.

Using a situation adapted from Piaget, Inhelder and Szeminska (1960), Hutton (1978) tested 11 year olds who were a new intake into a comprehensive school and covered the whole range of ability. She showed them the following congruent figures (same size and shape) which were made out of card and coloured and cut as shown. They were displayed on the board.

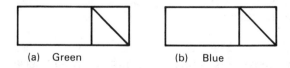

Fig. 2.27   (a)   Green   (b)   Blue

One girl was doubtful that the blue and green shapes were congruent but was soon satisfied that this was the case once the blue pieces had been placed on top of the corresponding green ones. The blue pieces were then rearranged and presented with the green shapes in the following manner:

Fig. 2.28   (a)   Green   (b)   Blue

The children were then asked to comment as to whether they thought the blue was the same, larger than or smaller than the green. This was further clarified by suggesting that if both were to be painted, would the same amount of card have to be covered on the blue figure as on the green, or more, or less. The children were also encouraged to give reasons for their judgments.

This procedure was also followed for two circles of the same size, one of which was rearranged as follows:

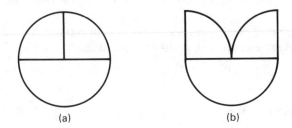

Fig. 2.29   (a)   (b)

Hutton classified the children's responses into three types:

1. The first type was where children conserved area for both situations i.e. the rectangle and circle rearrangements. They judged the amount of card to be the same for both configurations as shown in figure 2.28 and also for both configurations as shown in figure 2.29. Thirty-one of the 48 children made this judgment.
2. The second type of response was one in which area was conserved in one situation but not in the other i.e. the rearranged parts of the rectangle in figure 2.28 were judged to involve the same amount of space as the original configuration, but the rearranged parts of the circle in figure 2.29 were judged to require more or less covering than the comparison circle. Alternatively area may have been conserved in the circle situation but not with the rectangle. Eight of the 48 children responded with this type of judgment.
3. The third category of response was where neither rearrangement of parts was considered to preserve the amount of space to be covered. Nine children were unable to conserve area in either situation.

Over a third of these 11 year olds were unable to conserve area consistently. Of the 31 who appeared to be conservers only 22 were able to give a 'logical reason' for their judgment, as for example,

    ... because they are just the same pieces only put in a different position.

The non-conservers made comments such as:

> Bigger because you've spreaded the blue one out more.
> Bigger because it's almost like a triangle. (Referring to figure 2.28). (Hutton, 1978)

As part of the CSMS Project, Hart (1981), gave secondary children two questions involving conservation of area. One was very similar to the rearranged rectangle situation of figure 2.28 and area was conserved by 80 per cent of 12 year olds, 85 per cent of 13 year olds and 85 per cent of 14 year olds in the CSMS study. The other item, which was adapted from one of Piaget's items concerning the placing of houses on a field (Piaget, Inhelder and Szeminska, 1960), was the following.

This picture shows two squares of tin which are the same size:

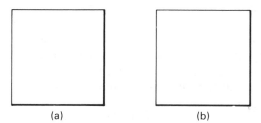

Fig. 2.30   (a)   (b)

A machine makes 8 equal holes in each tin square:

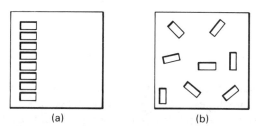

Fig. 2.31   (a)   (b)

The two squares were considered to now have the same amount of tin by 80 per cent of the 12 and 13 year olds and by 82 per cent of the 14 year olds.

Taking both the items of conservation of area into account, CSMS found that over a quarter of their secondary pupils did not consistently conserve area across the two situations.

This percentage also seems to be in line with the results of Rothwell Hughes (1979) at primary level. Children in the middle 50 per cent of the IQ range were tested on a variety of different conservation tasks and found that around 20 per cent of 7½ year olds in this group could consistently conserve. For a similar group of children aged about 10 years, the percentage of consistent conservers was around 50 per cent, although in many of the separate tasks the success rate reached 80 to 90 per cent.

2.2.2.2 *How Do Children Measure Area?*

Ward in his study of 10 year olds used two parallel questions concerning area which were administered to two different groups of children. One was worded as:

> If you had a leaf describe how you would find out how big it is? (Draw a diagram if it helps.)

The other as:

> Describe how you could find the area of a leaf. (If it helps, draw a diagram.) (Ward, 1979)

It was found that when the word 'area' was used then there was a slightly greater tendency for children to describe their approach in terms of multiplying two dimensions; width and length or top and sides.

The phraseology which did not include the word 'area' led to a greater tendency to approach measurement by means of a single dimension, e.g. width or length or, illustrated by lines on a sketch.

About 10 per cent of the children, for both formats, indicated measurement in terms of attention to the perimeter such as measuring with string around the edge. Since this was the case even when 'area' was specifically mentioned there again appears to be confusion over the two concepts, or at least the labels for these i.e. the words 'area' and 'perimeter'.

Around 20 per cent of the children in each of the groups described the measurement of the leaf by reference to a grid of squares mentioning the counting of squares, together with some mention of the need to take account of partly filled squares, and square units of measure.

2.2.2.3   Children's Understanding of Area in Terms of Filling Space

We have seen that measurement involves the reiteration of a particular unit of measure throughout the extent of whatever it is that is being measured. In a concrete sense this means that no unit should overlap another and no gaps should be left. For area the standard unit of measure is the square, particularly for measuring the area of figures typical of much of school mathematics. However any shape that may be used to fill a space can provide a means of measurement. The reiteration of this unit is seen in terms of either a single unit repeatedly positioned until all space has been accounted for, or alternatively, in terms of the space being covered by a grid in which one shape is repeated many times.

Most of the activities children engage in concern these approaches. For instance Ward (1979) asked his 10 year olds to determine how many small triangles like this

Fig. 2.32
(a)

Fig. 2.33

would fit exactly into the larger triangle: (b)

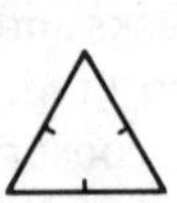

44 per cent were correct in their response of four, but 25 per cent said three. A further 5 per cent correctly drew in the lines on the larger triangle illustrating the subdivisions but neglected to then state the number of the units they had shown.

Quite probably the children who had difficulty with this item required far more experience with such activities as the tessellations described in Section 1. These would help facilitate the development of the notion of filling space by means of the repetition of a unit; this being prerequisite to any operational sense of the measurement process.

Rothwell Hughes (1979) adapted two of his area measurement tasks from those given by Piaget et al. (1960), and found very similar results from a sample of British primary school children to those obtained in Geneva thirty years earlier, in spite of his expectation that the more concrete introduction to area now used in many British primary schools would have improved the situation.

His first task simply required children to find how many white plastic (unit) squares were equivalent to an irregular red shape, the answer being a whole number.

Fig. 2.34

This was done correctly by 73 per cent of average ability 7½ year olds and 91 per cent of a similar group of children aged 9 yrs 11 months. (All samples were taken from the middle 50 per cent of children.)

The second task required children to compare the areas of two irregular shapes involving half squares, using a sheet of acetate on which a square grid was printed.

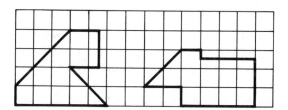

Fig. 2.35

This time the percentage correct ranged from about 10 per cent of average ability 7½ year olds to about 50 per cent of average ability children aged about 10 years, the remainder either ignoring the presence of half-units or counting carelessly.

The first APU Primary Survey (1980a) also found that 70 per cent of the 11 year olds they tested were successful in determining how many of the triangles

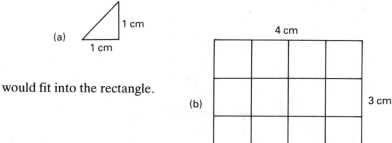

would fit into the rectangle.

Fig. 2.36

But only 57 per cent were successful with the following item:

Put a ring round each of the two shapes which have equal areas.

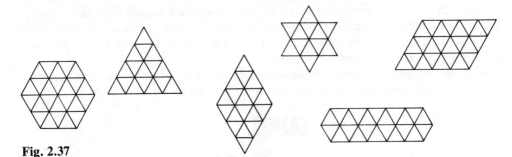

Fig. 2.37

The first APU Secondary Survey (1980b) with 15 year olds included a similar item to this involving shapes built up of equilateral triangles and achieved a success rate of just over 70 per cent. Just under 70 per cent of the 15 year olds could cope with the measurement of area of shapes drawn on large squared grids which were filled with squares and half-squares.

However, when the shape of the quadrilateral was more complicated and had to be considered as being made up of triangles and rectangles, and not as just built up from squares and half-squares, the facility for finding the area was around 40%.

The National Survey in the USA (see Carpenter et al., 1980) found that 28 per cent of their 9 year olds were able to determine the area of a rectangle which had been divided into square units whereas by 13 years of age the success rate for this had risen to around 70 per cent.

Hart (1981), in the CSMS measurement study involving 12, 13 and 14 year olds, found 80 per cent of the pupils able to determine area by counting whole and half squares but this rate of performance dropped to 57 per cent when quarter squares were involved as in the following exercise.

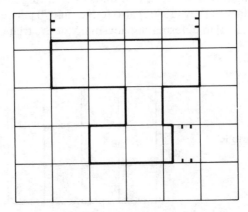

Fig. 2.38

As will become clear in the next section, the success with which children can conceptualise the measurement of area in situations where the space in question is *not* visually filled with the necessary units, is far lower than in the previous situations where the filling of space is readily visible. In other words children can more readily measure area when the units of measure are fully visible. However, even so, about a fifth and often a much higher proportion of children, even in the secondary school, are still unable to handle this concrete approach to the measurement of area where the units of measure are discrete and countable.

As we have seen, much of school work on area is concerned with dividing the area in question into squares and counting these – this being particularly the case at the primary level. Ultimately area has to be considered in terms of square units and so these activities are necessary experiences. However as Hutton (1978) points out they do not foster a notion of space as containing an infinite number of points. As we saw earlier this notion is the final Piagetian stage in becoming fully operational with the concept of measurement. Although such a sophisticated notion may well be beyond the grasp of many low attainers it is nonetheless necessary for them to attempt to achieve some inkling of this idea if they are to appreciate measurement in terms of its continuous and approximating nature. Otherwise they are confined to the bounds of the inaccurate view that measurement is always discrete and precise because the vast majority of their mathematical experiences in school promote this unrealistic picture.

Hutton suggests that an activity for estimating area would be to use paper printed with a square grid of dots. She suggests that as children approach 10 or 11 years of age they may be ready to investigate this aspect of space by using different grades of grid in which more and more dots are used per square unit of space. This provides the basis for a sequence which may then be continued in the imagination leading to some sort of notion of space containing an infinite number of points.

Activities leading up to this dot exercise could begin with any repeated pattern. Hutton suggests wallpaper as useful material for early estimates of the area of shapes: 'see how many rosebuds fit in etc.' She goes on to describe an activity based on her work with first year top stream secondary modern children which may be adapted for the less able older child. This particular exercise concerns an empirical lead into the notion of π and the formula for the area of a circle.

Each child is given the diagram as shown in Fig. 2.39. The child counts how many spots lie in the circle taking account of fractional quantities (roughly ¼ and ½ dots). Then he counts how many dots lie in one of the small squares. Comparing the answers leads to the result that the circle contains about $3\frac{1}{7}$ times as many dots as a small square is equivalent to the radius of the circle multiplied by itself, since the radius is the side of the small square. So the number of dots in the circle which indicates the approximate area of the circle is about $3\frac{1}{7}$ times the number of dots in the small square of which the area is equal to the square of the radius. Hence the area of the circle is about $3\frac{1}{7}$ times the square of the radius.

This relationship can be further investigated using finer and finer grades of dots allowing progressively closer estimates to the value of π . Again this helps develop the notion of space containing an infinite number of points as well as of the approximating nature of measurement.

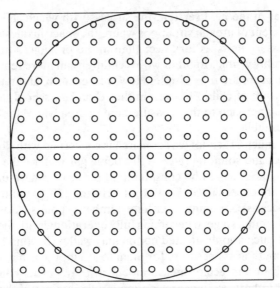

Fig. 2.39   Adapted from Hutton (1978)

2.2.2.4   *How Difficult do Children Find Typical Classroom Area Problems?*

Many of the educationally designed problems involving the measurement of area are such that only the minimum of linear measurements are given and the space involved is not already partitioned into square units. As was mentioned earlier in this section, very little research has been undertaken to discover how children make the transition from the visual representation of units of measure of area to coping with situations where these are not readily available.

Ward (1979) found 35 per cent of his 10 year olds who were able to give the area of this rectangle.

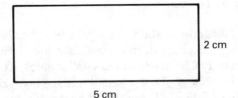

Fig. 2.40   5 cm

This 35 per cent included answers which gave inappropriate units such as 10, 10 cm and 10 cm^3. It is interesting to note again that 32 per cent of the children provided such answers as 14, 14 sq cm, 14 cm, apparently confusing perimeter and area.

In a similar but less conventional item Rothwell Hughes (1979) presented primary school children with a ruler and two plastic shapes, a square of area 9 cm^2 (3 cm x 3 cm) and a rectangle of area 8 cm^2 (4 cm x 2 cm), and asked them to use the ruler to find out which had more surface, or alternatively, which would need more paint. Again the results were poorer than might have been expected, with 1 per cent of 7½ year olds up to

14 per cent of children aged about 10 years succeeding. (The samples in each case were drawn from children in the middle 50 per cent of the ability range.) The major strategies leading to error were to measure one of the sides only (55 per cent of 7½ year olds and 18 per cent of 9 to 10 year olds) or to measure the perimeter (7 per cent of 7½ year olds up to 30 per cent of 9 to 10 year olds).

The NAEP Survey in the USA (Carpenter et al., 1980) found only half of their 13 year olds able to calculate the area of a rectangle presented in a similar format to the one in the diagram above which was used by Ward. Only 4 per cent of these 13 year olds could find the area of a more complex figure such as right-angled triangle and 12 per cent for a square given one of its sides.

Throughout the national surveys in both the USA and Britain there was generally a very poor performance for finding areas of figures particularly where a grasp of the geometry of the figures was required.

The following table giving results from the NAEP study in America illustrates this.

Basic area exercise	*% correct*		
	9 yrs	13 yrs	17 yrs
Area of rectangle given two dimensions	4%	51%	74%
Area of a square given one side	–	12%	42%
Area of a parallelogram	–	–	19%
Area of a right-angled triangle	–	4%	18%

Table adapted from Carpenter et al. (1980).

The authors point out that from their results of this NAEP survey 23 per cent of 13 year olds seemingly confused area with perimeter. Also only 16 per cent of their 17 year olds could find the area of a region made up of two rectangles.

The APU Primary Survey (1980a) included an item involving the finding of the area of a

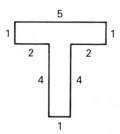

Fig. 2.41

shape which could be broken into two rectangles. All the dimensions were given for each side and essentially the figure broke down into a 5 x 1 and a 4 x 1 rectangle but only a quarter of the 11 year olds successfully found the whole area.

Thirty-seven per cent of the APU's 11 year olds were correct with the following item:

Fig. 2.42

3 cm

What is the area of this square? ..... cm^2

However 15 per cent of the pupils replied with 12 cm^2, again apparently calculating the perimeter.

The 15 year olds in the APU's Secondary Survey (1980b) were given a similar item which involved two steps. They had to find the area of a square whose perimeter was 12 cm. Only 35 per cent were successful.

Both the APU's Primary and Secondary Surveys included the following exercise:

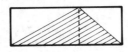

Fig. 2.43

(The dotted line was omitted on the secondary version).

They were required to find the area of the shaded triangle given that the area of the rectangle was 20 square centimetres.

Forty-seven per cent of the 11 year olds were successful and fifty per cent of the 15 year olds. Possibly the older children would have been more successful had the dotted line been present to guide them into attending to the geometry of the figure. Many children probably require much more experience with the spatial aspects of such shapes before they are prepared to cope with the measurement of area. Such a set of practical activities are, for instance, described by Bell et al., (1975). These involve the same tasks as used by Rothwell Hughes (1979) in his research project, and the associated equipment can be purchased as a classroom set from Taskmaster Ltd.

2.2.3   Confusion Between Area and Perimeter

It has become apparent throughout the last few sections that children often become confused between area and perimeter. Unfortunately their first structured encounters with these concepts occur within the context of measurement and often with the presentation of formulae. They may not have sufficient opportunity to explore practically the spatial foundations of these two ideas and the relationships between them.

The first APU Secondary Survey involved a practical test which was concerned with the relationship betwen perimeter and area. The commentary was:

> The testers' reports indicated that this topic was one of the most difficult to administer, especially to less able pupils. Many pupils' ideas about area and perimeter were confused and this may have been as much of a linguistic as a mathematical problem, but the number of pupils who failed to obtain the perimeter of a rectangle at the first time of asking suggests that mathematical concepts are deeply involved in the confusion.

They cite the case of a 15 year old boy.

His response to the initial question, 'What does the word perimeter mean?' was very hesitant – he found it difficult to find suitable nouns.

'Is ... a ... outside ...' He traced an outline of a shape on the cloth, then sat back and smiled;

'... hard to explain. It's a ... like what do you call ... circumference of a circle ...' He tried to distinguish between a circle and a rectangle but in so doing confused the idea of area with perimeter.

'It's the outer area of a ... not area ... surface area ... of what do you call it? ... rectangle ... or ... square ...' His last phrase trailed off almost indistinguishably.

The first APU Secondary Report goes on to give a very detailed account of the interview with this boy which is far too lengthy to recount in full here. However it is an extremely valuable piece of work which illustates a situation where assessment, understanding and learning are all richly interwoven. A few extracts are given below.

The tester showed him a sheet of centimetre squared paper upon which was drawn a 7 cm x 3 cm rectangle. When asked for the perimeter he calculated 20 and with prompting said 20 cm. He found the area easily, again having to be reminded about the units of measure i.e. 21 cm² and not 21 cm. He was then asked to draw a different rectangle but still with a perimeter of 20 cm. His first attempt was the same rectangle in a different orientation which he did not immediately recognise as being congruent. He was then encouraged to try again to think of a different rectangle but with a perimeter of 20 cm.

He momentarily confused the total previously established for the area with that for perimeter and queried whether the tester did mean 20 cm, and not 21. He then fell into puzzled thought.... then counted ten squares along, marking the extremities with dots.

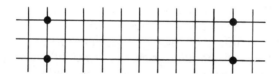

Fig. 2.44

He then seemingly realised that he had drawn a rectangle with an area of 20 sq cm and again queries 'perimeter or area'.

He made further attempts, but still based on the original 7 x 3 model and finally gave up. He was

clearly puzzled as to how these measurements could be changed and still retain the same perimeter.

When asked how he had originally worked out the perimeter he replied,

double the length and double these two and add them together. So the tester prompted that all he had to do was think of two numbers, which when he doubled the one and doubled the other, would add up to 20 cm. After this with a lot of thought and muttered countings, the pupil drew a rectangle 6 cm by 4 cm. The tester then asked what the area of this rectangle was, and (he) carefully recounted the squares before answering correctly 24 sq cm.

This 15 year old had obviously not readily appreciated that figures with the same perimeter could have different areas.

Lunzer (1968) recounts the results of a similar experiment done in collaboration with Piaget. A closed loop of string, 20cm long, was arranged on a geoboard so as to enclose a 5cm x 5cm square, which was then rearranged so as to in turn give rectangles of 6cm x 4cm, 7cm x 3cm, 8cm x 2cm and 9cm x 1cm, and each time the child was asked whether the cows would now have the same amount of grass to eat (area), and whether the farmer would have the same distance to walk round the field (perimeter). None of the 9 year olds would believe that the area changed if the perimeter did not; some of the 10 and 11 year olds would do so, but in the case of the extreme comparisons only. By the age of 13 a few realised that the area changed each time, although most of these only reached this generalisation after seeing many of the different shaped rectangles, and all admitted that it changed in the extreme cases, at least. By the age of 15 most of the children accepted the non-conservation of area.

In another experiment Lunzer started with a 12cm x 12cm square, and chopped off a larger triangle each time, replacing it in the other end so as to conserve the area but lengthen the perimeter e.g.

Fig. 2.45

Again the questions were posed in the context of a field of cows. This time the results were similar but slightly better, with most of the 13 year olds and upwards recognising the non-conservation of perimeter at least in the most extreme cases, although most of the 9 and 10 year olds still thought that the perimeter remained the same if the area did not change. Again this confirms that low attainers are likely to have especial difficulty distinguishing between the behaviour of these two measures.

In 'Checking Up II' (Nuffield Mathematics Teaching Project, 1972) an activity is presented that aims to help teachers gain insight into the child's development of perimeter and area concepts. The suggestion is to use four 20cm lengths of card and drawing pins. The strips are arranged to form a square ABCD with C and D pinned firmly to a board throughout. A and B are moved to different positions each being pinned into place so that a series of parallelograms emerge with different areas but all with the same perimeter

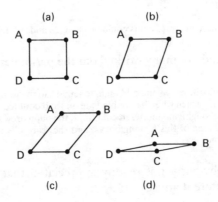

Fig. 2.46   (c)   (d)

The 'check-up' gives step by step procedures for questioning the child and helping him in his performance of such activities.

Holcomb (1980) advocates the use of the geoboard and dot paper for developing notions of perimeter and area. He claims that 6 and 7 year olds who have had plenty of experiences with geoboards and dot paper involving activities concerned with perimeter, area and symmetry, experience fewer confusions later on. He suggests that children could be asked to construct a figure which has a perimeter of, say, 8 units on a geoboard with a rubber band. Children can then compare their figures and discuss the similarities and differences. Another exercise could then be for each child to construct six different figures with a perimeter of 12 units and transpose these onto dot paper for more permanent recording and ease of comparison. For example here are eight possibilities.

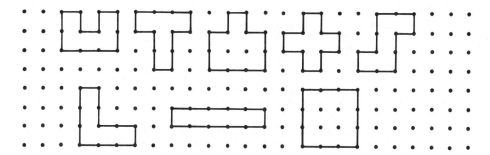

Fig. 2.47

Discussion may also centre around comparisons of the areas of these figures. Similarly children can construct different figures which enclose a given area and transpose these onto dot paper and discuss their findings.

It is probably preferable to introduce children to these activities by using more meaningful vocabulary such as, 'the fence around a garden', for perimeter and 'the garden' for area before presenting the more formal mathematical terminology. Again, suggestions are given in 'Checking Up II' (Nuffield Mathematics Teaching Project, 1972).

2.3 THE MEASUREMENT OF ANGLES AND THE APPLICATION OF MEASUREMENT IN GEOMETRICAL CONSTRUCTIONS

There appears to be a marked lack of research literature concerning angular measurement and any insight into the difficulties children experience with this. Likewise for their ability in performing geometrical constructions which involves a practical application of spatial understanding and certain measurement concepts, as well as the facility to use the instruments of measure at their disposal. The only sources referred to in this section are the APU Primary and Secondary Surveys (1980), together with a Scottish secondary survey (Giles, 1977) for which the sample was smaller but still representative of the full ability range.

2.3.1 The Child's Understanding of the Measurement of Angles

2.3.1.1   The Concept of an Angle

As part of an APU 'practical test' given in an individual interview situation, 15 year olds were asked about the nature of an angle:

2.62   Finally pupils were asked to explain in words what an angle is. 4 per cent gave responses judged to be acceptable and examples of these are given below:

'The degree of turn between two intersecting lines.'
'Amount of rotation between two lines where they meet.'

2.63   Around 30 per cent defined an angle as the distance, area or space between two lines and examples of such responses are:

'Distance measured between two lines.'
'The gap in between where two lines meet.'
'Area between two lines.'
The spacing between two lines which meet at a fixed point.'

These were not considered acceptable since no mention was made of how the space between the lines is measured. It can be seen from the written tests ... that some 20 per cent of pupils judged the size of angles on irrelevant features such as the size of the arc which labels the angle or the length of the bounding lines.

2.64   A further 10 per cent simply said an angle was where two lines meet and others said an angle was a corner. Nearly 10 per cent defined an angle in terms of degrees, for example:

'Number of degrees between two straight lines.'
'So many degrees, like a circle is 360°.'
'An amount of degrees.'

2.65   Almost one quarter did not respond to this question and many pupils were surprised at being asked to explain this concept of an angle. Indeed one pupil commented, 'People talk about angles but do not mention what they are'.

(APU, 1981b)

In a written item which tested more directly whether children had grasped the concept of an angle, the results were much better, although over 40 per cent of 11 year olds and over 20 per cent of 15 year olds were misled by other features:

41(P)   Which of these angles is the largest?   ....

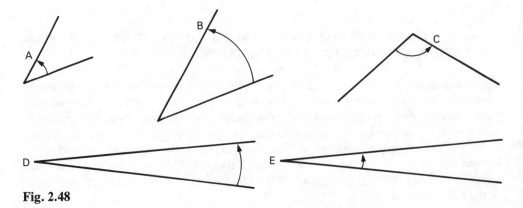

Fig. 2.48

	Selected by	
Response	11 year olds	15 year olds
A	1%	1%
B	21%	13%
*C	56%	75%
D	12%	5%
E	4%	2%
Others	3%	0%
Omitted by	4%	3%

Similarly when the Scottish 13 year olds were given a very similar set of five angles, the arms being of various lengths, about 65 per cent could put them in order of size, with little difference between classes using an individualised approach in a mixed ability organisation and those more traditionally taught (Giles, 1977).

2.3.1.2   Estimating and Measuring Angles

A summary is given in Table 2.1 of APU practical testing results (1979 survey) concerning the estimation and measurement of angles:

Item	11 year olds	15 year olds
'What are angles measured in?'	54%	–
Estimation of the size of an angle shown in a diagram		
(a) 90°	over 60%*	85%
(b) 33° (25°–40° accepted)	under 20%*	46%
(c) 106° (95°–115° accepted)	–	70%
(d) 235° (210°–260° accepted)	–	42%
Measurement of a given angle		
(a) acute	42%	72%
(b) obtuse	58%	69%
(c) reflex	–	41%
(accuracy to within ± 1°)		
Estimated drawing of an angle with pencil and ruler only		
(a) 70°	–	71%
(b) 170°	–	62%
(accuracy to within ± 10°)		
Accurate drawing (using protractor) of angle of 55°	–	73%**
(accuracy to within ± 1°)		

Table 2.1   Summary of APU practical testing (1979) results on estimation and measurement of angles (data obtained from APU, 1980a; 1981a; 1981b).

*approximate data only, from unreleased written items in 1978 survey
**data from 1979 written test item

2.3.1.3   Relations between Angles

The first APU Primary Survey (1980a) also included items which involved angular properties of triangles and angles on a straight line in terms of degree measurement. They
found in particular that children had difficulty in recognising basic angular relationships
in more complex situations. The indications are that this is more of a visualisation rather
than a measurement problem. The report quotes the example of the sum of adjacent
angles on a straight line appearing in three different situations, as illustrated in the
following figures.

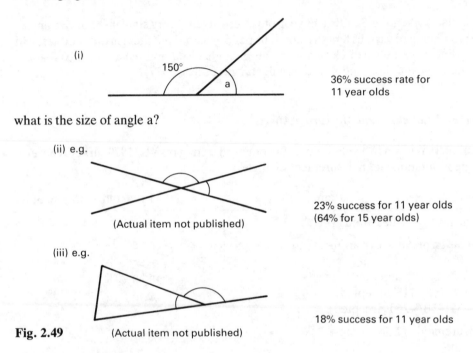

what is the size of angle a?

APU also confirmed this trend with two more items.

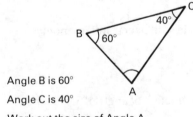

Fig. 2.50   Work out the size of Angle A.

In Figure 2.50, 44 per cent of 11 year olds found angle A correctly but in another unpublished item concerning the sum of the angles of a triangle in a more complex diagram
there was 35 per cent success.

Similar results were found with the Scottish sample of 13 year olds. On a multiple-choice question with a complex diagram (see Fig. 2.51) the percentage who could select the correct size of one of a pair of corresponding angles, the other being marked, was scarcely above the level expressed by chance.

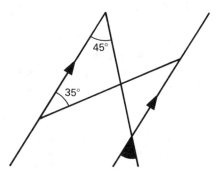

Fig. 2.51

The size of the shaded angle is:
(A) 35°
(B) 45°
(C) 80°
(D) the size cannot be worked out

Almost the same was true when children were asked to say how many of the following sets of angles could be the angles of a triangle:

(90°, 90°, 90°), (10°, 20°, 60°), (40°, 60°, 80°)
(100°, 200°, 60°), (20°, 30°, 60°, 70°), (80°, 80°, 20°).

Only children in the top 10 per cent of the IQ range scored significantly above the chance level for this item.

In the case of the 15 year olds in the second APU survey (APU, 1981b), although 75 per cent could deduce the value of the third angle in a triangle in a written item, only 30 per cent of those who were asked to estimate the size of each of the angles in a given triangle in a practical test bothered to check that their estimates totalled to roughly 180°.

These sorts of items do not really give much insight into the child's grasp of angular measurement. They may only reflect his ability to remember formulae and perform the necessary calculation, as was the case with many 'area' items quoted in 2.2.2. Again this format of question is detracting from the approximate nature of measurement as it really only represents the hypothetical state of precision and exactitude.

When a formula is not readily available and a problem requires a deeper understanding of the way in which angles are measured, then it seems that children are not so successful. For example the first APU Secondary Survey (APU 1980b) found only 30 per cent of 15 year olds able to calculate the number of degrees in turning from North to South-West (presumably in a specified direction although this is not reported). Furthermore only around 15 per cent of the 15 year olds were able to use a ruler and protractor successfully in answering other items on bearings, involving the drawing of accurate figures.

2.3.2   Measurement in the Construction of Geometric Figures

The first APU Secondary Survey (1980b) included practical tests involving the practical application of measurement in the context of constructing geometric figures. For these tests each 15 year old had at his disposal a ruler, pencil, compasses, and three different protractors.

It was found that over half of the pupils were able to draw a square accurately by making all the necessary measurements. Although many others did not use the instruments available for constructing the angles, their visual estimates were found to be accurate when checked. For example Stephanie was asked to draw a square with sides 6cm.

> Stephanie drew the square making each side 6cm but without measuring any of the angles. The tester asked her to check her drawing. Stephanie drew both diagonals and then said, 'Do I have to use a protractor or can I just put them as right angles?' (She was referring to the angles between the diagonals.) The tester repeated that it was a check that she required. Stephanie then proceeded to measure with a protractor the angles between the diagonals and the angles at the corners of the square, marking each as a right angle. She then measured each side with the ruler and painstakingly drew arrows on each length and wrote in the measurements.

This girl did not appreciate the necessity of measuring the angles at the corners of the square in order to construct them. This may have been because she was initially under the mistaken impression that provided all four sides were equal in length and the diagonals intersected at right angles then the figure was a square. (These are the properties of any rhombus-diamond.) It therefore appeared that her performance was mainly marred by the lack of understanding of the properties of a square – a spatial problem – rather than her measuring abilities, although it is not certain whether or not she would have been able to construct the angles accurately with either compasses or protractor.

Fig. 2.52

The pupils' next task was to draw an inscribed circle inside an 8cm square. The square was already drawn. Again about 55 per cent could do this accurately. Very few pupils in fact employed the simplest method of finding the cente of the square, i.e. by intersecting the diagonals. Stephanie again relied partly on a visual estimate for finding the centre:

> Stephanie picked up the compasses and ruler and set the compasses to 4cm. She then placed the ruler across the middle (judged visually) of the square and marked the 4cm point. With the point of the compasses on this point she then drew the circle and said, It's a bit out at the bottom.

Stephanie did not display an understanding of the need to coordinate measurements in two dimensions in order to find the centre of the square. Following some discussion with the tester she indicated that finding the point of intersection of the two diagonals of the square would be a more accurate method for finding the centre.

On another task 15 year olds were asked to mark a point on a sheet of paper which was equidistant from two given points A and B. About 75 per cent could do this, most choosing the one point midway between A and B. Only a few drew the locus of the point unprompted.

Fig. 2.53

•
A

•
B

With prompting, such as

Can you find a different point which is equidistant from A and B?, Now can you find another?

A total of around 54 per cent of the pupils eventually drew the line which was the locus of the point. However less than half of them could describe why all the points on the locus were the same distance from A as from B. The acceptable explanations which were given generally involved such reasons as, symmetry, congruency and isosceles triangles.

Another task was one in which pupils were given the lengths of the three sides of a triangle which they were to construct using ruler and compasses. Only about 40 per cent were successful. Many seemed to be unfamiliar with the method of using intersecting arcs for finding the third vertex of the triangle.

Below is a table of the results of the APU Practical tests (1978) on constructions. Boys were consistently more successful than girls in their initial attempts particularly for drawing an inscribed circle in a given square. It was generally concluded that this sex difference was probably due to the effects of many boys studying technical drawing.

Topic on geometrical constructions: results.
(Table from APU Secondary Survey Report No. 1 1980b).

		Unaided Success		
		All	Boys	Girls
1.	What is a square?	44%	49%	38%
2.	Draw, as accurately as you can, a square with sides 6cm	56%	61%	50%
3.	Here is a square with sides 8cm. Draw, as accurately as you can, a circle which is inside this square and just touches each side.	55%	70%	40%
4.	Mark, as accurately as you can, a point which is the same distance from A as it is from B.	75%	82%	67%
5.	Which scale of the ruler would you use to check that the distances are equal?	82%	88%	74%
6.	Why?	83%	86%	79%
7.	Check that the two distances are equal.	85%	87%	81%
8.	Mark, as accurately as you can, a different point which is the same distance from A as it is from B.	61%	64%	55%
9.	Check that the two distances are equal.	77%	83%	69%

	Unaided Success		
	All	*Boys*	*Girls*

10. Draw a line so that every point along it is
the same distance from A as it is from B.   54%   59%   47%

....

13. How do you know that all points on the line
you have drawn are the same distance from
A as from B?   28%   31%   25%
14. Draw a triangle so that the sides are
7cm, 3cm, 5cm.   42%   46%   37%

Unfortunately no further details are given for items numbered 5, 6, 7 and 9.

2.4 MASS, WEIGHT AND VOLUME

2.4.1 Mass and Weight

2.4.1.1 Terminology

It is difficult at an elementary level to distinguish between the concepts of *mass* and *weight*. It is only when a child begins to appreciate why a piece of rock which feels light on the moon will appear to feel much heavier when brought back to earth, and yet the number of kilograms it will balance against will be identical in both cases, that any real distinction between mass and weight can be made. The *mass* of the rock, measured by the number of kilograms against which it balances, remains invariant, whereas the *weight*, or *heaviness*, measured by the gravitational force (in newtons) acting on it depends on its position in space.

In the early stages the child's concept of mass/weight will depend on the sensation of 'heaviness', which is essentially a property of *weight*, and yet he will gradually be introduced to the accurate measurement of this quantity by the use of a balance, probably using, first, objects and, later, standard units like the gram and kilogram, which technically measure the object's *mass*. Since the child is as yet unable to understand the distinction, the teacher is virtually forced to keep to one word or the other in order to avoid confusing the child, and yet there is a genuine dilemma as to which to choose. At home the child is likely to 'weigh out' flour, probably in pounds and ounces, using direct-reading scales, and may therefore find it hard to make any connection between this activity and using a scale-pan balance at school to find how many plastic cubes, or gram masses, balance against a heap of fir cones. However in the teaching of science, the word mass is used consistently in all cases except to refer specifically to the notion of gravitational force.

Both popular primary mathematics programmes and researchers differ in the convention preferred. In this book it has been decided to adopt the convention which uses the words *'weight'* and *'weighing'* in the context of basic work for the reason that the idea the young, or low-attaining, child is likely to be working with is essentially that of 'heaviness', and hence 'weight'. (The concept of mass is in fact a very sophisticated one, and is not likely to be differentiated until much later, if at all, and only by recourse to formal theories of physics.)

It is clear from research that children overwhelmingly endorse the use of 'weight' rather than 'mass', for whatever reason.

The APU report, of 11 year olds:

> When presented with a 20g mass and asked to identify it, 60 per cent of the pupils called it 'a weight' or, more specifically, 'a 20g weight'. Another 25 per cent, possibly misunderstanding the question, replied that it was 20 grams. No pupil called it 'a 20g mass' and only 2 pupils of the 400 tested referred to it as 'a mass'. (APU, 1981a)

In fact the real problem for the child is to sort out the difference between the concepts of weight and volume. For example, a balloon may be *bigger* than a brick but it is not necessarily *heavier*.

2.4.1.2   The Sequence of Concept Development

Piaget (1952) suggests that a more primitive notion than either those of weight or volume is the concept of *substance,* or *quantity* (which is often confusingly translated as 'mass'). This is simply the ability to compare two amounts of the *same material.* For example, if two balls of plasticine appear in every way identical, then there is the same amount of *substance* in them, and if an extra bit is added to one, then that must now have 'more' than the other. This concept is thus a combination of weight and volume; the balls 'have the same amount' because they look the same *and* feel the same.

Before the child can become fully operational with the measurement of the weight of the lump of plasticine, as measured on a balance or scale-pan, he has to be able to conserve *substance* i.e. he must appreciate that whatever shape the lump of plasticine is made into – be it a ball, sausage-shape, pancake etc. – provided none has been added or taken away, it is still the same amount of 'stuff'· According to the Piagetian School the child is then able to proceed to the development of the idea that the weight of this 'stuff' will also remain invariant despite any change in shape.

Piaget's researches showed that children attain conservation of 'substance' at about the age of 7 on average, conservation of weight at about the age of 9, and conservation of volume at about the age of 12 (Piaget, 1952; 1974). Similar results were found in England by Lovell and Ogilvie (1960, 1961a) and in the United States by Elkind (1961a, 1961b). The usual Piagetian 'plasticine ball and sausage experiment' referred to earlier in which the child is asked whether there is more, less, or the same amount of plasticine when the shape of the ball is rolled out from a ball to a sausage, was also used by Beard (1963) with English children. She found conservation of substance was attained by 33 per cent of 5 year olds, rising to 58 per cent of 9 year olds. However she also found that the questions became harder when she broke the plasticine into small pieces, and again repeated the question. Smedslund (1961–2) and Uzgiris (1964) however report that children more easily conserve discontinuous quantities (when the quantity can be broken into discrete pieces than continuous quantities. Uzgiris states also that there was some variation in response according to the type of material used.

Beard's results suggest that low attainers have not generally grasped the idea of invariance of substance by the age of 9, and since there is considerable evidence that this concept is likely to precede those of invariance of weight and volume, there would seem to be a good case for restricting the treatment of these topics for low attainers in the primary school to practical activities and questioning aimed at development of invariance concepts. However as Smedslund pointed out, a grasp of conservation does not depend on practical experience alone but involves children in resolving a mental conflict between visual evidence and logical thinking.

2.4.1.3 Conservation of Weight

Rothwell Hughes (1979) carried out a comprehensive study concerning the concept of weight among children aged 7½ to 10 years in British primary schools. He found that almost all of them (i.e. 92 per cent of 7 year olds and 97 per cent of 9 to 10 year olds) could correctly compare two objects of the *same shape and size* and decide which was heavier, or lighter, by feeling them. Also 88 per cent of 7½ year olds rising to 95 per cent of 9 to 10 year olds understood how a balance worked i.e. that the pan holding the heavier weight would go down but if the weights were the same the arm would remain horizontal. However his sample for these tests was confined to the middle 50 per cent of the ability range, so that the children below this level may have proved less successful.

In the case of conservation of weight, three different sets of materials were used. One was the traditional Piagetian ball of plasticine transformed into a long thin sausage. The second consisted of two sets of nesting Russian dolls; the children were asked whether they thought a nested set would weigh more, less, or the same as the similar set of separate dolls. The third type of material used was Lego blocks; a 'brick' made up from smaller blocks could be separated and recombined into a long thin shape, thus giving a situation similar to the 'ball and sausage', but using discrete unit (blocks) rather than a continuous piece of plasticine.

For the plasticine 'ball and sausage', the results were:

Percentage of children who thought that the weight remained constant although the shape was changed. (Data from Rothwell Hughes, 1979.)

		IQ Band		
		Lowest 25%	*Middle 50%*	*Top 25%*
age	7 yrs 5 months	26%	57%	73%
	9 yrs 11 months	59%	86%	92%

The corresponding figures for the Russian dolls and the Lego were broadly similar, with slightly more children conserving in the case of the Lego bricks and slightly fewer in the case of the Russian dolls, again reinforcing that a change in the context of the question can produce a difference in the child's response.

The above figures suggest that many low attainers in the primary school think that the weight of a lump of material will change if its shape is changed, and hence do not have a very precise concept of weight.

2.4.1.4   Practical Studies Involving Balances

In addition to the items concerned with conservation of weight, Rothwell Hughes (1979) also asked children to perform a number of tasks which involved both practical weighing on a balance and the use of consistent logical reasoning. Typical of these was a task in which the child was asked to find the heaviest box out of three differently coloured boxes of equal volume, using a balance. The results were:

Percentage of children able to find the heaviest of three weights using a balance. (Data from Rothwell Hughes, 1979.)

		IQ Band		
		Lowest 25%	*Middle 50%*	*Top 25%*
age	7 yrs 5 months	9%	31%	57%
	9 yrs 11 months	46%	72%	83%

These results are rather lower than those for the conservation task, which suggests that the ability to conserve may well be effectively a prerequisite for solving simple practical problems in the area of weight. It is worth noting that a minority of low-attaining 10 year olds were able to solve what might appear to be a relatively straight forward practical problem involving weighing. When the task became slightly more complex, the percentages were reduced still further. For instance one task involved four blocks of equal volume, of which two (the green and the black) were of equal weight, one (the blue) was of twice this weight, and the fourth (the brown) was of intermediate weight. Children were asked first to find whether the green and the black blocks were of equal weight, and were then asked to use the balance to find which of the other two blocks weighed twice as much as the black one.

In this case although the percentage success for the lowest 25 per cent of children (by IQ) is not given, the success of the middle 50 per cent of children ranges only from 1 per cent of those aged 7 years 5 months to 20 per cent of those aged 9 years 11 months. Although this task is more a test of logic than of the ability to weigh the results do suggest that children need practice in applying their knowledge to the solution of practical problems.

It is suggested that in fact these, and many other practical tasks in the field of area, weight and volume used by Rothwell Hughes in his study can form the basis of a practical teaching programme; Bell et al. (1975) have provided a teachers' guide and the equipment is available from Taskmaster Ltd for purchase by schools.

The APU also conducted practical tests with 11 and 15 year olds to investigate their understanding of weight through the use of a balance. There were three main tasks.

(a)   Ordering Blocks by Weight
Each *15 year old* pupil was given five similar looking wooden blocks labelled A, B, C,

D and E, and a balance with two pans and no calibrated scale. The pupils were first to find the heaviest block, which was accomplished, unaided, by 95 per cent of the boys and by 82 per cent of the girls. They were then required to order them from heaviest to lightest. This was achieved by 72 per cent of the boys and by 67 per cent of the girls (APU, 1980b).

It is noteworthy, then, that one third of the 15 year old girls were unable, by themselves, to order five blocks of wood according to weight. And similarly over a quarter of the 15 year old boys did not succeed.

In the case of the *11-year olds*, 82 per cent of boys and 76 per cent of girls could find the heaviest of three blocks. Some of the sources of confusion are illustrated in two excerpts from interviews:

> Places each in one tray, one at a time – nothing in the other.
>
> Pupil: 'Do we have some weights?'
> Tester: 'No, just the scales.'
>
> The pupil then balances X against Y.
>
> Pupil: 'X is heavier than Y.'
> Tester: 'So which is heaviest?'
> Pupil: 'Z is heaviest.'
>
> Pupil puts Z in one pan, nothing in other.
> Removes Z, replaces with X.
> Removes X, replaces with Y.
>
> Pupil (indicating X): 'I think this.'
> Tester: 'Why?'
> Pupil: 'Because it stays down.'
> Tester: 'What do the others do?'
> Pupil: 'Stay down, but not as much.'
>
> Tester (pointing to other pan) 'Could you use this side at all?'
>
> Pupil now 'weighs' each in other pan as above.
>
> Tester: 'What would happen if you put one in each pan?'
>
> Pupil now weighs Z and X in turn against Y.
>
> Pupil: 'Z is heaviest because it goes down more than this one (X).'   (APU, 1981a)

(b) Making a Lump of Plasticine Weighing 20g

Again using the balance, each pupil was then required to make a lump of plasticine weighing 20g given a heavier amount of plasticine and a 20g weight/mass. Almost all the *15 year old* pupils were successful in doing this with the use of the balance. However only 80 to 85% of the *11 year olds* could do this (APU, 1981a; 1980b). Some 11 year olds failed to conserve, for example:

> The plasticine was heavier than the 20g mass and the pupil was attempting to adjust it.
>
> Tried squeezing all the plasticine and balanced it against the 20g mass (3 times).
>
> Tester: 'What are you doing?'
> Pupil: 'Squeezing it.'
> Tester: 'Why?'
> Pupil: 'To make it heavier.'
> Tester: 'Why do you want it to be heavier?'
> Pupil: No response.

Tester: 'What will you have to do?'
Pupil   (apparently realising her mistake) 'Take some off.'

Eventually balanced but once again tried squeezing to make it lighter.

(APU, 1981a)

The tester removed the residual plasticine and withheld the 20g weight/mass and asked each *15 year old* pupil – who was equipped with his 20g lump of plasticine and the balance – to make a lump of plasticine *a quarter* as heavy as the metal 20g weight/mass. The *11 year olds* were asked to make a lump *half* as heavy as the weight.

In the case of the 11 year olds, only 30 per cent spontaneously used the balance to ensure that the halves of their 20g lump were of equal weight, although most of the further 20 per cent who halved it visually used the balance when they were asked to check their result. Many of the 20 per cent who did not succeed even after prompting were not able to use their knowledge that the lump of plasticine now weighed 20 grams.

In the case of the 15 year olds, 40 per cent of the boys and 28 per cent of the girls were successful, unprompted, in making a lump of plasticine a quarter as heavy as the 20g weight/mass using the balance to check. However again a further group used visual means, which often turned out to be reasonably accurate.

(c) Finding the Weight of a Peg

Each *15 year old* pupil was equipped with a balance, a bag of pegs and a 20g weight/mass and asked to explain how he would find the mass/weight of one of these pegs using the apparatus. The following responses occurred:

i.   About 25 per cent stated they would see how many pegs weighed 20g but could go no further.
ii. 10 per cent of the pupils proceeded to explain that having found how many pegs weighed 20g they would then divide that number by 20. (This would give how many pegs weigh 1g – not the weight of one peg.)
iii. In all, 64 per cent of the boys and 47 per cent of the girls could give a satisfactory *method* of finding the weight of one peg.

All pupils were then asked to perform the task. 53 per cent of the boys and 34 per cent of the girls were able to do so without help. With prompts these percentages rose to 65 per cent and 48 per cent respectively.

About 36 pegs weighed 20g and answers were given as follows:

Just under 30% did the division (e.g. $20 \div 36$).

Just under 20% gave the weight in fractional form e.g. $\frac{20}{35}$

About 5% gave an acceptable approximation e.g. about ½g.

When asked how a more accurate measure of the mass of a peg could be found with this balance, just over 10% of the pupils suggested repeated weighings of 20g of pegs and taking the average or using a heavier weight

(APU, 1980b).

About half the *11 year olds* were successful in a similar but easier task in which each plastic shape actually weighed 5g, so that it was only necessary to put four together to balance the 20g weight/mass. Some children were reluctant even to do this:

20g weight in one pan. One shape added to other pan.

Pupil: 'It weighs about 5g.'
Tester: 'How do you know?'
Pupil: 'It doesn't weigh as much as the 20g. It feels lighter than the 10g.'
Tester: 'It might be 3g or 6g.'
Pupil: 'If it were 3g it would be as light as a feather.'
Tester: 'Think of a way of finding out what it weighs exactly.'
Pupil: 'Can you use all the shapes?'
Tester: 'Yes.' (More shapes added to the pan)
Pupil: '4 weigh 20g.'
Tester: 'I want to know the weight of one.'

Pupil does not reply.

Tester: 'If 4 weigh 20g, one weighs?'
Pupil: '5g.'

(APU, 1981a)

The APU found that this topic involving balances produced, in the first survey, the greatest discrepancy of any of the practical tasks between the performance of 15 year old boys and girls. They deduced from various comments made by the pupils that performance with practical tasks concerning mass and weight may well have been biased in favour of those studying science, especially physics. Nonetheless it was reported that most pupils showed an enthusiastic approach to the tasks.

2.4.1.5   A Case Study of Children with an Incomplete Understanding of the Concept of Weight

A case study described by Dichmont (1972) illustrates very clearly the difficulties experienced by children in their grasp of mass/weight concepts. He writes about two boys aged 8 and 9 years who were equipped with a balance, a book and plasticine. They were posed a problem written on a work card:

Balance a book with a lump of plasticine. Now can you balance half a book?

They managed the first part involving balancing the book with a lump of plasticine. Despite the ambiguity of the wording of the second part of the task it was decided that cutting the book in half was not what was required! Their subsequent struggles with the problem highlight many of the difficulties mentioned above.

Their first method towards a solution involved laying three rulers side by side so that they transferred the rulers to one scale plate with the plasticine on the other. Dichmont thought perhaps they were experimenting with the notion that the rulers which covered half the book would be the same weight as half the book. Then one boy measured the book saying,

It's 16 across. Half 16 is 8 isn't it? But if we put it on we haven't split it in half, have we?

They then discussed making a book of the same weight which they could then split in half, but they weren't sure how to make the spine of a book!

They then decided to make a model of the book out of wood. Dichmont comments that,

They seemed happy to assume that two things with the same surface area and thickness (i.e. volume) would have the same weight, though one was made of paper, the other of wood. They lacked, and needed a working knowledge of the concept of density not just the simple notion of weight.

Dichmont intervened at this point, offering them a sheaf of exercise paper which they could use as a model book mentioning that he did not want the paper cut.

Some time later the boys had a pile of the paper which balanced the book but were unable to halve it. They tried folding it in half but then realised that it would not change the weight. The next day the boys returned for the apparatus, one of them having thought in the meantime of cutting the lump of plasticine in half. Dichmont recounts,

They split the lump of plasticine roughly into two and put one piece on either scale plate. The two 'halves' don't balance. To make them equal in weight, they take bits away from the heavier 'half', but fail to add these to the lighter side. (They don't conserve their total weight of plasticine.)

Dichmont suggests they now see if the two lumps still balance the book. The book of course is now heavier because they have dispensed with some of the original plasticine – although they do not appreciate this. The boys then add extra bits to each lump of plasticine by eye until the book is balanced again. They then remove the book and place one lump of plasticine on each pan. They still do not balance. One of the boys then swaps the lumps over, thinking this might work. Then he adds spare plasticine to the lighter lump – again failing to conserve – and finds that the book is now lighter than the two lumps. The 9 year old then begins to realise that adding or removing spare plasticine is not the answer.

They then experiment with placing the lumps in different positions on the scale pans, trying to find a point of balance. They then decide to alter the fine adjuster of the balance thus indicating their lack of understanding of how the balance itself works. They try sticking one lump underneath the pan to see if that works. They try altering the shape of the heavier lump to a more regular sphere – again evidence of not conserving.

By the third day Dichmont decides to show them the solution by taking bits off the heavier lump and adding them to the lighter one until the pans balance, saying,

Didn't you think of doing that?

The 9 year old said,

Yes: we tried it but it didn't work so we tried something else.

Dichmont states,

I suddenly realised that 'my' solution was for me the logical thing to do, whereas for them it was just one among many possible moves. I hadn't appreciated ... how totally confusing it is to be faced with some 'raw' facts; I mean facts not yet secured within a framework of expectations and assumptions.

In other words these boys were tackling a problem which was conceptually beyond their grasp. They did not appreciate the invariance of weight and displayed confused notions of 'substance', weight and volume. They did not appreciate the approximating nature of measurement in terms of repeatedly 'taking a bit of plasticine off here and putting it on there' until they approached a point of balance. They did not have a working knowledge of the instrument of measurement at their disposal, which is probably hard to grasp when they do not fully understand the nature of what it is that is being measured by this instrument.

It is hard to tell from their performance on this task how far they had grasped the notions of 'substance' and the invariance of 'substance' alone i.e. devoid of the context of weight. It may be that they could appreciate that a lump of plasticine is still the same amount of plasticine if it is shaped into a more regular ball but cannot appreciate that it will still weigh the same. Obviously the teacher should go back and assess the depth of their understanding of these matters and devise instruction accordingly.

2.4.2   Volume and Capacity

2.4.2.1   *Different Aspects of the Concepts of Volume and Capacity*

Kerslake points out that there also exists a certain amount of confusion between the two concepts, volume and capacity. Furthermore, as can be seen in the section concerning the measurement of weight, these two ideas are often linked with the notion of weight. Not only do these confusions take place in the classroom, for Kerslake gives the example of having seen a bottle of sauce, the label of which read,

> Net Volume = 225g.

Mathematically speaking *capacity* concerns,

> The ability of hollow containers to hold something ... filling boxes with tins, ... using liquids or free flowing materials like sand ... filling a cup with tea. (Kerslake, 1976)

The standard units of measure for capacity are millilitres and litres (or, in the Imperial system, pints and gallons).

Volume according to Kerslake is used in two senses,

i.   *Internal volume* of a hollow which is the same as capacity although the units of measure are usually expected to be in, 'solid sounding units like cubic centimetres'.
ii. *External volume,* in the sense of the amount of space taken up by an object e.g. a brick, in relation to other objects, i.e. occupied volume of space.

(This classification is somewhat different to that used by Piaget, Inhelder and Szeminska (1960)).

Kerslake emphasises that most practical day to day experiences of volume are concerned with internal volume/capacity and filling or partial filling of hollow shapes, and not with external or occupied volume. She cites a case of where a group of primary teachers from many different countries could not give an accurate estimate of the volume of their own bodies – their estimates ranged from 1½ cubic feet to 700 cubic feet.

She claims that concern with occupied volume is largely restricted to school mathematics exercises for finding the volume of a solid such as a brick or a cone for its own sake and is not a very realistic activity outside of the classroom. She further argues that this tends to lead to a situation where the child becomes preoccupied with calculating volume in terms of a, 'search for an appropriate formula and a numerical operation' without much understanding of how the formula is derived or any sound conceptual grasp of volume. This is often evidenced by the confusion shown over the formulae for the surface area and volume of a cylinder. Furthermore she feels that there is a marked lack of concrete experience of occupied volume since many practical activities are restricted to filling hollow spaces and thus centre on internal volume/capacity.

The parts of the text which follow are each concerned with a particular aspect of volume: 'external volume', 'internal volume', 'liquid volume and capacity', and 'displaced volume'.

2.4.2.2   What Difficulties do Children Experience with Occupied Volume?

The first step in forming the concept of the external volume of a solid is the conservation of the quantity of *substance*, as tested in Piaget's 'ball and sausage' experiment referred to on page 123. (Results for a recent British sample, as obtained by Rothwell Hughes, 1979, are given on page 124.)

Rothwell Hughes (1979) investigated some basic notions of occupied volume of solid shapes among children of primary school age. He used the alternative forms of wording 'Do they take up the same amount of room?' and 'I want you to imagine that these are blocks of chocolate. Do they have the same amount of chocolate?.' (The latter, he felt, evokes an answer in terms of volume, whereas others have used it as a test for the concept of substance or weight.) There was in fact little difference between the two forms, although the latter was generally found slightly easier.

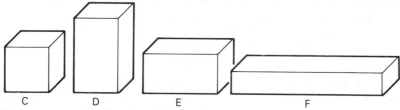

C   D   E   F

Fig. 2.54

In one example he asked children to pick out, from a set of four blocks, all the pairs of blocks which were of the same volume. Two of the blocks (D and E) were of identical dimensions but arranged in different orientations, and out of a sample of children drawn from the middle 50 per cent of the IQ range, 85 per cent of 7½ year olds and 93 per cent of children aged 9 yrs 11 months could pick out these as having the same volume. Block F was also the same volume as D and E, being half the height of E but twice the length, a fact which could easily have been ascertained by matching one against the other. However in this case only 12 per cent of the average 7½ year old children, and 29 per cent of those aged 9 yrs 11 months, selected such a pair of blocks, even when asked if there were any further pairs with equal volumes.

However when the children were presented directly with a similar pair of blocks to this latter pair, but where the second one was already divided into two

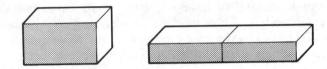

Fig. 2.55

and could thus be dismantled and reassembled to exactly match the other, the percentages who recognised that the volumes were equal rose to 60 per cent of average 7½ year olds and 84 per cent of average children aged 9 years 11 months.

On other similar items testing conservation of volume in which one or both blocks were dismantled and re-formed into a different shape, and children were asked whether one now took up more room than the other, the success rates were between 45 and 55 per cent of 7½ year olds, and between 65 and 80 per cent of the older group, according to context. However for a group of low attaining children drawn from the lowest quarter of the ability range, the success rate for one of these items was only 18 per cent of 7½ year olds and 48 per cent of children aged 9 years 11 months.

When it came to the meaurement of volume, Rothwell Hughes reports that when, in the pilot test, children were asked to find the volume of a block given a ruler, very few children even at 11 years had any notion of how to proceed.

However when the question was translated into more concrete terms and children were asked to determine which of two solid blocks contained more wood, having been provided with a set of 'unit' cubes, 12 per cent of average 7½ year olds and 28 per cent of children aged 9 years 11 months were able to succeed. For the lowest 25 per cent of the ability range in these age groups the corresponding figures were 3 per cent and 18 per cent respectively.

Thus although the majority of children of junior school age have some concept of the occupied volume of solids, in that they can conserve, only a much smaller number have the notion that volumes can be compared using the repetition of a unit cube.

These results accord reasonably well with those of Piaget et al. (1960) since he judged that children at the stage of concrete operations (starting in his view typically at aged 6 to 8 and continuing to 11 to 12) can conserve but that their comparisons of volume are qualitative rather than quantitative. His experiment involved children being shown a block, 4 units high on a base of a 3 x 3 grid and asked to construct a 'new house, with exactly as much room as the old, on a new island' with the base on the new island shown as a 2 x 2 grid. A typical response of this age group was given by Bar (9 years 6 months), who built a house 8 units high on the 2 x 2 base giving a volume of 32 units against the original 36.

> 'I've made it twice the height, because this one (the original model) is less high but it's wider, and that one is thinner and narrower but it's higher'.

This shows the realisation that it is necessary to adjust the height to conserve volume, but not the ability to measure the extent to which it should be adjusted, which Piaget suggested arose typically at aged 11 to 12.

One of the major difficulties experienced by children in activities such as the last one reported which involved thinking of solids as built of unit cubes, is that many of the cubic units cannot be seen, such as those in the centre of a cuboid. One of the items given to secondary school children in the CSMS project (Hart, 1981) which was adapted from the Piagetian 'islands' item referred to earlier illustrates these points.

This block is made by putting small cubes together:

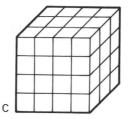

Fig. 2.56i

How many cubes make this block if there are no gaps inside?
All the cubes from the block are put in a pile.

Fig. 2.56ii

I am now going to use *all* these cubes from the block to build a 'skyscraper' so that the bottom floor is four cubes.

Fig. 2.56iii

How many cubes high would this 'skyscraper' be from the ground?......

It was found that the percentage of children for each age group who gave the answer 24 cubes making the block – i.e. seemingly just counting those which they could see – was as follows:

12 year olds	13 year olds	14 year olds
10.7%	13.5%	12.6%

From 40 per cent to 50 per cent were able to say how many cubes high the 'skyscraper' would be. However it was found that a further 10 per cent for each year although giving an incorrect answer for the 'skyscraper' part nonetheless gave an answer which was one quarter of the number of cubes they had given for the first part of the question.

One of the conclusions suggested by these results is that 40 to 50 per cent of these age groups in the secondary school are unable to cope with a very straightforward question on volume which involves measurement only in the sense of a repetition of unit cubes.

Hart goes on to report that,

> The volume of the block is of course obtainable by multiplication but some third year children on interview obtained their answers by addition, 'a layer of twelve and another layer and another'. Having obtained the answer 36, the last part can again be seen as an addition of fours until 36 is reached. Piaget states that the realisation that two volumes are equal if the products of their respective elements (dimensions) are the same is indicative of early formal (thought).

The indications are, as mentioned at the beginning of this chapter, that many children of early secondary age have not reached such an advanced conceptual level and still operate at a concrete stage of development.

The NAEP Elementary Survey in America (see Carpenter et al., 1980) found that only,

> 7% of 9 year olds and 24% of 13 year olds could find the volume of a rectangular solid cut into unit cubes. 46% of 9 year olds and 36% of 13 year olds simply counted the faces of the cubes shown in the picture or found the surface area of the solid.

However the first APU Primary Survey (1980a) of 11 year olds found that,

> Around 80% gave the correct number of unit cubes missing from an uncompleted larger cube, and around 60% for the number missing from an uncompleted cuboid.

As these items were unpublished it is not clear how easily the children could visualise the situation in that the positions of the missing cubes could or could not be readily seen.

Finding the volume of a solid where only the linear dimensions are given does not necessarily call upon the child's conceptual grasp of volume. He may just apply the 'length x breadth x height' formula and proceed to display his computational skills. However Ward (1979) managed to overcome this to a large extent by posing the following question to 10 year olds

How many small cubes would make this block?

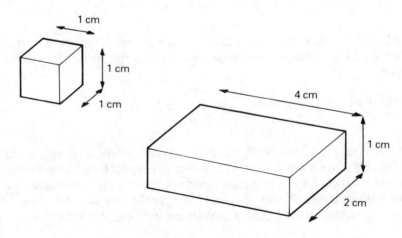

Fig. 2.57

Forty-one per cent of 2300 ten year olds were correct. The height being one centimetre probably facilitated performance since attention needed only to be focussed on the other two dimensions.

The first APU Primary and Secondary Surveys (APU 1980a,b) did not release details of their written items on volume save to relate that when the three dimensions of a cuboid were given and these were whole numbers, one-third of the 11 year olds could find the volume and three-quarters of the 15 year olds could do so.

However the National American Survey (NAEP, 1980) gives figures of 17 per cent of 13 year olds and 39 per cent of 17 year olds as being able to find the volume of a rectangular solid. 42 per cent of the 13 year olds and 18 per cent of the 17 year olds appeared to have added the three dimensions instead of multiplying, again illustrating how difficult this was for pupils to grasp.

2.4.2.3 The Child's Grasp of Internal Volume

Kerslake's (1976) comments, quoted on page 130, suggest that it might be expected that children would find the idea of internal volume i.e. 'how much does that box hold?' easier than that of external volume i.e. 'how much space does that object occupy?'. However there appears to be little relevant research in this area.

The CSMS project (Hart, 1980, 1981) devised two items to try to assess whether children could appreciate the invariance of internal volume. The first was as follows,

I have 24 cubes which will just fit into a box A, leaving no spaces.

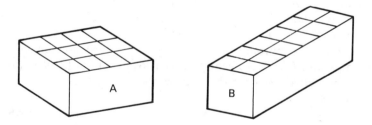

Fig. 2.58

The same 24 cubes will also just fit into box B, leaving no spaces.

Tick the statement which is true about the *volume of air space* in the two boxes when the cubes are taken out.

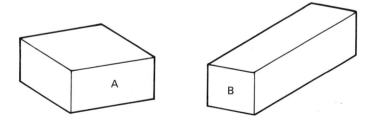

Fig. 2.59

1. Box A has more air space ....
2. Box B has more air space ....
3. Box A and B have the same amount of air space ....
4. You cannot tell if one has more air space or not ....

The percentages of pupils for each year who correctly judged the air space for the two boxes to be the same were:

12 year olds	13 year olds	14 year olds
56%	65%	68%

The second CSMS question concerned a piece of plasticine which was placed inside a box, first as a single lump and later having been cut into three separate pieces which were shown spaced out inside the box. Children were asked, in both cases, what 'volume of air-space' was left in the box. The per centage of children who gave the same answer in both cases were almost identical to the percentages for the first CSMS question quoted above.

The percentages in the CSMS items concerning the conservation of 'external' or 'occupied' volume of solid blocks were again very similar, so that there is no evidence in this case that 'internal' volume is any easier to conserve than 'occupied' volume.

The first APU Primary Survey assessed children's performance in finding the internal volume of a rectangular box of given dimensions by asking how many cuboids of a given size would fit into it. This was an unpublished item in which the unit block had one dimension equivalent to one of the dimensions of the larger box: therefore, only two of the three dimensions of the larger box were needed. 45 per cent of the 11 year olds were successful. However, on a similar problem, also unpublished, where all three dimensions had to be considered, performance dropped to about 25 per cent success.

This kind of problem highlights the way in which the idealised situation presented in the classroom differs significantly from an actual practical problem. For example, consider a comparable CSMS problem (to which 46 per cent of 12 year olds and 66 per cent of 14 year olds obtained the 'correct' answer):

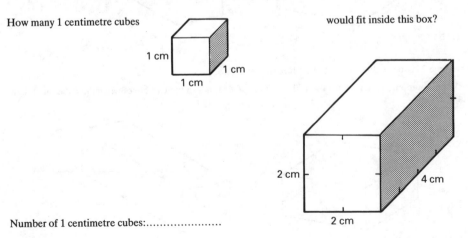

How many 1 centimetre cubes   would fit inside this box?

1 cm
1 cm
1 cm

2 cm
4 cm
2 cm

Number of 1 centimetre cubes:......................

Fig. 2.60

To arrive at an answer of 16 cubic centimetres certain assumptions have to be made, namely:

i.   Either you ignore the thickness of the box or you assume the dimensions are internal dimensions.
ii.   The centimetre cubes are so well made that they slide together 'perfectly' without taking up the slightest bit of extra space.

Thus again the impression is conveyed to the child that the measurement of volume is exact.

Furthermore the very same diagrams are often given for solids where consideration is centred around 'occupied' space; so the child is never really given the opportunity to differentiate clearly between the two types of volume nor to consider the different implications involved in the measurement of each type.

2.4.2.4   *Liquid Volume and Capacity*

Since the *capacity* of a container is simply the *volume of liquid* it will hold, (i.e. *capacity* is the 'liquid equivalent' of the concept of '*internal volume*', as discussed in the previous section) the two aspects of liquid volume and capacity are here considered together. Logically, there is no real need to differentiate between the volume measurements of liquids and solids, although as a matter of convention they are measured in different units. (In the metric system liquid volume and capacity are measured in litres, although conversion is not difficult since 1000 cubic centimetres = 1 litre.)

There are however psychological differences due to the fact that liquids have no fundamental 'shape'. For example, a child can determine the volume of a rectangular solid directly, by building the shape with centimetre cubes and simply counting (provided the length of the sides are whole numbers of centimetres). This can only be repeated with certain quantities of liquid, and then only if it is poured into a suitable rectangular container first, and this process requires the notion of conservation, i.e. that the volume of liquid remains the same regardless of the shape of the container.

On the other hand, at a more complex level, it is no easy matter to estimate the volume of a *complicated* solid shape, like a chair, a bicycle or a person. However, the capacity of an irregular vase or teapot can easily be determined by first filling it and then transferring the liquid to a jug or measuring cylinder which is already calibrated in appropriate units.

One of Piaget's most famous experiments is concerned with the conservation of liquid volume (Piaget, 1952); it is the liquid equivalent of the 'ball and sausage' task, concerned with the conservation of solid 'substance', which is referred to on page 123. The basic experiment can be represented as shown in Fig. 2.61.

There are variations e.g. A_2 poured into two smaller beakers. The question asked of the child, after he had initially agreed to the fact that he (with A_1) and his friend (with A_2) would have the same amount of drink, was, once A_2 was poured into L whether he and his friend still had the same amount to drink, or whether one of them had more. In some cases the child was asked to predict what level the liquid would reach in L before the 'orange juice' was poured in.

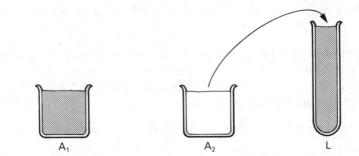

Fig. 2.61 A_1 A_2 L

Piaget showed that younger children in general tend to judge quantity only on the basis of one dimension, usually height e.g. Madeleine (Sim), aged 5, said:

> M: 'There's more orangeade (in L) because it's higher.'
> Interviewer: 'Is there more to drink, or does it just look as if there is?'
> M: 'There's more to drink.'

Older children, on average around the age of seven, although maybe much later for low attainers, would reason that there must necessarily be the same amount of liquid, and explain the higher level as did Aes (aged 6½ years):

> Aes: 'Oh yes! It's the same; it seems as if there's less (in A_1) because it's bigger (i.e. wider),
>   but it's the same....'
> Experimenter: 'And if I pour yours into 4 glasses?'
> Aes: 'It'll still be the same.'

Rothwell Hughes (1979) repeated Piaget's experiment in Britain and found results shown in the following table.

Percentage of Conservers in Various Ability Groups (by IQ) at different ages

	Lowest 25%	*Middle 50%*	*Top 25%*
7 yrs 5 mths	6%	43%	76%
9 yrs 11 mths	38%	82%	89%

When the liquid in A_1 was divided between 4 squat cylinders, and that in A_2 was divided between 5 tall thin cylinders, the percentage of conservers decreased; 35 per cent of the middle group of younger children now conserved against the 43 per cent in the first experiment, although the percentage for the middle group of older children was very similar.

Some experimenters claim to have had success in teaching young children to conserve. (These training experiments are summarised by Modgil and Modgil, 1976).

However Carey (1979) claimed that by no means all young 'conservers' are consistent; for instance when the container L was replaced by a trick container L^1 (see Fig. 2.62) with the result that the level in L^1 after the contents of A_2 had been poured into it was the same as that in A_1, the majority of the 'conservers' showed no surprise, and 90 per cent of them still maintained that there was the same amount to drink in A_1 and L^1. This means that although children may appear to conserve liquid volume, they do not necessarily have a complete grasp of the situation.

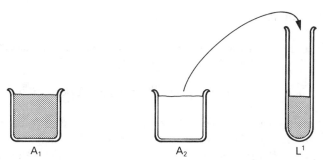

Fig. 2.62   A₁   A₂   L¹

The next step, after conservation, would appear to be the ability to compare in qualitative terms the capacities of two differently shaped containers. Rothwell Hughes (1979) asked children to do this, given an uncalibrated measuring cylinder of greater volume than either container, a pen for recording heights, a funnel and a large bowl. He then asked them to use the same method to compare the size of one of the containers with that of a third container. Only 14 per cent of the children in the middle 50 per cent of the ability range of the 7½ year old group were able to successfully judge the larger container in both cases, and only 46 per cent of the middle group of the children aged 9 yrs 11 months.

It was arranged that these children would have obtained two comparisons of the form 'A contains more than B' and 'B contains more than C', and they were reminded of their conclusions by the tester before being asked whether they could decide which contained more, A or C, without measuring. Only 5 per cent of the middle-ability 7½ year olds and 32 per cent of the 9 to 10 year olds were able to make this final *transitive* step.

The next stage, after qualitative comparison, would appear to be the measurement of liquid volume, and hence the capacity of containers, in standard units. Ward posed the following situation to 10 year olds. It was an open-ended question which was useful for investigating the different interpretations given to it by the children, reflecting some of their ideas and conceptual development in the volume/capacity area of measurement.

This is a fish tank nearly full of water. If you had the job of finding out just how much water there was, how would you do it? (Ward, 1979)

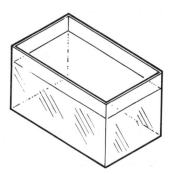

Fig. 2.63

4 per cent took 'how much' to mean 'how heavy' and described procedures involving weighing the tank plus water and weighing the tank empty and then subtracting to find the weight of the water.

19 per cent interpreted the situation in terms of capacity:

'You would get a milk bottle and fill it with water from the fish tank. Count how many times you did it until the fish tank is empty.' (Ward, 1979)

6 per cent gave such an answer in terms of litres whilst 11 per cent mentioned pints or gallons.

4 per cent of the ten year olds tackled the situation in terms of volume i.e. measuring the dimensions of the tank. Often it was not clear whether the height was for the tank or the water level. Few children made any mention of accounting for the unfilled space at the top of the tank.

11 per cent of the children described a 'dipstick' method – either placing a stick into the water and seeing how much was wet or measuring with a ruler up the side of the tank. It was suggested that this category of answer may have resulted from a particular primary mathematics scheme which uses the case of a petrol tanker driver using a calibrated dipstick, to introduce this concept. Alternatively the children might, like Piaget's non-conservers, have used height as a measure of volume. These children have certainly not understood that the other two dimensions need to be taken into account.

The first APU Secondary Survey (APU 1980b), in an unpublished item, required 15 year olds to judge which of three containers could have the most water added to it. All the containers had a circular cross-section and all were partly filled to the same height at a point where the cross-sections were equal. However, one container was diverging, one was cylindrical and the other converging. Thirty per cent made the correct choice. APU states that in a similar item on the Primary Survey, but requiring the vessel *containing* the most water to be chosen, seventy per cent of 11 year olds were correct. No further details were given.

It is not easy to compare the difficulty of grasping the notion of 'the volume of a liquid' with that of 'the volume of a solid', since few of the items used are directly comparable. However, in cases where they are, for instance when Rothwell Hughes tested conservation of solid and liquid 'substance' on the same sample, the liquid conservation was found to be somewhat more difficult (see the tables on pages 132 and 138).

Ward's result concerning the tank of water, referred to above, certainly suggests that only a minority of primary school children have really grasped the idea of liquid volume.

2.4.2.5 *What Difficulties do Children Experience with the 'Displaced' Aspect of Volume?*

Section 2.4.2.2 (page 131) was concerned with the notion of 'external' volume in the sense of: 'Which has more stuff in it?' or 'Which takes up more room?'. The latter question would appear to have a slightly different connotation, since it considers an object in relation to its surroundings. However as reported on page 131 Rothwell Hughes (1979) found that the second form of the question was answered almost as well as the first.

This relationship between objects and their surroundings can be made more precise by putting each of a pair of objects into a container, either empty or filled with water, and asking the child: 'Which takes up more room in the box?'

Rothwell Hughes (1979) used two empty perspex boxes into each of which he placed a block. When the blocks were of the same heights but of different thicknesses, 86 per cent of children in the middle 50 per cent of the ability range of 7½ year olds, and 95 per cent of a similar group aged 9 years 11 months, correctly judged that the thicker one took up more space. However when two blocks were demonstrated to have equal dimensions, but one was put in standing up and the other on its side,

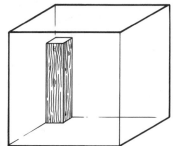 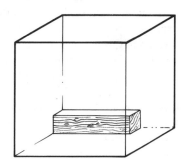

Fig. 2.64

only 47 per cent of the middle 7½ year olds and 75 per cent of the middle 9 to 10 year olds agreed that they took up the same amount of space, the majority of the remainder focusing on the height. These percentages were generally rather less than for the 'conservation of substance' tasks.

Piaget and Inhelder (1974) also found that the notion of 'occupied volume' (i.e. the amount of space taken up) was acquired later than that of 'amount of stuff', but that even when children agreed that two objects took up the same amount of space in a box, they did not always agree that the 'amount of space left' was the same.

Another Piagetian task involved a plasticine ball which was rolled out into a 'sausage' and dropped into a narrow glass of water. Children were asked whether the 'sausage' would take up 'as much room in the water and make the water rise' to the same level as would the original ball. This notion that volume of an object is equivalent to the volume of liquid it displaces when immersed in a container of water is known as 'displaced volume', and it has generally been found to be the most difficult qualitative aspect of volume to acquire.

Piaget and Inhelder's figures show no child under age 7 as deciding that the 'ball' and 'sausage' would make the water rise the same amount but 12 per cent of 7 year olds, 32 per cent of 9 year olds, and 82 per cent of 11 year olds produced this conclusion. Beard (1963), working here in the U.K., also found 31 per cent of 'conservers' among 9 year olds.

As in other cases, the percentages of successful children could be reduced by introducing complicating factors into the experiment e.g. using two identical metal cylinders but allowing one to sink to the bottom of a jar of water and suspending the other half way up (although fully submerged). This is described in Nuffield Checking Up III (Nuffield Mathematics Teaching Project, 1973), in which it is suggested that the child may either say

'the one that's only halfway down the jar can't push the water up as far as the other one that's gone right to the bottom',

thus showing an ill-defined grasp of the concept of volume, or

'The one at the bottom weighs more than the one in the middle because it's sunk – you're holding part of the weight of the other one,'

thus confusing weight with volume.

Using an almost identical task to this, Rothwell Hughes found very low success rates, ranging from 7 per cent of middle ability 7½ year olds to 16 per cent of middle ability children aged 9 years 11 months. The remainder expected the object to displace more water as it sank deeper, generally for the same reasons as those suggested earlier.

Also described in Checking Up III is an experiment due to Inhelder and Vinh-Bang in which they use cylinders of identical shape and size but made of different metals, and therefore of perceptibly different weights. Only 7 per cent of 8 year olds, 15 per cent of 10 year olds and 37 per cent of 11 year olds could give a reasonable explanation of why the water levels were the same even when the weights were obviously different. This task was repeated by Beard (1963) here in the U.K., who obtained correct predictions and reasons from 9 per cent of 9 year olds, but more surprisingly from 19 per cent of 7 year olds!

Apparently paradoxical results were also obtained by Rothwell Hughes (1979), who found with the same experiment that the success rate varied from 15 per cent of 7½ year olds, down to 9 per cent of children aged about 9 years, and up again to 26 per cent of children aged 9 years 11 months, with almost all the remainder centring on weight as the relevant attribute and hence predicting that the heavier object would displace more water. (Again these figures relate to children in the middle 50 per cent of the ability range.)

However when two objects of the same weight and different volume were used, some children changed their minds and decided that volume rather than weight was the relevant attribute, thus predicting correctly that the larger object would displace more water (31 per cent of 7½ year olds, 24 per cent of 9 year olds, and 34 per cent of children aged 9 years 11 months, all middle ability, succeeded with the task).

This again confirms the fact that children judge according to appearances: if the objects are seen to differ significantly in any way it will be predicted that they will displace different amounts of water.

The research results reported here have a very important implication for teaching, since they suggest that children at the top of the junior school and the lower end of the secondary school may be confused by methods of measuring irregular volumes which depend on displacement. The link between the volume of the object, the amount of rise in the water level, and the volume of liquid collected in a separate container as a result, is likely to be at best a very tenuous one for the majority of children of this age.

2.4.2.6   *Some Suggestions for a Scheme of Work Concerning Volume and Capacity*

Kerslake (1976) offers suggestions for a scheme of work concerning volume and capacity which she considers suitable for roughly the 5 to 11 year old age range. Although her suggestions are too lengthy to be reproduced here, they can be summarised as follows:

A.   Work involving comparisons of the capacity of different containers. Ordering in terms of this capacity and filling and measuring capacity in terms of 'informal' units such as spoonsful of water to a bottle, number of marbles which fit into a bag.

B. Collecting containers of similar capacity, e.g. wine bottles, paint tins which hold one litre.

C. Dividing a quantity, such as a litre, into equal proportions leading to experience of ½ litre (5dl or 500 ml) and ¼ litre (2.5dl or 250 ml).

D. Making objects of a particular capacity and volume, e.g. a cubic metre frame.

E. Making different shapes of the same volume using identical units, e.g. different shapes from five bricks.

F. Looking at the volume of solids using, say, plasticine and moulding it into a 1000 cubic centimetre cube. Dividing it up and checking by weighing perhaps.

G. Using different substances to fill identical beakers with the same volume and comparing the weights.

H. Activities involving the relationship between surface area and volume, on an informal level, e.g. make different rectangular blocks from eight bricks. Cover each large block with sticky coloured paper. Will each rectangular block require the same amount of paper? Do they each take up the same amount of space?

I. Comparing cost and capacity, e.g. best buys for washing-up liquid, miles to a gallon for different models of cars.

J. Introduction to standard units by making cubes of different sizes. Finding the volume of an irregular solid e.g. potato, by displacement of water.

K. More sophisticated measurement of surface area to volume as in H but using standard units rather than visual approximation.

This scheme of work covers virtually all the difficulties children experience with volume and capacity that have been mentioned in this chapter. It would help to clarify their confusions with occupied space, internal volume and capacity, volume and weight, and volume and surface area. It mainly concerns practical activities which are necessary for children to develop an appreciation of the approximate and imprecise nature of measurement.

2.5 TIME AND MONEY

2.5.1 Time

2.5.1.1 The Distinction between the 'Telling of Time' and the Concept of Time

As Lovell points out:

> That a child can tell the time on the clock does not necessarily imply that he has a concept of time. Telling the time is no more and no less than dial-reading. One may read a dial without having any concept of what it is that the dial registers.
> Children may be quite capable of working the four rules of number in relation to days, hours, minutes and seconds, without having a concept of time in which instants and intervals on the time-continuum are co-ordinated in the mind. (Lovell, 1966)

But although children *can* be trained to 'tell the time' and to calculate with times before they have a corresponding concept of time, this may prove an uphill task. For instance, in the parallel book to this volume, an example is given:

Ian, a 6 year old in an infant school, had an exercise book in which his teacher had stamped some clock faces and a work card which said: 'half-past 8', 'half-past 11', etc. He had drawn hands of equal length on the faces to represent these times and for the first one had put:

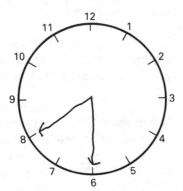

Fig. 2.65

with the hour hand on the eight.

Ian had no idea what time it was (it was about 10 a.m.) and, although he knew that dinner time was at twelve o'clock, when asked whether it was before or after dinner he was not able to reply. When the question was discussed more carefully, he realized what he was being asked, and said: 'Oh, I see. Dinner's coming.'

(Denvir, Stolz and Brown, 1982)

Indeed, Kerslake (1975) emphasises the fact that a preoccupation with activities aimed at teaching children to 'tell the time' can often be detrimental to their development of the concept and measurement of time. For instance, in a situation where the teacher is altering the hands of a clock for a class to practise telling the time, the concept of the passage of time is greatly distorted. For example, the teacher may adjust the hands to make it 'half past two' and then a few seconds later presents the children with a situation which shows 'a quarter past ten'.

Similarly if a child is presented with a page of clock faces all showing different times he may not grasp that at any point in time the measurement of time as shown on a clock face can only be one particular setting of the hands. Nor will the child necessarily appreciate the manner in which the hands move, in particular the relationships between the speeds of rotation of the second, minute and hour hands.

2.5.1.2 How Does an Understanding of the Concept of Time Develop?

Piaget's conclusions concerning the concept of time (Piaget, 1969) are summarised in the Nuffield Project publication 'Checking Up III':

Before children acquire the concept of time, they must grasp two important facts:

(i)   that there are series of events which occur in a temporal order, and
(ii) that between these events there are intervals whose duration must be appreciated.

Some evidence of the development of these two aspects of *order* and *duration* can be obtained by observing children's understanding, and, later, spontaneous use, of words which relate to the passing of time.

Ames (1946) undertook a comprehensive study of the development of the concept of time among young American children of 'superior intelligence', gathering the data by long-term observation of children's spontaneous language in nursery school, together with some more systematic interviewing. She arrived at the results shown in the following table, in which is shown those accomplishments which were achieved by a majority of her sample at a particular age.

Although the ages given are clearly very inappropriate for low attainers (even within her 'superior' sample, Ames gives evidence of wide differences in progress), the information indicates a probable sequence of development.

Stages of development reached by a majority of children of 'Superior Intelligence' in the concept of time

18 months

The 18-months-old child lives in the immediate present and has little if any sense of the past and future. He cannot wait. No time words are used by him, but he responds to the word "Now" and in his time psychology is interested in the "now."

There is some slight sense of timing: he may roll a ball and wait for it to stop before he pursues it. The sight of juice and crackers may bring him to the table.

21 months

The child still lives chiefly in the present. His chief time word is "Now." Projection into the future begins to come in. He will wait in response to "In a minute." There is an improving sense of timing: two children may rock in rhythm; or the child may sit at the table and wait for juice.

24 months

An important advance takes place at this age. Though the child still lives very much in the present, several words which denote *future* time (especially "gonna" and "in a minute") become part of his own spoken vocabulary. He will wait in response to such words as "wait," "soon," and "pretty soon."

He now has several different words to indicate present time: "now," "today," "aw day," "dis day." He uses no specific words implying past time, but is beginning to use the past tense of the verb, often inaccurately. The 2-year-old cannot answer questions involving time concepts, but he comprehends simple sequences as in the adult's promise, "Have clay after juice."

30 months

Though the child's vocabulary of time words may still be very limited (comprising probably not more than twenty or so different words), a definite advance takes place at this age in that he now uses freely words implying past, present and future time, having numerous different words for each. Thus the finer divisions of time— "morning" and "afternoon"—have been added to "day" to indicate present time. The future may be indicated by "some day," "tomorrow," and several others.

There are fewer different words for the past than for the future. Past time is usually designated as "last night." Though "tomorrow" is used, the word "yesterday" has not yet appeared.

Altogether, his time expressions sound quite versatile at this age, in spite of the smallness of his time vocabulary. He uses freely names of the days of the week.

36 months

More different new time words come in between 30 and 36 months than in any similar interval. Most of the more common basic time words are now in the child's vocabulary. Past, present and future are all referred to though there are still more different words for expressing the future than the past. There is now nearly as much spontaneous verbalization about the past and future as about the present.

Expressions of duration—such as "all the time," "all day," "for two weeks" come in at this age.

There is a pretense of telling time, and spontaneous use of clock time phrases usually inaccurate. There is much use of the word "time" alone or in combination. Thus, "What time?", or "It's time," as well as "lunchtime," "puzzletime" etc. Phrases beginning with "when" are very common.

The child of three can answer a few questions involving time concepts. He can tell how old he is, when he goes to bed (in terms of some other activity) and what he will do tomorrow, at Christmas, in the winter.

At 3 years, and at no other age conspicuously, the child may perseveratively answer all questions about time with some one inappropriate clock time or some one number (as "fifty-nine").

42 months
Expressions indicating past, present and future time are now used in spontaneous conversation to an equal extent. Past and future tenses are used accurately.

There is not so much an increase in number of time words used at this age, as in the refinements of use. The whole week," "in the meantime." Or such phrases as "Two things at once." Also there come in at this age many different new ways of expressing sequence.

There is not so much an increase in number of time words used at this age, as in the refinements of use. The child says, "It's almost time," "a nice long time." He expresses habitual action, as "On Fridays."

With increasing complexity of time expressions comes a confusion characteristic of 42 months. The child frequently refers to future happenings as in the past. Thus "I'm not going to take a nap yesterday."

Ability to answer questions about time is not much greater than at 3 years.

48 months
Past, present and future all continue to be used freely and about equally. Many new time words or expressions are added at this age. Particularly is the word "month" used in different contexts. Also such broad time concepts as "next summer," and "last summer," are used accurately.

By this age the child seems to have a reasonably clear understanding of when events of the day take place in relation to each other.

5 years
Data are limited from this age on. Present data indicates that in addition to the preceding concepts, which constitute a free verbal handling of the more common aspects of time, the child can tell what day it is, can name the days of the week in correct order, and can tell what day follows Sunday. He can also project forward to tell how old he will be at his next birthday.

6 years
At this age comes an understanding of the four seasons, and an increasing knowledge of duration.

7 years
Now the child can tell the season, the month and the specific clock hours (including how many minutes of or past the hour). The larger concept of what year it is, is still beyond him.

8 years
At this age the child can handle well extremes of time. He can not only tell the clock time but can also tell what year it is and what day of the month. He also indicates an understanding of the more generalized concept of his ability to answer the question: "What does *Time* mean?"

To give an example of how advanced these age norms are, Ames found that the majority of her bright sample were able to say whether it was morning or afternoon at 4 years of age, whereas Bradley (1948), in a more representative sample, found that this distinction was not achieved until 6 years, on average.

Bradley also found that facility in handling words connected with the calendar, like weeks, months and years, was not achieved until 8 years of age, with more general understanding of duration of time, and the distinction between time and space (as discussed in the next section), not until after 8 years of age. One of the problems with Ames' data is that, as Lovell (1966) comments: 'the use of the words does not ensure that he has a grasp of the concept'. Thus knowing his age does not necessarily mean that the child has a concept of age, as is demonstrated in the following text. However, it is worth noting, that even in Ames' bright group, a majority were not able to tell the time until they were 7 years old, which was after they had developed an intuitive sense of time and an appropriate vocabulary.

2.5.1.3   What are the Difficulties Children have with the Concept of Time?

Piaget (1969) undertook a number of experiments concerning the two aspects of the concepts of time which he identified, namely, *order of events* and *duration of intervals*. Many of these were replicated in England by Lovell and Slater (1960). In many cases these illustrated particular confusions which were not resolved until the child reached the age of 7 or 8 years or more.

Piaget (1969) investigated the idea of *order of events* by showing children a sequence of pictures of an event which they had previously observed (a bottle being filled up, and, in another experiment, an object falling) and asking the children to arrange the pictures in order. He observed that children were not able to do this until the age of 7 or 8 on average, but 15 per cent of his sample could not do it at the age of 8. He suggested that the ordering of a time sequence requires the ability to reverse an operation, and that thus, like the ordering of objects by length, it is not achieved until the start of the concrete operations period. The implication is that many low attainers in the junior school may not have a firm grasp of this idea of a temporal sequence of events, and that the activity of ordering pictures representing such a sequence, for instance in a well-known story, or in an action which they first observe for themselves, is therefore a useful teaching approach.

As for the *duration of time-intervals*, Greenes says that with respect to time measurement some low attaining children have little or no concept of duration.

> Duration implies an understanding of 'now', 'later', 'before', 'sometime' and constancy of motion, a recognition that time is steady flow, not related to the amount of work completed....children are frequently confused by temporal terminology in the form of time comparisons. This may be due partially to the fact that in the English language many of the same terms are used to describe spatial (measurement) and temporal ideas. We use 'long' to refer to length, and 'long' when referring to time. (Greenes, 1979)

In Section 1.9 on page 49, it was noted that children experience difficulty in differentiating between time and space. A specific example of this is given by Piaget (1969).

> Hes (aged 4 years, 5 months) was shown two dolls, one blue and one yellow, who 'walked' along parallel paths.

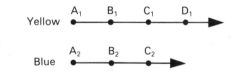

Fig. 2.66

The experimenter started the yellow and blue dolls off simultaneously at A_1, A_2 respectively. The yellow one was moved much faster than the blue one, so that by the time the yellow one had reached D_1, and stopped, the blue one had only reached B_2. However the blue one was then moved on to C_2 while the yellow one remained at D_1. Thus the blue one had been moving for twice as long, but had not travelled as far as the yellow one.

Experimenter:	'Did they stop at the same time?'
Hes:	'No.'
E:	'Which one stopped first?'
H:	'The blue one.'

E:	'Which moved for longer?'
H:	'The yellow one.'
E:	'When this (yellow) one stopped it was lunchtime, so did this (blue) one stop before or after lunch?'
H:	'Before.'
E:	'Let's do it again.' (Repeats experiment)
H:	'The yellow one stopped first, the blue one was still moving so the yellow one went on longer.'
E:	'But did one stop before the other?'
H:	'The blue one.'

(Piaget, 1969)

Lovell and Slater (1960) with a sample of British primary school children, used a simpler version of this task, in which the two dolls started and finished at the same instants, signalled by clicks, but travelled different distances. In this case the percentage of correct replies given to questions concerning the time for which the dolls travelled ranged from 13 per cent, for the 5 to 6 year olds, to only 43 per cent for the 9 to 10 year olds, although the sample contained only 'average' and 'above average' children.

Although there appears to be a confusion between 'longer' as referring to either distance or time in the above interview with Hes, Piaget demonstrated that this was not the only difficulty by repeating the same type of experiment in different situations. For instance when the experimenter and the child (Per, aged 4½) started off moving across the room together at the same time, and reached the other wall simultaneously, the result was:

Experimenter:	'Did we take the same time?'
Per:	'No. I took longer because I was running.'

Similarly when two taps, running at the same rate but into different sized bottles, were turned on, and later off, simultaneously, Per denied this:

Experimenter:	'Did we stop them at the same time?'
Per:	'No, because this bottle (the larger) is not full (like the other).'

(The experimenter repeats the action, this time counting 1... 2... 3...)

E:	'Did we stop them at the same time?'
Per:	'No.'

(Piaget, 1969)

(Lovell and Slater, 1960, also replicated this experiment with their sample of British primary school chilren of 'average and above average' ability. Only in the 8 to 9 year old age group did half of the children interviewed correctly reason that the two bottles had been filling for the same time, although almost all of the 9 to 10 age group were able to reply in this way.)

Thus the child tends to judge duration of time according to visual perceptions (distance, capacity) or motor perceptions (speed).

Another instance of the failure of children to distinguish between space and time occurred in a study by Sturt (1925) in which children were asked first which day of the week it was, and then which day of the week it was at a nearby town. One third of the 8 year olds, and 15 per cent of the 10 year olds, gave different answers to the two questions.

Piaget (1969) also gives evidence that children of under eight years on average also tend to identify age with height. For example Vet (aged seven years, ten months) gave the following responses:

Experimenter:	'When you are old, will Florian (a younger brother) still be younger than you?'
Vet:	'No, not always.'
E:	'Does your Father grow older every year?'
V:	'No, he remains the same.'
E:	'And you?'
V:	'Me, I keep growing bigger.'
E:	'When people are grown-up, do they get older?'
V:	'People grow bigger and then for a longer time they remain the same and then quite suddenly they become old'

(Piaget, 1969)

Kerslake (1975) considers it advisable to give children 'timing' activities to develop their concept of duration. Although the child is usually first confronted with duration in terms of hours, half-hours, and quarter-hours it is probably more meaningful to begin with seconds and minutes. Sand-clocks, or a stop-watch, are very useful as measuring instruments for activities of short duration. Seconds and minutes are much more accessible to the child as units of measure than is the hour because he is far more able to give his concentration to short intervals of time. This also greatly facilitates his ability to estimate and judge periods of time which are readily within his grasp.

2.5.1.4 What are the Problems with Telling the Time?

Of course the child has to learn to tell the time but it must be remembered that the traditional clock face although it may say 5 o'clock does not tell you whether it is morning, evening or which day of the week it is. Thus, unless the clock or watch is equipped with extra displays, like the day and date, the child still requires some concept of the flow of time and the relevant vocabulary.

The child also needs to understand the continuous nature of time, so that by the time he has worked out that it is 'a quarter past five,' it is no longer exactly that time but some while later. Certainly the consideration of the accuracy of timepieces – whether clocks or watches show the same time – whether one is faster or slower than another – and the idea of the standard of Greenwich Mean Time against which these are measured, give opportunities for developing the approximating and continuous nature of the measurement of time, but little if any literature is available on this.

Nor is it known whether the widespread use of digital clocks and watches will assist children's conception of time as a continuous quantity. However digital watches are easier for telling the time than traditional ones, since, as Harris (1981) points out:

To use statements like '14.45' you need to be able to read and understand numbers up to 60; you need to know there are 60 minutes in an hour, ...

Nevertheless in order to convert between such a time and a standard spoken form (in this case, 'a quarter to three') requires considerably more knowledge and number relationships.

Kerslake (1975) contrasts the ease of reading a gas or electricity meter, which is not normally taught until the late secondary school if at all, and which the gas and electricity boards do not trust customers to do without guidance, with the difficulty of telling the time from a traditional clock face:

> Compare this (meter) with the two hands of a clock, moving at different rates over a single scale with digits 1–12, which have to be interpreted differently, depending on the size of the hand. So a hand pointing at a seven may call for the response 'seven', or 'twenty-five to' or 'thirty-five past'. A hand pointing to a nine is sometimes to be interpreted as 'quarter-to' in addition to 'nine', 'forty-five, 'fifteen'. To add to the confusion, some watches have no digits at all marked on them, and the child will see clocks with Roman numerals, twenty-four hour clocks and digital clocks. With all this, eight-year olds are expected to be able to tell the time, while adults are not expected to read meters.
>
> (Kerslake, 1975)

Often the difficulties children experience in telling the time from the traditional clock faces result, according to Greenes (1979), from a child's poor spatial ability. This is particularly the case if he has difficulties with directionality in terms of differentiating between left and right and up and down leading to problems in distinguishing between 6 and 9; three o'clock and nine o'clock; 11:45 and 12:15 etc.

Springer (1952) gives a breakdown of the relative sources of error among a small sample of 4 to 6 year olds in the United States:

Types of error in reading a clock

	Percentages of all errors		
Type of error	4 yrs	5 yrs	6 yrs
Minutes incorrect, hour correct	6%	27%	44%
Hour identified with long hand	11%	21%	29%
Counting incorrect	7%	26%	5%
Inaccurate identification of numbers	13%	8%	0%
Others – including guesses	63%	18%	22%

Many 5 year olds counted incorrectly, failing to coordinate saying the numbers and pointing to them. Most of the inaccurate identification of numbers was due to confusion between, for instance, 6 and 9, 8 and 0, 5 and 2, 11 and 2, and difficulties with 10, 11 and 12.

In all, 77 per cent of 4 year olds had *no* answers correct, compared with 28 per cent of 5 year olds and 10 per cent of 6 year olds. Sample results are given in the following table:

Accuracy in 'telling time' (data from Springer, 1952)

	Percentages giving correct responses		
Time	4 yrs (n = 22)	5 yrs (n = 36)	6 yrs (n = 31)
8.00	14%	53%	74%
8.30	0%	11%	32%
8.15	0%	8%	13%

Setting a clock to a particular time was generally found by Springer to be much more difficult than reading a given time.

Kerslake (1975) considers it best to relate clock reading to activities which have some meaning in the child's life. For instance the times of television programmes, when he has to leave home for school, bedtime etc., even though these are not necessarily on the hour or half hour activities. In fact they are more likely to be what she calls 'awkward' times, but providing the child can relate to them in a meaningful way they are the best starting point for telling the time. For example, Springer (1952) found that 50 per cent of 5 year olds and 84 per cent of 6 year olds were able to give a reasonably accurate answer to the questions 'What time do you leave school?' Ames' bright children also became familiar around the age of 5 to 6 years with the times at which they got up, went to school, had supper, and went to bed.

2.5.1.5 The Tabulation of Time

Often children come across exercises involving the reading of some sort of tabulation of time. This may be in the form of timetables for public transport, or calendars etc. There appears to be little research on how children cope with, say, the symbolic representation of the 24 hour clock as it appears on such timetables (e.g. 14.37), or on their understanding of the concepts of duration as presented in this format. However familiarity with digital watches and clocks especially 24 hour displays is likely to help here, once the initial problems of translation are overcome. In the main items used in research test children's ability to read tables and draw certain conclusions from the information without really tapping their depth of understanding of the measurement of time which underlies these tabulated forms.

Ward (1979) showed ten year olds a table giving the dates of birth of five children as follows:

Ann	14th June 1964
Brian	20th April 1963
Simon	23rd March 1964
Mark	12th August 1963
Ellen	2nd May 1964

Who is the oldest?

59 per cent correctly chose Brian but 21 per cent chose Simon: possibly because they only attended to the month of the year – March coming before the rest – rather than taking into account the actual year of birth as well. Alternatively they may have centred on the date of the month – 23 being the highest number therefore he is the oldest! Obviously much further probing is necessary to determine to what extent the actual concept of time is lacking and to what extent the symbolisation of this is misunderstood.

When asked for the youngest child, only 37 per cent were correct.

The first APU Primary and Secondary Surveys (APU 1980a,b) with 11 and 15 year olds respectively, included a similar item involving the dates of birth of five boys. The results are given in the following table:

Results for APU item on 'ages'

	11 year olds	15 year olds
Correctly selecting the oldest	55%	80%
Correctly selecting the youngest	40%	70%

When the 15 year olds in the APU survey were given a date of birth only 30 per cent could calculate the age in years and complete months. How far this reflects an inadequate grasp of time, or their inability to perform the necessary calculation, is not known and this is where far more research is essential in order to identify the real nature of the problems experienced by children.

The first APU Primary Survey (APU 1980a) included an item which involved reading a calendar for the month of February and drawing conclusions on the basis of this concerning matters not directly evident from the actual information which was presented.

		February			
Sunday		5	12	19	26
Monday		6	13	20	27
Tuesday		7	14	21	28
Wednesday	1	8	15	22	
Thursday	2	9	16	23	
Friday	3	10	17	24	
Saturday	4	11	18	25	

This is a calendar for the month of February. On what day of the week is
a) 1st February? ....
b) 1st March? ....
c) Is this calendar for a leap year? ....

APU report that 92 per cent of 11 year olds could answer part (a). 78 per cent showed understanding that 1st March follows 28th February, thus displaying some appreciation of the structure of the calendar. Fifty-nine per cent indicated that the calendar was not for a leap year, nearly all justifying their answers with some reference to 29th February as an 'extra day'.

The first APU Secondary Survey (APU, 1980b) presented their 15 year olds with a

small section of a railway timetable (and) 80 per cent could read the time of arrival of a specified train, but less than half could calculate the journey time when the subtraction required "borrowing".

Again this type of item does not really test the child's understanding of the measurement of time, but rather his ability to read a timetable and perform a calculation.

2.5.2 Money

2.5.2.1 *Money and Measurement*

The 'hotness' of a room can be measured using a scale of temperature (e.g. 17°C); the duration of an action can be measured using a scale of time (e.g. 3.5 seconds); the amount of water in a jug can be measured using a scale of capacity (e.g. 2.3d1). Thus all systems of measurement provide a method of translating the extent of some physical property into a number of standard units, where the units themselves are of arbitrary size.

The monetary system has much in common with other measurement systems, since it provides a method of indicating the 'economic value' of something into a number of standard units (e.g. £27.50), and again the units chosen are essentially arbitrary. (For instance the £ sterling could be replaced by dollars, pesetas, and so on.) Because of these similarities, money is included with other types of measurement in this book, as in other places.

However it should be borne in mind that there are several important differences between the monetary system and other systems of measurement, for example:

a. There is in general, *no process of measuring* 'economic value' which is analogous to, for example, measuring area by counting the number of square centimetre units, or measuring temperature using a thermometer, either with a scale or with a digital display. 'Economic value', or 'price' is normally determined by a person or group of persons, and hence the usual way to find out a price is to read a notice, or table, or simply to ask someone (There are occasions when cost appears to be 'measured', for example in selling goods by mass/weight; once the 'unit price' is determined then the scale which measures mass/weight can be adapted to 'measure' the cost of the goods. For example a scale used in butchers' shops or supermarkets may give a digital display of both mass/weight and cost, as does a petrol pump.)

b. Money is *essentially discrete;* there is no unit smaller than ½p in the British currency system. This means that the idea of increasing approximation, by the use of decimals or by subdividing the unit of measure, as discussed in Section 2.1.2 on page 83, is generally inappropriate. It may be legitimate to work out that if a tube of ten sweets costs 32p then a single sweet costs 3.2p, but it is not feasible to pay this amount, nor is there any sense in which it is an approximate answer which could be made accurate with more exact measuring instruments.

c. Money involves a *system of exchange using physical tokens.* It is true that 1 centimetre is equivalent to 10 millimetres, but we do not represent these units by 'tokens' in the same way that we deem a certain piece of metal to represent 10p, and another to represent 20p.

In spite of these differences, money is the system of measurement most frequently encountered in everyday life, as is shown in the following text.

2.5.2.2 *Money in Everyday Life*

In order to investigate which arithmetical processes were most often used in employment and in everyday life, Thompson (1962) surveyed a group of very low-attaining children soon after they left school. He found:

> The greatest single use of arithmetic, both in leisure time and during working hours found by the investigation was that associated with money in shopping activities. By using a liberal interpretation of the word 'shopping' (dancing, pictures, etc.) 88% of all arithmetic was confined to buying either goods or services of one kind or another. (Thompson, 1962)

Most problems were oral, except for a simple form of budgeting and completion of time sheets. The language of number was frequently required, but actual number work was limited to reading and understanding of numbers and the use of addition and counting. Thompson also investigated the areas of hire purchase and insurance policies. He found that insurance policies were undertaken with no real knowledge of the term of the policy or even of the final amount involved. With hire purchase, while some of the subjects had a rather hazy idea that articles did cost more when purchased by this system, they were completely ignorant of the actual amounts involved. Others had no idea at all that any H.P. charge was in fact made, and simply accepted the shop assistant's assurance that it was the cheapest they would buy anywhere.

Thompson concluded from his research that if very low attaining children were to get a proper understanding of how to deal with money (which was most essential if they really were to become occupationally and socially adequate after leaving school) the work at school had, above all else, to be practical. He saw it of paramount importance to give guidance to older children on issues of hire purchase and insurance as well as familiarity with coins, vocabulary and immediate application of monetary skills.

It is sometimes assumed that children simply acquire the ability to handle money from their everyday life experiences, without requiring much specific instruction in school. However this is not necessarily the case. Stringer (1979), reporting on the effectiveness of an 'adult numeracy' television series (*Make it Count*), found:

> There are indeed many adults in Britain who have the greatest difficulty with even such apparently simply matters as adding up money, checking their change in shops or working out the cost of five gallons of petrol. Yet these adults are not just the unintelligent or the uneducated. They come from many walks of life, and some are very highly educated indeed...

There is some evidence that children do indeed pick up some knowledge from outside school. For instance Rea and Reys (1971), in the United States, found that of children starting school at five years old, over 75 per cent could already identify common coins, over 50 per cent could identify common notes, but less than 25 per cent were able to give change. Nevertheless these figures appear very high and may suggest that the sample was not representative.

The 11 year olds tested in Britain in the first APU primary survey (APU, 1980a) were, in fact, still far from completely competent:

> Items testing ideas about notation and coinage generally gained facilities of 70% or more, but when halfpence were involved the success rate dropped to just below 50%.

> A group of items was concerned with applications set in everyday contexts, such as buying, bills and savings. The facility values ranged from 85% down to about 15%....

(No further details of these items are given.)

One reason why some children do not necessarily 'pick up' money ideas is that, although they may be sent shopping, this does not necessarily require any decision making. For instance Kelly (1967) gave a questionnaire to a group of very low-attaining children and discovered that although many of the children went shopping, they usually went to the corner shop with a written note and the goods purchased were sent out on a weekly bill, or alternatively the money was wrapped in a piece of paper on which the required articles had been written down, and the change was returned to the child, often wrapped in the same piece of paper. Many of the children said that they never attempted to find out if they had been given the right amount of change, but just accepted it.

This suggests that it is necessary to ensure that children can not only perform arithmetical manipulations using money which are often presented in textbooks, but that they can actually use money in a real-life context, making decisions about what to buy, what coins to offer and what to expect in change.

Various aspects of the use of money are discussed in the following text. It should be noted that, although the ability to work with money is obviously essential to adult competence, there is very little relevant research in this area.

2.5.2.3   What Perceptual and Mathematical Skills are Needed?

Gibson (1981) identifies three levels of perceptual and mathematical skills required in handling money:

(a)   coin recognition; ;
(b)   equivalence;
(c)   practical situations.

This classification will be followed here, with each aspect discussed in turn.

a.   *Coin-recognition*   The problems some low-attaining (or young) children find in identifying coins is illustrated by a conversation recorded by Gibson:

> Gary (shown 10p and 5p pieces) Yes, they match... no they don't. Both got hair, grey, they've got numbers on. One's 10 and one's 5. (Gibson, 1981).

(It is made clear that the last sentence was said while the boy was actually reading the numbers 10 and 5 on the coins.)

Here Gary was attending to unimportant features (e.g. that they both had the Queen's 'hair' on one side), while missing the important feature of relative size. It also shows that at least some children become familiar with written numerals before they learn to recognise coins on sight.

Gibson found that by the second year in primary school (i.e. aged 6 to 7 years) children assessed as 'of average ability' could identify common coins and notes, whereas this was not true of the average 5 to 6 year olds. (However the samples were very small.) Similarly with low-attaining children of primary age, all those whose intellectual competence was assessed to be equivalent to that of an average 6 year old or above succeeded. However a few older children who were generally functioning below the 6 year old level could identify the coins and notes, which suggests that plentiful experience with using money may be the over-riding factor in what is essentially a perceptual task.

b.   *Equivalence*   This involves the understanding of the relative value of the coins, and is thus dependent on a grasp of the number system itself, and in particular of the relationships between the numbers ½, 1, 2, 5, 10, 20, 100, etc. (in the British system).

Each coin or note has to be appreciated in terms of

  i. its *relative* value (e.g. a 10p coin is worth more than a 2p coin but less than a 20p coin).
  ii. its *unit value* (e.g. 10p coin is worth the same as ten 1p coins, taking these as the unit of currency)
  iii. other *equivalences* (e.g. a 10p coin is worth the same as two 5p coins, or five 2p coins, or one 5p coin and five 1p coins, or half as much as a 20p coin, etc.)

Thus an appreciation of equivalence incorporates many of the basic notions of number, as discussed in Section 3; in particular it involves conservation, counting, ordering, addition, doubling and halving and number bonds.

A choice has therefore to be made between

  a. waiting until the child has developed a particular number relationship before introducing this in the setting of money,
  b. introducing ideas in number by using money as a concrete embodiment, and
  c. running the teaching in parallel in the two areas.

In Section 2.1.8.3 on page 94, it was suggested that measurement, in particular that of length, area and volume, might be an appropriate way of *introducing* number concepts in a concrete way. Thyer and Maggs however point out that:

> ...money is much more difficult to teach than length or capacity, for example, because the relationship between coins cannot readily be seen. Ten decimetre rods can be put end to end against a metre stick to show equivalence, and ten decilitre measures of water poured into a litre measure, but we cannot show that ten pennies have the same value as a tenpenny bit. Pennies and tenpenny bits are of different metals and the exchange rate between them has to be accepted. It is true that 2 halfpennies have the same mass as a 1 penny, and 2 pennies the same mass as a twopenny bit, but the teacher would need an extremely sensitive balance to show these relationships. (Thyer and Maggs, 1971)

This suggests that perhaps, at least at the early stage, money may be a less appropriate introduction to number than other measures.

Gibson compared the performance of both average and low-attaining children on individually administered tests of number and money, including some items which were designed to be parallel e.g.

> Here are three cars. Each has 4 tyres. How many tyres are there altogether?
>
> I want to give 4 pence to each of these three friends of mine. How much do I need altogether?
> (Gibson, 1981)

(Equipment was available to the child e.g. cars, a pile of pennies, pictures of the friends.)

In almost all cases the 'number' items was found to be slightly easier than the money item, although in general children either succeeded with both or failed with both.

This again suggests that money may not necessarily be a good medium for introducing elementary number ideas. (In Section 2.5.2.4 on page 159 the relationship between money, place value and the decimal system is considered.)

c. *Practical situations* Tackling practical problems both assists children to cope in everyday situations of buying and selling and is likely to increase their grasp of the underlying number relations. For instance the following questions are critical in most 'shopping' situations:

'Do I have enough money to buy this (or these)?'
'Which of these can I buy for the money I have?'
'What coins shall I offer in order to pay?'
'What change should I get?'

Evidence from various studies of children's abilities to deal with such questions is given below.

Gibson (1981) showed that average 7 to 8 year olds were generally able to answer simple questions of this type, involving amounts under 50p although average children 6 to 7 years or below were generally unsuccessful. (Again, the samples were small.)

Results on practical test on money for 11 and 15 year olds
(from data in APU, 1980a and b)

			Unaided sucess %	
			15 year olds	11 year olds
1.	Pupil has 50p; tester 4 x 10p, 1 x 5p, 3 x 1p	All	98	82
	If you want to pay me 10p, show me	Boys	99	82
	how to do it.	Girls	97	91
2.	If you want to pay me 17p, show me	All	99	93
	how to do it.	Boys	99	88
		Girls	99	96
3.	Pupil now has 2 x 10p, 3 x 1p			
	Tester has 1 x 50p, 2 x 10p, 1 x 5p	All	87	64
	If you want to pay me another 17p,	Boys	92	60
	show me how to do it.	Girls	83	66
4.	First set of coins received and pupil now given 1 x 50p, 4 x 10p, 1 x 5p, 3 x 2p, and tester has £1 note.			
	Suppose I want to pay you 32½p. I	All	91	
	give you £1 note. Now give me the	Boys	91	not asked
	right change.	Girls	91	
5.	Explain what you were doing.	All	94	
		Boys	95	not asked
		Girls	93	

The tests were individually administered, with coins and labelled goods available, so as to simulate a real situation. Some examples of the questions used are given below:

> I went into a shop and I wanted to buy some sweets which cost 2 pence, but I only had a 5p piece. Could I buy any of them? How many could I buy?
>
> I want to buy this bag of sweets for 37p but I only have a 50p piece. Can I buy the sweets? How much change should I get?
>
> These sweets cost 30 pence. Can you show me how much you would give the shopkeeper if you wanted to buy them?
>
> Here's 10 pence. If we spent four pence on a bag of sweets, how much would we have left?
>
> I wanted to buy 2 of these sweets (which cost 6 pence) and one orange (which costs 7 pence). Show me how much I will have to pay for them.
> If I give the shopkeeper two 10p pieces, how much change will he have to give me?

The difference between this type of practical test situation, and those in which children are given the problem without any access to objects or coins, needs to be stressed. Pencil-and-paper tests do not necessarily indicate children's ability to cope in an actual shopping context.

Another practical test in which children were allowed to handle coins and discuss the problem with the tester was given by the APU to 11 and 15 year olds.

The item and results were as shown in the table on page 157.

This suggests that given a fairly straightforward practical situation, almost all 15 year olds and most 11 year olds are able to cope reasonably well. It is reported that 70 per cent of 15 year olds found the answer to question 4 by counting on from 32½ to 100, while 20 per cent subtracted 32½ from 100.

These encouraging results on practical tests contrast with poorer results on written items. For example Ward asked a large sample of 10 year olds the following question as part of a much longer test:

> How much is left from £2 after buying 4 books costing 30p each? (Ward, 1979)

Sixty-five per cent of 10 year olds obtained the correct answer. The question was ranked by teachers as among the 5 most important out of 45 items, the other 4 highly-ranked questions including one other on money and three formal computation 'sums'.

Similarly the first APU primary written survey which tested 11 year olds found that only:

> half of the pupils could correctly determine the total cost of four articles, two at 60p each, and the others at 50p and 75p respectively (APU, 1980a).

The 'money' section of the written tests in the first APU secondary survey of 15 year olds contained rather more complex items relating to 'everyday' uses of money, but again the results were disappointing:

> All the items in this section had facilities less than 50 per cent and involved calculations with more than one step. For instance, a facility of just over 40 per cent was gained on an item which asked for the new weekly

wage after £50 per week had been increased by 6 per cent and £1 per week. Around 40 per cent of the pupils could calculate the monthly payment when told that £85 was to be made up by a deposit of £10 and twelve monthly payments. However, other items of this kind, e.g. calculating an electricity bill from two meter readings, with a standing charge and price per unit, gained lower facilities of around 25 per cent. (APU, 1980b)

In this last set of examples it was not clear where the difficulties lay. It seems possible that the two-step nature of the items may have caused problems, or simply the arithmetical calculations, rather than necessarily anything to do with understanding the system of money itself.

2.5.2.4 Money, Place Value and Decimals

It is often suggested that decimals should be introduced using the monetary system, partly because money provides a concrete embodiment of a decimal system, and partly because it is the context in which the majority of people will most frequently use a decimal system. For example, HMI found in the Primary Surveys (DES, 1978) that:

In a few seven year old classes children were introduced to the notion of decimals, normally associated with the recording of amounts of money or metric measures. It was more common for such teaching to be introduced at a later stage, within the programme of about three-quarters of nine year old classes and almost all eleven year old classes. (DES, 1978)

However Brown lists some reasons why the money system is often not operated as a complete decimal system:

(i)   it is commonly utilised on the basis of experience (of) and with reference to 'concrete' objects i.e. coins and notes;
(ii)   there are effectively two units, pence and pounds. This means that £1.42 is read as 'one pound and forty-two pence' rather than as 'one point four two pounds';
(iii)   £6.4 has no clear meaning, at least to children;
(iv)   £6.42½ is allowable notation.

(Brown, 1981b)

Brown found that when asked for an example of where the number 6.4 might be encountered in real life, around 10 per cent of 12 to 15 year olds translated this into money, with around 5 per cent interpreting it as 'six pounds and forty pence' and the remaining 5 per cent interpreting it as 'six pounds and four pence'. Some of the latter group wrote down £6.4p, which supports the second point made above about the use of two units, pound and pence.

A further confusion was that about half of the 12 to 15 year olds sample used by Brown confused 'tens' and 'tenths' at some stage. Thus in the 'pure' number 0.75, the 7 might be identified as 7 'tens', and the 5 as 5 'units'. (Similarly in 0.752 the 7 is said to be 7 'hundreds' or ' hundredths', the 5 to be 5 'tens' or 'tenths', and the 2 to be 2 'units'.) The use of decimals in the money system may have caused this, for in 0.75, the seven *can* legitimately be seen as 7 tens (of pence) as well as 7 tenths (of a pound) and the 5 similarly as 5 units (of pence) as well as 5 hundredths (of a pound).

This all suggests that money may not be a very appropriate context for introducing ideas about decimals, and that, again, other measures such as length may be more appropriate, although they do also suffer from the problem that more than one unit is involved, e.g. 1.42 metres is thought of as one metre and 42cm, with no conception of 42cm as 0.42 of a metre.

In a more limited way 10p and 1p coins may be used to introduce basic concepts of place value up to 100 i.e. 52p means five 10p pieces and two 1p pieces. As a second stage after 'grouping activities' this may be a useful approach to place value. All is well as long as examples are confined to the numbers up to 100, but problems occur with the introduction of the next place, as the form 142p is rarely encountered as an alternative for £1.42.

2.5.2.5   Teaching about Money

Gibson notes that some primary schemes include exercises such as

2 pence + 1 pence =

Simon bought 3 sweets at 2p each and had 1p left. How much did he have to spend?
     $(3 \times 2) + 1 = 7$
He had 7p.

(Gibson, 1981)

She stresses that these may be useful in teaching about number, but are little use in the teaching of how to handle money, since they are very far removed from real situations. For example in the unlikely event that children were faced with a problem such as the second one above in everyday life, they would generally solve it mentally and would rarely record it formally as $(3 \times 2) + 1 = 7$.

A quotation from Langdon highlights the fact that real-life problems are often solved by radically different methods from those found in mathematics textbooks:

When I asked a colleague, he said that he often multiplies decimals to help him in spending. When he put petrol in his car, for exampie, he multiplied 4 x 78.9. Did he really? Ah, no, he said. He served himself 4 gallons and the dial showed £3.15½. But he continued, suppose he only had £3 and he wanted to know if he could afford 4 gallons? Really? It's more likely that he would serve £2 worth of petrol and keep the rest for incidental expenses if that's all the money he had. (Langdon, 1979)

Langdon uses this example and others to stress that, especially in the teaching of low attainers, realistic and honest strategies are very important.

Practical and realistic exercises are suggested by Cawley (1974) and Bowers (1981). Cawley asks numerous questions based on a newspaper, for example:

Find the best-paid job advertised.... Large food adverts: make a specific grocery list up and compare the cost from various supermarkets: use the adverts to plan the cheapest enjoyable meal.... Holiday adverts: which of the Spanish holidays cost least per day? (Cawley, 1974)

Bowers (1981) describes an exercise on 'planning a day's outing', which involved a group of low-attaining fourth year secondary girls in a trip to Victoria Station to collect information. This led to a determination to actually go on a day trip to Boulogne, which included working out the cheapest way to travel, fixing up all the arrangements, calculations as to what hour the coach should leave London to catch the boat, deciding how much French currency to take, and so on. The trip was a great success from every point of view, not least because it gave the girls confidence in their own ability.

Gibson (1981) suggests the use of games to teach basic skills such as coin recognition and early ideas of equivalence. She devised and tested several simple games based on 'snap', 'dominoes', 'snakes and ladders', and the use of dice and spinners. For example, in 'dominoes' (or 'snap') a 5p coin (mounted on a card) was 'matched' by a similarly mounted pair of 2p coins and a 1p coin; in a game using a spinner the players each tried to be the first to make up from their own coins the amount shown on the spinner (e.g. 13p).

Sears also reports favourably on the use of games for consolidating skills needed in handling money with 11 year old low attainers.

Both games are for two players. In each case, dice and a "bank" of plastic money are needed; the first game also requires two counters. The games can be set out quite simply on duplicated sheets and given to the children.

Game One.

The main aim here was to help the children become skilled at counting out different sums of money from any given coins.

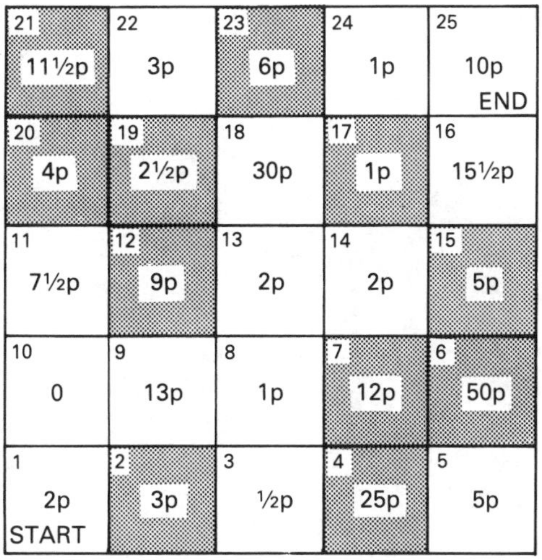

21 11½p	22 3p	23 6p	24 1p	25 10p END
20 4p	19 2½p	18 30p	17 1p	16 15½p
11 7½p	12 9p	13 2p	14 2p	15 5p
10 0	9 13p	8 1p	7 12p	6 50p
1 2p START	2 3p	3 ½p	4 25p	5 5p

Fig. 2.67

1. Take £1 each, in coins, from the bank.
2. Take turns to throw the dice. Move the number of places shown by the dice.
3. If you land on a white square, take the money shown on the square from the bank.
4. If you land on a shaded square, pay the money on the square to the bank.
5. The winner is the one with most money when you have got to the end.

From simply counting out different amounts of money, the next step was to develop skill in making up given amounts from a number of coins.

	PLAYER 1	PLAYER 2
3½p		
33p		
51p		
24½p		
11p		
17p		
27p		

Fig. 2.68

Game Two.
1. Throw the dice. This tells you how many coins you may take from the bank.
2. If you can make up the value shown in the box take the coins. Draw the coins you have used.
3. Now the other player has a turn.
4. Do the same for the other amounts. Put your results in the table.
5. The winner is the player with most money at the end of the game.

(Sears, 1974)

Both authors report considerable enthusiasm in the playing of the games, and easy adaptation of them to different levels of competence by changing the sums of money, or the types of equivalence, involved.

References for Section 2

Assessment of Performance Unit (APU) – see Department of Education and Science.

Ames, L.B. (1946) The Development of the Sense of Time in the Young Child, *Journal of Genetic Psychology,* **68,** 97–125.

Beard, R.M. (1963) The Order of Concept Development: Studies in Two Fields, I and II, *Educational Review,* **15,** 105–117 and 228–237.

Bell, D; Rothwell Hughes, E; Rogers, J; (1975) *Area, Weight and Volume: Monitoring and Encouraging Children's: Conceptual Development* Sunbury-on-Thames: Nelson, for the Schools Council.

Bowers, J. (1981) Excursion to Boulogne. *Struggle: Mathematics for Low Attainers – ILEA,* **4,** 12–16.

Bradley, N.C. (1948) The Growth of the Knowledge of Time in Children of School Age, *British Journal of Psychology,* **38,** 67–78.

Brookes, W.M. (1970) Magnitudes, Measurement and Children, In *Mathematical Reflections.* Association of Teachers of Mathematics. Cambridge University Press.

Brown, M. (1981b) Place Value and Decimals, In *Children's Understanding of Mathematics: 11–16* (Ed) Hart, K. John Murray.

Carey, S. (1979) Cognitive Competence. In *Cognitive Development in the School Years* (Ed) Floyd, A. Croom Helm for the Open University.

Carpenter, T.P. and Osborne, A.R. (1976) Needed Research on Teaching and Learning Measure. In *Number and Measurement: Papers for a Research Workshop* (Ed) Lesh, R.A. Ohio: Eric/SMEAC Center, Ohio State University.

Carpenter, T.P. (1976) Analysis and Synthesis of Existing Research on Measurement In *Number and Measurement: Papers for a Research Workshop* (Ed) Lesh, R.A. Ohio: Eric/SMEAC Center, Ohio State University.

Carpenter, T.P. et al. (1980) Results and Implications of the second NAEP Mathematics Assessments: Elementary School. *Arithmetic Teacher,* **27(8),** 44–47.

Carpenter, T.P. et al. (1980) Results of the Second NAEP Mathematics Assessment: Secondary School, *Mathematics Teacher,* **73(5),** 329–338.

Cawley, N. (1974) Mathematics from a Newspaper. *Remedial Education,* **9 (1),** 29–33.

CSMS (Concepts in Secondary Maths and Science). See Hart, K.M. (Ed, 1981).

Denvir, B; Stolz, C; Brown, M. (1982) *Low Attainers in Mathematics 5–16: Policies and Practices in School* (Schools Council Working Paper 72) London: Methuen Educational Ltd. for the Schools Council.

Department of Education and Science (DES), 1978 *Primary Education in England: a Survey by HMI.* London: HMSO.

Department of Education and Science, APU – Assessment of Performance Unit (1980a) *Mathematical Development, Primary Survey Report No. 1.* HMSO.

Department of Education and Science, APU – Assessment of Performance Unit (1980b) *Mathematical Development, Secondary Survey Report No. 1.* HMSO.

Department of Education and Science, APU – Assessment of Performance Unit (1981a) *Mathematical Development, Primary Survey Report No. 2.* HMSO.

Department of Education and Science, APU – Assessment of Performance Unit (1981b) *Mathematical Development, Secondary Survey Report No. 2.* HMSO.

Dichmont, J. (1972) Balancing: A Conversation. *Mathematics Teaching, 59*, 27–29.

Dienes, Z.P. (1959) The Growth of Mathematical Concepts in Children through Experience, *Educational Research, 2*, 9–28.

Elkind, D. (1961a) The Development of Quantitative Thinking: A Systematic Replication of Piaget's Studies. *Journal of Genetic Psychology, 98, (1)*, 37–46.

Elkind, D. (1961b) Children's Discovery of the Conservation of Mass, Weight and Volume: Piaget Replication Study II. *Journal of Genetic Psychology, 98, (2)*, 279–287.

Fogelman, K.R. (1970) *Piagetian Tests for the Primary School.* Windsor: NFER Publishing Co. Ltd.

Galperin, P.Y. and Georgiev, L.S. (1969) The Formation of Elementary Mathematics Notions. In *Soviet Studies in the Psychology of Learning and Teaching Mathematics, Vol. 1: The Learning of Mathematical Concepts* (Ed) Kilpatrick, J.E; Wirszup, I. – School Mathematics Group, Stanford University.

Gibson, O.E. (1981) *A Study of the Ability of Children with Spina Bifida to Handle Money – Ph. D. thesis.* University of London.

Giles, G. (1977) *School Mathematics under Examination 2: A Comparison of the Cognitive Effects of Individualised Learning and Conventional Teaching.* University of Sterling.

Greenes, C.E. (1979) The Learning Disabled Child in Mathematics. *Focus-on Learning Problems in Mathematics (Framingham, Massachusetts), 1 (1).*

Harris, M. (1981) *Dissecting a Clockface* Struggle: Mathematics for Low Attainers, *4*, 26–28.

Hart, K.M. (1978) Mistakes in Mathematics. *Mathematics Teaching, 85*, 38–40.

Hart, K.M. (1980) *Secondary School Children's Understanding of Mathematics – Research Monograph* (A report of the Mathematics Component of the Concepts in Secondary Mathematics and Science Programme). Chelsea College, University of London.

Hart, K.M. (1981) Measurement. In *Children's Understanding of Mathematics: 11–16* (Ed) Hart, K.M. London: John Murray (Publishers) Ltd.

Holcomb, J. (1980) Using Geoboards in the Primary Grades. *Arithmetic Teacher, 27 (8)*, 22–25.

Hutton, J. (1978) Memoirs of a Maths Teacher. *Mathematics Teacher, 82*, 8–14.

Inhelder, B; Sinclair, H. & Bovet, M. (1974) *Learning and the Development of Cognition.* London: Routledge and Kegan Paul.

Kelly, M.B. (1967) *An enquiry into the Ability of ESN Children to Handle Money in Practical Situations, in relation to their Mathematical Understanding – M.Ed. thesis.* University of Manchester.

Kerslake, D. (1975) Taking Time Out. *Mathematics Teaching, 73*, 8–10.

Kerslake, D. (1976) Volume and Capacity. *Mathematics Teaching, 77*, 14–15.

Langdon, N. (1979) Strategies for Spending Money *Struggle: Mathematics for Low Attainers – ILEA, 1*, 25–27.

Lovell, K. (1966) *The Growth of Basic Mathematical and Scientific Concepts in Children – 5th edition.* Sevenoaks: Hodder & Stoughton Educational.

Lovell, K. & Ogilvie, E. (1960) A Study of the Conservation of Weight in the Junior School Child. *British Journal of Educational Psychology, 31(2)*, 138–144.

Lovell, K. & Ogilvie, E. (1961a) The Growth of the Concept of Volume in Junior School Children. *Journal of Child Psychology and Psychiatry, 1(1)*, 191–202.

Lovell, K. & Slater, A. (1960) The Growth of the Concept of Time: a Comparative Study. *Journal of Child Psychology and Psychiatry,* **1**, 179–190.

Lunzer, E.A. (1968) Formal Reasoning. In *Development in Human Learning* (Ed) Lunzer, E.A. and Morris, J.F. New York: American Elsevier.

Modgil, S. & Modgil, C. (1976) *Piagetian Research Compilation and Commentary. Vol. 7: Training Techniques.* Windsor: NFER.

National Assessment of Educational Progress (1980) *Mathematics Technical Report: Summary Volume,* Denver Colorado: NAEP.

Nuffield Mathematics Teaching Project (1972) *Checking Up II.* John Murray and W. & R. Chambers for the Nuffield Foundation.

Nuffield Mathematics Teaching Project (1973) *Checking Up III.* John Murray and W. & R. Chambers for the Nuffield Foundation.

Osborne, A.R. (1976) The Mathematical and Psychological Foundations of Measure. In *Number and Measurement – Papers for a Research Workshop* (Ed) Lesh, R.A. Ohio: ERIC/SMEAC Center, Ohio State University.

Piaget, J. (1952) *The Child's Conception of Number.* London: Routledge and Kegan Paul.

Piaget, J. (1969) *The Child's Conception of Time.* London: Routledge and Kegan Paul.

Piaget, J. (1974) *The Child and Reality: Problems of Genetic Psychology.* London: Frederick Muller Ltd.

Piaget, J. & Inhelder, B. (1974) *The Child's Construction of Quantities: Conservation and Atomism.* London: Routledge and Kegan Paul.

Piaget, J; Inhelder, B; Szeminska, A. (1960) *The Child's Conception of Geometry.* London: Routledge and Kegan Paul.

Rea, R. & Reys, R. (1971) Competencies of Entering Kindergarten Children in Geometry, Number, Money and Measurement. *School Science and Mathematics,* **71**, 389–402.

Rothwell Hughes, E. (1979) *Conceptual Powers of Children an Approach through Mathematics and Science – Schools Council Research Studies.* London: Macmillan, for the Schools Council.

Sears, H. (1974) Topics: Money Games, *Mathematics Teaching,* **68**, 44.

Shayer, M; Küchemann, D; Wylam, H. (1976) The Distribution of Piagetian Stages of Thinking in British Middle and Secondary School Children. *British Journal of Educational Psychology*, **46**, 164–173.

Shayer, M. & Wylam, H. (1978) The Distribution of Piagetian Stages of Thinking in British Middle and Secondary School Children II: 14 to 16 year-olds and Sex Differentials. *British Journal of Educational Psychology,* **18**, 62–70.

Smedslund, J. (1961–2) The Acquisition of Conservation of Substance and Weight in Children I–IV. *Scandinavian Journal of Psychology,* **2**, 11–20; 71–84; 85–87; 153–155; 156–160; 203–210. **3**, 69–77.

Springer, D. (1952) Development in Young Children of an Understanding of Time and the Clock. *Journal of Genetic Psychology,* **80**, 83–96.

Stringer, D. (1979) *Make it Count – a Study.* London: Independent Broadcasting Authority (Fellowship Scheme).

Sturt, M. (1925) *The Psychology of Time.* London: Routledge and Kegan Paul Ltd.

Thompson, G.E. (1962) What Arithmetic shall we Teach our Educationally Subnormal Children? *Special Education,* **51**, 3.

Thyer, D. & Maggs, J. (1971) *Teaching Mathematics to Young Children*. Eastbourne: Holt, Rinehart and Winston Ltd.

Uzgiris, I.C. (1964) Situational Generality of Conservation. *Child Development,* **35, (3),** 831–841.

Ward, M. (1979) *Mathematics and the 10-year-old – Schools Council Working Paper 61.* Evans/Methuen for the Schools Council.

Wheeler, D. (1975) Humanising Mathematical Education. *Mathematics Teaching,* **71,** 4–9.

SECTION 3: *Number*

SECTION 3: *Number*

3.1 EARLY STAGES IN THE DEVELOPMENT OF NUMBER IDEAS

3.1.1 Introduction: the Complexity of Number Ideas

To most adults, the knowledge and use of the first nine natural numbers (one, two, three... up to nine) appears to be a very simple and straightforward business. Yet an average child takes around five years, from about the age of two to the age of seven, to learn to handle such numbers consistently and to apply them to a variety of everyday situations. And the period would become even longer if the use of number operations were included.

The fact that basic number development proceeds so slowly may seem surprising, especially in comparison with the relatively prodigious speed at which the young child appears to pick up language.

Before going into details of children's mathematical development it seems worth examining some of the reasons why the process should be so slow compared with language.

One use of number is to specify the size of a collection of objects. (This is known as the *cardinal* aspect of number.) It may be true that the young child of two years learns to distinguish between two toy cars and three toy cars merely by looking at the group of cars, in just the same way as he learns to use his visual perception to distinguish between a car and a bus, or between a red car and a green car. However this will not get him very far in distinguishing a collection of eight cars from one of nine cars; here he can no longer rely on perceptual discrimination but will need to master the skill of accurate counting, and this is itself a considerable accomplishment.

It might on first sight be thought that learning to count was merely a matter of reciting a string of words, just like learning a nursery rhyme, which is a feat that young children can master surprisingly early. However as we shall see, the art of counting involves a number of additional features, such as only pointing to one object at a time, and keeping track of which objects have already been counted.

This process of counting i.e. the assigning of a number to a particular object which forms one of a sequence of objects, is known as the *ordinal* aspect of number. There is however a final step, that of knowing that the number the child 'finishes on' in counting a collection can be used to represent the size ('manyness', 'numerosity') of the whole collection. This is the link between the *ordinal* and *cardinal* aspects of number.

And even when a child has got this far, he may not realise that he would arrive at the same number, in whatever order he counted the objects or however the objects were rearranged.

Thus the complete grasp of even small natural numbers involves forming several relationships, particularly, as Piaget (1952) emphasised, that between the *ordinal* aspect (assigning a number to denote the position of an object in a sequence) and the *cardinal* aspect (using a number to denote the size of a collection). It seems to be this complexity which is responsible for the delay before numbers are used consistently; however it should not be forgotten that children can, and do, effectively learn some aspects of number at a relatively early age.

Various researchers have endeavoured to identify 'stages' in the development of number. The result of one such attempt, by Schaeffer, Eggleston and Scott (1974) is now taken as a convenient framework for a more detailed description of the various steps involved. However it should be remembered that their identification of stages is tentative, and indeed is challenged by some researchers. It is also based on a very small sample, of 65 American children between 2 years and 5 years 11 months, which was not necessarily representative. Although it is a helpful guide it should not be taken as 'the ultimate truth'; indeed it is very possible that an alternative and more valid scheme of stages will soon emerge as there is at present much research activity in this area, especially in the United States.

3.1.2   Schaeffer's Stage One: Pre-counting Achievements

The criterion for Stage One was 'not able to count correctly collections of five or more objects'.

In the study by Schaeffer, Eggleston and Scott (1974), the 13 children who were identified as being at this stage ranged in age between 2 years exactly and 5 years exactly, with a mean age of 3 years 8 months.

Pattern recognition   There was considerable evidence that Stage One children could discriminate between small numbers on a basis of the perceptual pattern. In Schaeffer's study, 12 out of the 13 correctly recognised as 'two' an array of two men; in the same way 7 recognised 'three', and 6 'four'. Only 1 out of the 13 children counted correctly the arrays of two, three and four men. When a row of between one and four poker chips were presented to the Stage One children, and they were asked how many there were, 51 per cent of the answers were correct, with very few instances of evidence of counting recorded. When the children were asked to put a certain number of sweets into a cup this group were correct 84 per cent of the time when the number was one or two, but only 22 per cent of the time on numbers between three and seven. Similarly when they were asked to tap a drum a given number of times, these children were correct 51 per cent of the time for one or two taps, but only 6 per cent of the time for three or more taps. The average number of taps they counted when the interviewer tapped the drum rhythmically was 3.8, but the interviewer generally explained to the children that she wanted them to count by demonstrating 'one, two, three', so that effectively the children did not get beyond this. Thus it would appear that young children can recognise the number of objects in very small collections, certainly one and two and sometimes three and four, without counting them. The fact that they performed better on visual arrays than auditory ones suggests that there is a perceptual basis for this skill. Schaeffer suggests it relates to the fact that

...an array of one is a dot, an array of two forms a straight line, an array of three usually forms a triangle...

This evidence agrees with that of Descoeudres (1921) who found that among children aged three years six months, 67 per cent consistently and correctly discriminated visually between 1 and 2 object arrays, and the same percentage discriminated between 2 and 3 object arrays, but only 13 per cent discriminated between 3 and 4 object arrays.

Descoeudres called this the 'un, deux, trois, beaucoup' (one, two, three, a lot) syndrome. Gelman and Gallistel (1978) also found evidence of this, in that when presented with rows of varying numbers of stars, 25 per cent of three year olds gave the same guess for all collections over a certain size; some children in fact said 'three' for all collections from three to nineteen.

Gelman actually challenges Schaeffer's assertion that young children do not count in order to recognise the size of small collections. Gelman found that some two year olds could distinguish between an array of two and an array of three, with 8 out of 18 two year olds identifying 'two' as the number of objects in an array of two, and 4 out of 18 correctly identifying three. However, she noted that all of these children *could* count up to these numbers, although she does not give evidence that they actually did count on these occasions. But only 3 out of 18 two year olds did not give evidence of 'counting', in the sense of repeating a sequence of 'number words', in answer to an instruction to count. However one child only repeated the word 'three', one recycled 'one, two, three, one two, three, ...', and five more used idiosyncratic sequences e.g. DS (aged 2 years 6 months)

'How many on this (the two-item) plate?' *'Um mm, one, two.'*
'How many on this (the three-item) plate?' *'One, two, six!'*
'You want to do that again?' *'Ya, one, two, six.'*

Many two year olds, however, failed to stop after the correct number of 'count words'.

Gelman therefore suggests a different developmental sequence to that proposal by Schaeffer in that at the earliest stage children may well recognise very small numbers through 'counting', and that pattern recognition is a later development. The evidence on this hypothesis is not yet clear.

Judgments of relative size (numerosity)   Schaeffer suggests that the idea of 'more' first develops between the age of two and two and a half, when children begin to 'take more' or 'ask for more'. Donaldson and Wales (1970) showed that children aged three and a half years could determine which of two arrays composed of 1 to 5 objects had more objects, although Donaldson notes that three to five year olds generally think that the word 'less' means 'more' (Donaldson and Balfour, 1968).

Siegel (1978) also notes that 35 out of 45 three year olds learned to distinguish a larger from a smaller array in a non-verbal test, although only 17 out of 45 could consistently respond correctly when asked verbally which set was bigger or smaller. (Among four year olds the corresponding proportions were 56 out of 57 and 38 out of 57.) Gelman confirmed these results in that 24 out of 30 three year olds learned to distinguish between three and five objects, although the words 'winner' and 'loser' were used instead of 'big' and 'small'.

Schaeffer asked children verbally whether they would prefer to have two sweets or six sweets, and repeated this with various other combinations of numbers differing by either one or four. He found the Stage One children responded correctly on 64 per cent of these choices, which was significantly better than chance, provided one of the numbers was less than five and the other was four greater than the first (e.g. four and eight), but otherwise did not do better than would be predicted by chance.

Although Schaeffer's test was very different from the others and required knowledge of the order of the names of the counting numbers, rather than a judgment based on visual perception, it does suggest that children at this initial stage are able to distinguish which of two sets is larger or smaller provided at least one of the numbers is less than five, although they do not necessarily understand the word 'less' or 'smaller'.

These findings support Bryant's (1974) claim that young children develop 'relative codes' before 'absolute codes' i.e. they learn to distinguish which of two shapes is larger or smaller, which of two lines is longer or shorter, whether two lines are pointing in the same direction or in different directions, and so on, before they learn to judge absolute size, length, direction or numerosity.

However this ability to judge which of two collections is larger clearly depends on how the collections are arranged. When two equal collections were arranged in rows of different lengths as below:

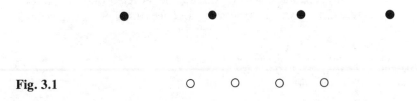

Fig. 3.1

children at age three and a half years, and beyond, tended to judge the longer one as the more numerous, even when there were only three or four objects in each set (Piaget, 1968; Bever, Mehler and Epstein, 1968; Gelman, 1972).

However Bryant (1974) showed that if the objects were spaced out identically to show 'one-one correspondence', or lack of it, as below:

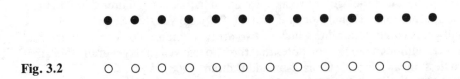

Fig. 3.2

three year old children could decide which row was the larger collection even when the distinction was between 19 and 20 objects.

Spontaneous use of one-to-one correspondence for comparing the size of two collections certainly occurs later than this. Wang, Resnick and Boozer (1971) found that only 24 per cent of a sample of American children aged between 4 years and 6 months and 6 years could consistently determine the larger of two sets, each containing less than ten objects, if they were not allowed to count them. In fact the success rate for the same task using counting was exactly the same, although the group of children who succeeded was different in each case. (With three sets from which the largest had to be chosen, the success rates dropped to 5 per cent for one-to-one correspondence and 0 per cent for counting.)

Summary of Stage One   Thus children at the first stage identified by Schaeffer could

a. recognise numbers up to two and sometimes three or four (probably by recognising a visual or auditory pattern although possibly by counting);
b. distinguish between larger and smaller collections in cases where at least one was less than five, both visually and verbally;
c. distinguish between larger and smaller collections of any size provided they were lined up so as to show one-to-one correspondence or lack of it.

On the other hand they could on no occasion count five or more objects.

Thus in Schaeffer's view Stage One children have grasped the cardinal aspect of number i.e. using a number to determine the size of a collection of objects, at least for very small collections, but do not yet have the ordinal aspect associated with attaching a sequence of numbers named to a sequence of objects.

3.1.3   Schaeffer's Stage Two: the Ordinal Aspect

Schaeffer used as his criterion for Stage Two that children could count at least one array of five or more correctly, but correctly applied the 'cardinality rule' less than half of the time. (The *cardinality rule* is the knowledge that the number on which the counting process finishes can be used to denote the size of the set.)

In Schaeffer's study the 20 children identified to be at Stage Two ranged in age from 2 years 9 months to 4 years 6 months, with a mean of 3 years 5 months. (It should be noted that the mean age is actually less than that for the Stage One children, presumably due to the small size of the sample and the presence of some relatively old children in Stage One. Thus even at this early age there is evidence of considerable variation in development between children of the same chronological age.)

The skills possessed by the Stage Two children are as follows:

Pattern recognition   Stage Two children can recognise small numbers as patterns, but they are more likely to count them than are Stage One or Stage Three children. Thus when they were shown arrays of one, two, three or four men, Stage Two children recognised the number on sight 46 per cent of the time and counted 34 per cent of the time. They also correctly identified the number of chips in a row containing between one and four on 89 per cent of trials, which was substantially more than the 51 per cent registered by Stage One children; many of these were without observable counting.

There are two alternative explanations for this; Schaeffer would say that the Stage Two child has now mastered counting and therefore although he does have the pattern recognition abilities of the Stage One child if in doubt he is likely to count. Gelman however prefers the hypothesis that children only learn to recognise a pattern once they have familiarised themselves with a number by repeatedly counting it. Thus she would say that the Stage Two children have come to recognise the patterns for small numbers (one to three) but not yet for larger ones.

Counting   Stage Two children counted correctly 71 per cent of the arrays containing between five and seven poker chips, in comparison to no correct counts from Stage One children. They also counted on average 8.3 of the interviewer's ten drum taps. These results suggest that Stage Two children understand the nature of the counting process, but are rather inaccurate in its performance. They have thus grasped two of the major counting principles proposed by Gelman, namely:

a. counting requires repetition of a string of number names in an identical order (the '*stable order principle*')
b. each number name must be matched to one and only one object (the '*one-to-one principle*')

The '*one-to-one principle*' may not be complied with for any of the following reasons:

i.   errors in *partitioning* the set of objects already counted from those as yet uncounted, resulting in double-counting or omissions
ii. errors in *assigning* number names, for example using the same one twice
iii. errors in *coordinating* names with objects e.g. reciting two number names while pointing to one object, or pointing to two objects while saying one name.

Taking each of these in turn, Gelman found that the '*stable order principle*' was followed, at least as far as three, by 9 out of 14 two year olds, even though some of them used idiosyncratic number name sequences. For 3 year olds, counting as far as five, the proportion was 17 out of 20; while the rule was complied with by all 4 year old children in the sample.

Gelman also noted the tendency of children to try to correct their number sequences until they corresponded to the 'official' one, e.g. AB (aged 3 years 6 months) attempting to count eight items:

> AB: One, two, three, four, eight, ten, eleben. No, try dat again.
> One, two, three, four, five, ten, eleben. No, try dat again.
> One! Two! Three-ee-four, five, ten eleben. No. [This pattern of self-correction continued for many attempts and ended with the following count.] One, two, three, four, five, six, seven, eleben! Whew!
> (Gelman and Gallistel, 1978, p.93)

The '*one-to-one principle*' however was the source of considerably more errors in all age groups. Gelman assesses that 11 out of 56 count sequences used by 2 year olds, even just for the numbers two or three, contained one-to-one errors. For 3, 4 and 5 year olds the percentage of children making one-to-one errors some of the time was as shown in the following table.

Percentage of children making one-to-one errors in counting collections of a given size:

		Age		
		3 yrs	4 yrs	5 yrs
	3	48	5	7
Size of	5	71	32	46
collection	7	62	47	53
	11	86	84	60
	19	95	68	67

Thus in counting only to five, even the 5 year olds were remarkably unreliable. Further support for this lack of reliability is given by the results obtained by Wang et al. (1971), quoted below.

When Gelman analysed the cause of the one-to-one errors, she found that they were equally split between partitioning and coordinating errors for small numbers (2 to 5), whereas beyond that, partitioning errors were more frequent (i.e. objects were counted twice or not at all). The coordination errors tended to occur at the end of the counting sequence.

It is worth noting that Gelman's objects were arranged in a row only on 50 per cent of these trials; on the remainder they were arranged in a random array. There is some evidence that arranging objects in a row not surprisingly reduces the partitioning errors as it is easier to keep track of those already counted. For example, Wang et al. (1971) report that 44 per cent of their sample of American children aged between 4 years 6 months and 6 years were consistently correct in counting rows of between six and ten pictured objects, whereas when the pictures were arranged randomly, the percentage dropped to 23 per cent. A similar result was obtained by Potter and Levy (1968). The same is true if objects are arranged in small clusters rather than spread randomly over a space (Beckwith and Restle, 1966). It seems likely that the ability to move objects around in order to group those already counted would further reduce partitioning errors, although Wang et al. noted that few children took advantage of this possibility in practice.

Gelman noted also that 98 per cent of children pointed when counting, presumably in a conscious or unconscious attempt to coordinate number names and objects.

Ginsburg gives a classic illustration of a failure in coordination:

At 4-11, they were engaged in counting the number of fingers on a hand... Rebecca counted hers and claimed 'I've got ten fingers'.

Deborah disagreed. She counted her own (ten) fingers and got 16.... Deborah believed that two identical collections (the twins' ten fingers) could have different numbers (ten and 16).

Rebecca said 'We don't get 16. Want me to count your fingers?' She then counted Deborah's fingers and got ten. Deborah counted both hands and again got the wrong result. Next she counted the fingers on one hand and got six. Rebecca counted the same fingers and got five. Rebecca showed Deborah what six really is – namely, five on one hand and one on the other.

Deborah held up the fingers of one hand and insisted, 'This is six too; this is six too!' (Ginsburg, 1977)

One possible reason given for the lack of coordination in young children is that the sheer effort of remembering the number names is so great that they cannot give full attention to coordinating them with the objects. Fuson (1980) quotes research by Case to show that the mean rate at which shapes on a card were counted increased from 840 milliseconds per object at age four, to 510 milliseconds per object at age 6, and to 210 milliseconds per object for adults. Case noted that this increase in operational efficiency should free the counter for other activities, like keeping more accurate check on the counting process.

Thus, returning to Schaeffer's Stage Two children, it would appear that they do recognise the basic principles of counting, but are not always able to carry them out very efficiently.

The 'Cardinality rule'   One of Schaeffer's criteria for Stage Two was that the child should fail on at least 50 per cent of occasions to use the 'cardinality rule' i.e. having carried out the counting process the child should not normally be able to apply the result of this to determine the size of a collection. Sometimes children appear to think that the answer to a 'How many?' question is the total counting sequence rather than a single number name e.g. CD (age 3 years 5 months), who was being tested with four objects that varied in colour:

How many are on that plate? *Um... look, this is red and so is this. There's one, two, three, four* (Gelman and Gallistel, 1978).

Schaeffer also points out that

Characteristically, when a Stage Two child was asked to count an array of six chips, he said correctly the numbers one to six while pointing to each chip in turn. When the array was covered and the child was asked how many chips were hidden, however, he recounted, sometimes pointing to the remembered position of hidden chips on the covering piece of cardboard, or named a number other than the last one he had counted. (Schaeffer et al. 1974)

Gelman's evidence supports that of Schaeffer, although her criteria are less rigid. She accepts either a repeat of the last number name at the end of a count, or the use of it at some later point to refer to the size of the collection. She notes that in almost all cases the cardinality rule is attained after the stable order and one-to-one principles, and also gives data which show that children's readiness to use the cardinality rule decreases as the size of the collection increases.

For example 60 per cent of 3 year olds use it with a collection of four objects, but only 30 per cent with seven objects; whereas for 5 year olds, 60 per cent apply it to collections of nine, and around 30 per cent for nineteen.

Another illustration of the failure to relate the counting process to the size of the collection is given by Schaeffer's experiments in which he asked children to put out a given number of sweets, and, later, to tap a given number of times on a drum. Although these work the opposite way round from the other illustrations – here the total is given and the child has to count to the required number, instead of having to count to obtain the total – the results are similar. For the sweets, the Stage Two children correctly placed 1 to 4 sweets in a cup on 75 per cent of the trials, but 5 to 7 on only 19 per cent. On the drum taps the percentages were lower – 48 per cent correct for 1 to 4 taps and only 10 per cent for 5 to 7. These are in spite of the fact that children at this stage could correctly *count* a collection of 5 to 7 objects 71 per cent of the time. Surprisingly the children spontaneously counted on only 10 per cent of the sweets trials, and 6 per cent of the tapping ones, which suggests again that they did not realise the connection between the counting process and the total size of the collection, but were attempting to make global estimates of the set size.

Irrelevance of order   In view of the above evidence it is perhaps not surprising that children of the same age as Stage Two children did not always understand that it did not matter at which object the count begins, but that the total would be the same regardless of the order in which the items were counted. Gelman judged that most 3 year olds did not understand this point, although a few did, as did the majority of 4 year olds. She gives the example of DS (aged 3 years 4 months) who had correctly counted a row of five objects:

> 'Make this' (an object in the middle of the row) 'number one' (DS touches the object). 'Can you start counting here and make this number one? *One, one*' (two successive objects *'What's that?'* (Experimenter repeats the question. This time DS starts with the designated object and counts to the end of the row, leaving out the remaining objects) 'Can you make this be two? *Two.*' (Gelman and Gallistel, 1978)

Gelman makes the point that although the instructions seem a little odd they were clearly understood by most 4 year olds.

It seems possible then that Stage Two children have not really grasped the cardinal aspect of number for collections larger than five, since they do not seem necessarily to have isolated the size of the collection as a stable characteristic of the collection. Gréco and Morf (1962) also note that children around this stage, and possibly later, are likely to re-count after a collection has been re-arranged.

Summary of Stage Two   Stage Two children, unlike Stage One children, seem to understand what is required in the process of counting. They can determine numbers from one to four by recognition or counting, but for larger numbers their counting becomes rather inaccurate, mostly due to errors in partitioning the items already counted, and in coordinating the speech with the pointing.

They have generally not connected the counting process with its result, which is that the final number represents the total size of the collection, and that this is invariant i.e. it does not depend on the order in which the objects are counted. Thus we can say that Stage Two children have grasped the ordinal aspect of number (i.e. assigning numbers to a sequence of objects in a counting process) and may understand the cardinal aspect for very small collections, but they have not as yet related the two aspects for numbers greater than four.

3.1.4   Schaeffer's Stage Three: the Cardinal Aspect

The criterion for Stage Three is that the child can apply the 'cardinality rule' i.e. he can use the counting process to arrive at the size of a collection, and consider the size itself to be a stable property of the collection. However Stage Three children, unlike those at Stage Four, do not recognise that, for instance, seven is necessarily more than six.

   The 15 Stage Three children in Schaeffer's study ranged in age from 3 years 3 months to 5 years 3 months, with a mean age of 4 years 2 months.

Pattern recognition   The Stage Three children were more likely than the Stage Two children to recognise the number of objects in a small collection (containing 1 to 4 objects) than to count them. In fact 86 per cent of these numbers were given correctly without counting, in comparison with 46 per cent for the Stage Two children. Gelman also reports that children are increasingly likely to identify the total size of the collection without counting as they get older (at least in the 3 to 5 year old age group), even when they are explicitly asked to count.

Counting   Although the Stage Two children seemed to have grasped the basic essentials of the counting process, the Stage Three children were more accurate, presumably because the process itself had become more automatic. Thus the collection of 5 to 7 chips were counted accurately 91 per cent of the time in comparison with the 71 per cent for Stage Two. Similarly all Stage Three children accurately counted 10 drum beats, whereas Stage Two children had only reached 8.3 on average.

Cardinality rule   When the children had counted a collection of between 5 to 7 chips, the collection was covered and the children were again asked how many chips were there. The Stage Three children correctly repeated the highest number counted (whether or not the count was accurate) in 99 per cent of cases. This contrasts with the Stage Two children who did not ever do so, preferring instead to re-count the memorised objects or to state a number other than the largest one reached.

   Thus only the Stage Three children seemed to have made the connection between the process of counting and its application in giving the number of objects in the whole collection.

   This is consistent with Gelman's results in which she quotes that some 3 year olds and most 4 year olds appear to realise that you can count a collection of objects in any order, and still reach the same result.

   For instance KG (age 4 years 10 months):

> Can you count and make this (an object in the middle of a nonlinear array) number one? *Yes* (KG demonstrates). Could this (another object) be number one? *Yes. All of them could.* Could it be number two? *Yes. All of them could be – six, four, and five and any number.* Why? (KG demonstrates by moving objects into different positions for a series of counts and showing that the number names are being reassigned as she does this.) (Gelman and Gallistel, 1978)

Further evidence is provided by Schaeffer's results when he asked children to put out a given number of sweets, or to tap a certain number of times on a drum. Here, in constrast to the Stage Two children, they counted as they performed the action and arrived at the correct numbers in 87 per cent and 75 per cent of cases, respectively.

Recognition of larger and smaller numbers   Although the Stage Three children seem reasonably adept at counting, they do not appear necessarily to connect the order of the counting numbers with the sizes of set they represent. Thus when asked whether they would prefer six or seven sweets, or any other pairs differing by one, ranging from four or five, to nine or ten, they identified the larger number exactly half of the time, which is what would be expected had the children guessed at random. (With smaller pairs like two and three, or with numbers differing by four like one and five, or six and ten, they were like the Stage Two children correct about 70 per cent of the time.)

Summary of Stage Three   Stage Three children were generally reasonably accurate counting up to ten, and had begun to connect the ordinal aspect of numbers used in assigning numbers in sequence to a series of objects with the cardinal aspect of assigning a number to represent the size of a collection. They could thus use a count to give a number to a collection, but could not use the reverse process of comparing the collections according to the order of their cardinal numbers in the counting sequence.

3.1.5   Schaeffer's Stage Four: Relative Size of Numbers

The criterion for Stage Four was the ability to recognise the larger of any pair of numbers up to ten.

In Schaeffer's study the 17 children at Stage Four ranged in age from 5 years exactly to 5 years 11 months, with a mean of 5 years 6 months. Their characteristics were very similar to the Stage Three children, except that their counting of numbers up to ten was accurate in each experiment at least 98 per cent of the time, and they were even more likely to recognise by sight numbers up to four. They also chose the larger of two numbers, where when asked how many sweets they would prefer, 92 per cent of the time, including 86 per cent of cases where both numbers were greater than three and differed by one.

Summary of Stage Four   Thus the Stage Four children seemed to have a fairly firm grasp of counting and its application to distinguishing between the relative sizes of two collections, at least when the two collections contained ten objects or less.

Thus from Schaeffer's evidence, it would seem likely that many, although not all, 5 year olds have a good working understanding of natural numbers up to ten, at least in an oral mode. That this precedes recognition of number symbols is demonstrated in part 3.4. It will be shown in part 3.3 that the oral knowledge appears to be enough to solve simple oral arithmetic problems. However that children's grasp of these numbers is not complete is illustrated in part 3.2 by reference to Piaget's classic experiments of the 1930s on both the cardinal aspect of number ('conservation') and the ordinal aspect ('seriation').

3.1.6   Implications for Teaching

Schaeffer's work has demonstrated that the major development in the pre-school years is not only the ability to count, but the ability to use the counting process in assessing the sizes of collections. Clearly the ability to count is not acquired spontaneously, but by imitating the actions of others. However the child has to construct for himself the connection between the recitation of arbitrary sounds and the size of a collection.

Parents and teachers can obviously assist by encouraging children simply to count, but further questioning, of the type involved in the experiments quoted, referring to the size of the collection, the order of the count, the relative size of two collections, and so on, may help the child to construct the important relationships which need to be made before the process of counting can be usefully applied.

One problem the young child has is in keeping control of a number of aspects while he is counting e.g. which objects he has already counted, remembering to say only one number for each object, and so on. This suggests that early counting experiences should be made as straightforward as possible, involving perhaps a row of solid objects which can be moved aside when counted. Only later when his counting has become more automatic should the child be expected to cope with a random array of fixed objects, as for instance occurs in the pages of early textbooks, e.g:

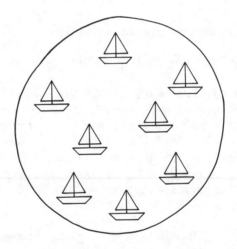

Fig. 3.3

Such a collection is often referred to as a 'set', which appears not unreasonable since it is a word which is easily read. However as Freudenthal (1973) points out, a collection such as that shown above is mathematically speaking not a set, since the objects are not distinguishable from each other. To avoid creating confusion later, it therefore seems worth playing down the 'set' notion at this stage, and simply asking 'How many boats?'.

Fuson (1980) quotes Case as having shown that considerable practice in the counting process not surprisingly enables children to apply it more automatically, and thus to concentrate on other aspects, like those required in accurate enumeration, or making the link between counting and the size of a collection.

The work of Schaeffer and Gelman, referred to previously, stresses the important role of counting in the child's early grasp of the ordinal and cardinal aspects of number. Brainerd (1979) suggests that the idea of 'one-to-one matching' between objects of one collection and those of another in order to compare the sizes of the collections is a relatively late accomplishment (unless the objects are set out to make the correspondence immediate as in the illustration given on page 172 by Bryant). The evidence of Wang et al. (1971), also referred to on pages 173 and 175, shows that one-to-one correspondence does not precede counting skills, but that both develop simultaneously.

Thus taken together these points may indicate that recent teaching schemes for young, and slow-learning children may give too much emphasis to making one-to-one correspondences between objects in sets, and consequently underplay the use of counting.

The availability of an effective 'stage structure' such as that given by Schaeffer should assist teachers to assess the stage reached by particular children and thus to provide suitable experiences to help them to master the stage which follows. In all the studies quoted, it was apparent that the range of attainment was very great, even at the age of 5 years.

3.2 Coordination of the Cardinal and Ordinal Aspects of Natural Number: Piaget's Experiments

The work of researchers such as Schaeffer and Gelman has shown that the average 5 year old child has come a long way in understanding and applying the idea of natural numbers. However during the 1930s, Piaget performed a series of classic experiments which illustrate that the process may not yet be complete.

3.2.1 Piaget's Experiments Relating to the Cardinal Aspect of Number: 'Conservation of number'

To explore the understanding of the cardinal aspect of number i.e. the use of natural numbers to denote the size of a collection, Piaget used the following situations:

a. Provoked correspondence between glasses and bottles

> Six little bottles (about one inch high, of the kind used in dolls' games) are put on the table, and the child is shown a set of glasses on a tray: 'Look at these little bottles. What shall we need if we want to drink? – Glasses – Well, there they are. Take off the tray just enough glasses, the same number as there are bottles, one for each bottle.' The child himself makes the correspondence, putting one glass in front of each bottle. If he takes too many or too few he is asked: 'Do you think they're the same?' until it is clear that he can do no more. Mistakes occur in fact only with children of the first stage (4 to 5 years). The correspondence can be made easier by getting the child to empty the bottles into the glasses, each bottle filling one glass. Once the correspondence is established, the six glasses are grouped together and the child is again asked: 'Are there as many glasses as bottles?' If he says 'no', then he is asked: 'Where are there more?' and 'Why are there more there?'. The glasses are then rearranged in a row and the bottles grouped together, the questions being repeated each time. (Piaget, 1952)

b. Provoked correspondence between flowers and vases.
c. Provoked correspondence between eggs and egg cups.
    (Each of the experiments (b) and (c) is essentially similar to the 'glasses and bottles' experiment described above.)

d. Provoked correspondence between pennies and objects. This is again similar except that the child is asked to exchange his pennies in turn for a collection of objects.
e. Spontaneous correspondences.
   This is also similar except that the child was presented with various shapes made of counters and asked to take out of a box the same number of counters as each figure contained. If the child matched the shape of the figure, one shape was distorted and the child was asked whether there were still the same number of counters.

In each of these experiments Piaget identified three stages of performance. (Unfortunately there is no research to show whether any connection can be made between the stages of Schaeffer and those of Piaget.)

Stage I: Global comparison without one-to-one correspondence or lasting equivalence
   These children appeared to simply try to find a collection which was globally 'similar', often in its arrangement, without trying to quantify by counting or by matching each object of the second collection with an object of the collection provided.
   All of the children quoted lie within the age range 3 years 6 months to 5 years 6 months, with the majority between 4 and 5 years old.
   An example of a Stage I response in the case of the flowers and vase experiment is that of Giri, aged 4 years 4 months:

> Giri put 13 flowers close together in a row opposite ten vases rather more spread out, although he had counted the vases from one to ten. Since the rows were the same length, he thought that the flowers and vases were '*the same*'. 'Then you can put the flowers into the vases? – *Yes*.' He did so, and found he had three flowers over. The flowers were taken out and bunched together in front of the vases. 'Is there the same number of vases and flowers? – *No* – Where are there more? – *There are more vases*... (Piaget, 1952)

A classroom example showing children beginning to appreciate the significance of one-to-one correspondence is provided by a teacher quoted in Biggs.

> We did a lot of number work at milk time in a group. At first we counted the number of children and the number of bottles, having first checked – 'one for Andrew, one for Peter' etc. We noted that the number was the same. Over the week they became aware that the numbers *must* be the same, or there would not be one each. Then I tried taking all the tops off, and, hiding them in my hand, asked how many they thought there would be. They had no idea! They were quite intrigued by this relationship between the number of children, bottles, tops and straws... (Biggs, 1970)

Stage II: One-to-one correspondence without lasting equivalence
   At Stage II the children managed to create a second collection equivalent in number to the one given to them, but judged that one of the collections became bigger than the other if it was more spread out.
   All the Stage II children quoted lie within the age range 4 years 1 month and 6 years 7 months, with the majority between five and six years old.
   An example of a Stage II response, to the glasses and bottles experiment, is that of Mul (aged 5 years 3 months) who made the correct correspondence between bottles and glasses, but only after initially he had selected a collection of eight glasses by eye to match the six bottles:

> Mul ... 'Were they the same? – *No, there were too many glasses.* – And are they the same now? – *Yes,*

they're the same. – (The glasses were then grouped together and the bottles spread out.) – Are they the same? – *No, because that's bigger.* – Can you count? – *yes.* – How many glasses are there? – *Six.* – And how many bottles? – *Six.* – So there are the same number of glasses and bottles? – *There are more when it's bigger.* (Piaget, 1952)

Another instance of this stage is given by Ginsburg, who was interviewing Jonathan, aged 4 years 6 months:

The interviewer next showed Jonathan two rows of candies, one designated as his and the other as the interviewer's. The rows were identical in appearance.

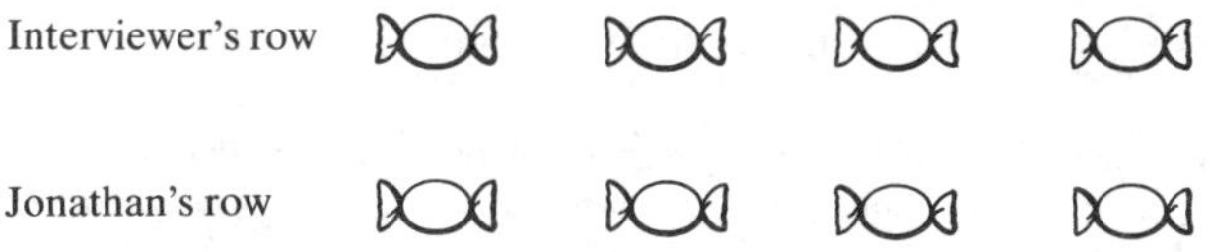

Fig. 3.4

After Jonathan agreed that the rows were the same, the interviewer said; 'Now watch what I do, I move yours out and I put one here.'
The interviewer added one to his own row and spread Jonathan's apart.

Fig. 3.5

Next the interviewer asked, 'Which one do you want to eat?'...Jonathan indicated that he would like to eat his row...

I: That one? Why do you want to eat that one?
J: 'Cause you put these out (meaning, because you spread them out)
I: Can you count them for me?
J: 1, 2, 3, 4
I: So you have 4, and how many do I have?
J: 1, 2, 3, 4, 5
I: So who has more?
J: Me
I: You? But this one has 4 and this one has 5.
J: You put yours like that and put mine right there.

(Ginsburg, 1977)

Thus at Stage II, the children have initially understood the cardinal aspect of number, in that they understand what is meant by two collections with the same number, and they can also spontaneously find some way of selecting a collection with the same number of elements as a given one, either by matching objects directly (one-to-one correspondence) or by counting (matching each collection separately with the sequence of natural numbers). This suggests that Piaget's Stage II children must be at least at Schaeffer's Stage Three, and that this may not be sufficient since Schaeffer's Stage Three children only had to match a collection of objects with a number, and not spontaneously count two collections.

However when one collection is spread out, the children feel the correspondence has been broken, since they insist on using a perceptual property, in this case length, as a criterion of which collection contains more objects. This is, as the previous quotation illustrates, independent of whether they have counted the collections.

Stage III: Lasting equivalence

At the final stage, children were not only able to produce a collection of objects equal in number to a given collection, they were also sure that the collections were equivalent however they were arranged.

All the Stage III children quoted in Piaget's work are at least 4 years and 11 months, and the majority are at least 5 years and 6 months.

An example of a Stage III response is given by Lau, aged 6 years and 2 months:

> Lau made six glasses correspond to six bottles. The glasses were then grouped together: 'Are they still the same? – *Yes, it's the same number of glasses. You've only put them close together, but it's still the same number* – And now, are there more bottles (grouped) or more glasses (spaced out)? *They're still the same. You've only put the bottles close together.* (Piaget, 1952)

Thus it is only at Stage III that children are clear that if two sets can be shown to be equivalent in number, this equivalence is not destroyed by re-arranging one set.

Piaget's 'conservation of number' experiments described have been much criticised; some claim they are too 'verbal' and hence children who have the concepts may be misled by the language used (e.g. Siegel, 1978). Others contend that the way the situation is arranged pre-disposes the child to answer incorrectly. For instance Donaldson repeated the experiments but told the child that the disturbance in the arrangement was caused by 'naughty teddy', and showed considerably increased levels of performance (Donaldson, 1978). Bryant (1974) reasonably claims that the classic conservation experiments are not good tests of grasp of the invariant property of number, since the rearranged collection is compared not with itself in a previous state, but with a second collection which has to be assumed to remain equivalent, and thus a further (transitive) step is required. He proves his point by demonstrating that 4 year olds were correct in identifying equivalence in 75 per cent of cases in which a single row was transformed, but in only 35 per cent of cases when a second row was used for comparison, as in the Piagetian conservation experiment.

The above criticisms would seem to be valid. Nevertheless Piaget's results themselves have been replicated many times (e.g. Elkind, 1964; Dodwell, 1962). What can we therefore conclude about what is demonstrated about children's understanding of the cardinal aspect of number?

First, it is clear that Piaget's Stage I children have at most a very vague notion of the cardinal aspect of number, since they have no valid method of assessing whether two sets are equivalent in size.

Second, it is clear that young children tend to use a perceptual cue, namely relative length, to help them to judge relative sizes of collections. Although this often gives the correct answer, it does not do so on all occasions.

Thus by employing this distraction, and in using an extra collection as a standard to judge whether or not one collection changes its size when its elements are re-arranged, Piaget can reasonably be accused of predisposing the child to answer incorrectly. Nevertheless, it cannot be contended that the child has a sound understanding of the use of natural numbers to compare sizes of collections unless he can cope with just such distractions, and indeed Piaget demonstrates that children are able to do so by around six years of age on average.

Children may thus have a reasonable grasp of the cardinal aspect of natural numbers, which may enable them to solve many practical problems, as is demonstrated in part 3.3, before their understanding is sure enough to enable them to cope with the so-called 'conservation' experiments. Children should not therefore be written off as 'non-conservers' if they fail on a Piagetian conservation task; they may well conserve in more favourable circumstances and may have a concept of number which is adequate for many basic number situations.

In fact many researchers have attempted to 'train' younger children to give correct responses in the conservation experiments. Although these have not always met with success, a number of them have produced very positive results, especially with 4 to 5 year olds. What is more interesting, is that this success has not been restricted to a single training method but has been demonstrated across a variety of methods. Some of these (e.g. Siegel, 1978; Gelman and Gallistel, 1978) have been behaviourist inspired and children have merely been given feedback on their success or failure, with no indication of the cause of the mistakes. Others have made more positive efforts to change the basis of children's judgments. For example, Bryant (1974) demonstrated to children that if they used a length cue to compare a set of 10 black and 9 white counters, then their answer changed each time the rows were compressed or spread out. However, using a one-to-one correspondence strategy produced the consistent result that the black set had more.

Thus many children who fail on traditional Piagetian conservation tests may not be in any sense 'incapable' of success, but merely may not have yet had cause to doubt that length is a reliable cue to the size of a collection. Whether or not it is actually worth 'training' children to conserve is less certain, especially when this failure does not otherwise appear to hold back their progress and is in any case likely to be remedied without such external intervention by the age of six or so.

3.2.2 Piaget's Experiments Relating to the Ordinal Aspect of Number: 'Seriation'

Piaget realised that his conservation experiments actually involved an ordinal aspect; in order to decide whether two collections were equivalent in size a child had to either count the objects, or to arrange both collections in parallel sequences so that the one-to-one relationship became clear.

However he also designed some experimental tasks which focused on the ordinal aspect. One problem which arises with these (the so-called 'seriation experiments') is that in most cases he incorporated a built-in perceptual sequence into his collection of objects. Thus the first task the child had was to order the object by some physical property (e.g. dolls by size, sticks by length etc.) into a sequence. The ability to do this efficiently is certainly evidence of a logical, and therefore mathematical, ability, which is relevant to a capacity to approach a task systematically and to check for consistency. However it is not clear that this ability is itself related to the idea of number, which requires a rather different assumption, namely that the order in which a collection of objects is counted or matched is irrelevant.

The importance in this context of the seriation experiments is that they are used to reveal whether a child can relate the ordinal aspect to the cardinal one i.e. whether he appreciates that when the seventh object in a sequence is reached, the collection previously counted is of size six, and the collection that will then have been counted is of size seven.

Thus Piaget gives as an example of part of a Stage III response that of Shen (aged 6 years 6 months), who had been introduced to a situation involving a collection of ten dolls of varying height, together with a collection of ten sticks, also varying in height.

> Which (doll) is the biggest? – *The last* (Doll 10) – We could call it the first as well, couldn't we? – *Yes* – And what about this one (Doll 9)? – *The second.* – And this one (Doll 8)? – *The third...* – If we say that a doll is fourth, how many are there in front of it? – *Three.* – And in front of the eighth? *Seven.* – How do you know – *I counted in my head how many were left...* – Which doll will this stick (Stick 5) go with? – *That one* (Doll 5) – Why? – *I counted in my head* (pointing at Dolls 10-5 and Sticks 10-5, which were not now arranged in order of size). (Piaget, 1952)

(The only children quoted who are said to be at Stage III are aged between 6 years 6 months and 7 years 7 months, whereas those at Stage II are between 5 years 5 months and 7 year olds. Thus they are rather older than the Stage III children quoted in the conservation experiments.)

Another Stage III response, this time in an experiment involving a number of rods of different length which the child has to arrange in a 'staircase', is quoted below by Ald (aged 6 years 6 months). The 19 rods were disarranged.

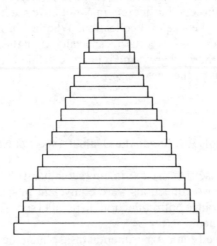

Fig. 3.6

> Look, the doll is here (Step 12). How many steps has it climbed?... *I'll have to arrange them as far as that* (he did so). *That makes 12* – How many steps are behind it? – (without counting) *11* – And how many has it still to climb? – (He counted the scattered sticks without arranging them and gave the correct answer) – *7.* (Piaget, 1952)

This is in contrast to the children at Stage II who reconstructed the *whole* staircase in order to answer the last two questions.

Perhaps an even clearer indication of the association between the ordinal and cardinal aspects is provided by an experiment similar to that above, but in which the sticks were replaced by rectangles, of equal width, the second rectangle being twice as long as the first, the third three times as long and so on.

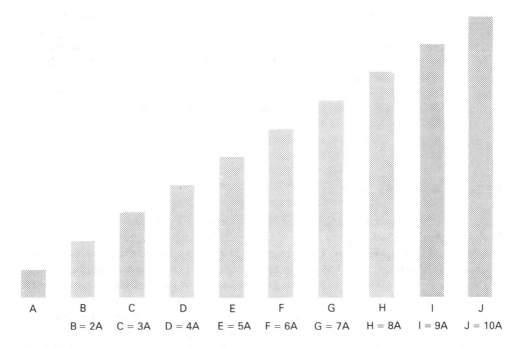

A	B	C	D	E	F	G	H	I	J
	B = 2A	C = 3A	D = 4A	E = 5A	F = 6A	G = 7A	H = 8A	I = 9A	J = 10A

Fig. 3.7   (reproduced from Copeland, 1974)

Here the contrast between a Stage II response, that of Bru, aged 5 years, and a Stage III response, that of Ald (the same boy as before), can be seen clearly:

> Bru 'Which is the first one? – (He pointed to A.) – And the second? (B) – If we cut up that one (B), how many little cards like this (A) can we make? – 2 – And out of this one (C) – 3 – And out of this one (D) – 4; etc, up to ten.' (But when without disturbing the series, we pointed to cards at random, he thought there were 4 in E, 6 in H, 3 in D, 7 in E, and so on. (Piaget, 1952)

> Ald (6.6) made the series without hesitating and was first asked

> How many like that one (A) can we make out of (B)? He replied: '3, *I knew I had to count that one* (A) *as well.*' He had therefore at once counted the positions. 'And with that one (F)? – 6, *because there are three there* (pointing to K, J and H) *that makes 10, 9, 8, then that one* (G) 7. When the series was disarranged, he reconstructed the staircase as far as the element in question and thus found: 5 for E etc. (Piaget, 1952)

The Stage III responses in each case illustrate the coordination between the order of the object in the sequence, and the collection of objects which come before it. In the latter experiment this is even clearer, since there is a correspondence between the order of the object in the sequence and the number of units of which it is composed.

Piaget thus concludes that in order to really understand and apply natural numbers, an appreciation of both ordinal and cardinal aspects are needed, together with the necessary relationship between them. He considers this to be acquired at the same time as many other logical operations develop, at roughly around 6 to 8 years for average children.

There has in fact been little informed challenge to this assertion of Piaget's, and there is therefore no reason at present to doubt that it is true. As can be seen from the responses given above, what marks Piaget's Stage III is often the ability to manipulate the numbers themselves in a flexible way to solve problems, for example, by counting on or backwards, rather than a total dependence on the presence and ordering of the objects involved. The beginnings of this can be seen in Schaeffer's Stage Four, in which children could reason that eight was necessarily more than seven, without requiring any visual array. Thus between the ages of 5 and 8 years, the average child is beginning to develop the ability to reason consistently by reference to numbers, rather than by reference to particular collections of objects.

Further illustrations of this process follow in part 3.3.

3.2.3 Implications for Teaching

Piaget's work in this area, like Schaeffer's, is important since it again illustrates the difficulties a child encounters en route to a full understanding of the basic idea of natural numbers, and their application. The experimental tasks can again be used by teachers to assess the child's progress and to give insight into his particular view of number.

There are however dangers in attaching too much importance to Piaget's criteria or to those of other workers. It has frequently been demonstrated that what appears to be a minor change in the design of the task can result in significant changes in the child's response. Hence the attainment of 'conservation' in its Piagetian definition should not be regarded as a great watershed, but only as a further small step towards a complete picture. There are certainly many problems involving number that a child who fails the test can adequately solve, some of which may help him eventually to arrive at the view of number as representing a stable property of a collection of objects. Nevertheless a child who has not responded in an adult fashion to the conservation and seriation tasks is unlikely to have abstracted a consistent concept of number, but still to be dependent on objects themselves, so it would appear to be unreasonable to expect such a child to manipulate numbers flexibly without reference to such objects.

3.3 EARLY STAGES IN THE DEVELOPMENT OF THE ABILITY TO ADD AND SUBTRACT

3.3.1 Concrete Understanding: 3 to 5 Year Olds

So far it has been demonstrated that most pre-school children have begun to develop a working grasp of small natural numbers, although this may not be complete until around the age of 7 on averge. Similarly, many children develop notions of addition and subtraction before they receive any formal instruction in school, although again the ideas are unlikely to be fully grasped until the age of seven or so.

Gelman and Gallistel, in a study referred to, in Section 3.1.2 on page 171, devised an experiment in which children learned to recognise a particular number of objects on a plate as a 'winner', and then substituted a plate containing a different number of objects in order to test their reactions. When a 'three-plate' was substituted for a 'two-plate' or vice versa, they found that children of 3, and sometimes even younger, could not only distinguish between the plates but could also describe the transformation which had taken place in terms of 'you put one on' or 'you took one off'. They could also reverse the transformation by spontaneously either removing one object or selecting one from a group offered to add to the collection.

However when the difference was greater than one – in fact three toy mice were substituted for five – although 26 of the 32 children aged 3 to 4 years indicated in discussion in some way that they thought that more than one was missing, by phrases such as 'They gone', 'Some came out', etc., only six of these were able to say that the number removed was two. And only four took exactly two mice out of a selection of four offered to restore the original position; others took one, three or four and proceeded by trial and error as did VB aged 4 years 4 months:

Win? *Wait! There's one, two, three.* Is that the plate that wins? *No.* Why? *Because it has three.* It has three? What happened? *Must have disappeared!* What? *The other mouses.* Where did they disappear from? *One was here and one was here.* How many now? *One, two, three.* How many at the beginning of the game? *There was one there, one there, one there, one there, and there.* How many? *Five – this one is three, not … but before it was five.* What would need to fix the game? *I'm not really sure because my brother is real big and he could tell.* What do you think he would need? *Well, I don't know…some things come back.* (Experimenter hands VB some objects, including four mice. VB puts all four mice on the plate.) *There. Now there's one, two, three, four, five, six, seven! No…I'll take these off and we'll see how many. One, two, three, four, five, no – one, two, three, four. There were five, right?* Right. *I'll put this one here and then we'll see how many there is now. One, two, three, four, five. Five! Five!* (Gelman and Gallistel, 1978)

Thus we can see that children of 3 and 4 have formed an intuitive notion of addition and subtraction in terms of the concrete actions of 'adding on' and 'taking away', recognise the effect of these operations in increasing or decreasing the size of the collection, and appreciate the inverse relationship of the two actions in that one 'undoes' the other. However they do not always have the ability to quantify the change and are generally dependent on repeated re-counting of the complete set.

But a study by Starkey and Gelman (1982) suggests that in simple cases even some 3 year olds are able to accurately solve addition and subtraction problems provided these are set in a concrete framework. In the task used the experimenter first asked the child to determine how many pennies were in the experimenter's open hand. These pennies were then covered and a second array of pennies was placed in the same hand so that the child could readily count them. At the same time the experimenter stated how many extra pennies he was putting in his hand. The child was then asked how many pennies were in the hand altogether. Since the two groups of pennies were never simultaneously visible, the child had to do more than merely count a collection of objects. The task was repeated with different numbers of pennies, and in some cases pennies were removed from, rather than added to, the original collection. The sample included 48 American pre-school children (16 3 year olds, 16 4 year olds and 16 5 year olds). The next table shows the number of children giving correct answers in a selection of the tasks. (The number given first in the expression indicates the initial number in the hand.)

Mean proportions of children in each age group giving correct answers in addition and subtraction problems (from data in Starkey and Gelman, 1982)

	2+1	5+1	14+1	4+2	2+4	2−1	5−1	15−1	6−2	6−4
3 year olds (N=16)	.73	.20	.13	.07	.07	.87	.27	.13	.07	.33
4 year olds (N=16)	1.00	.88	.31	.50	.25	1.00	.69	.19	.19	.31
5 year olds (N=16)	1.00	1.00	.62	.81	.56	.94	.94	.44	.56	.44

This suggests that 3 to 5 year olds are often able to obtain correct answers to simple concrete addition and subtraction problems. Many children were observed to solve the problems by counting even when some of the objects were hidden from view. Some appeared to be imagining the covered objects, while others used their fingers to represent them. A further group were able to simply use the sequence of number names, in some cases appearing to 'count on' (e.g. in 14+1 the answer 15 was sometimes produced rapidly). Fuson (1982) also notes the ability of some 3 year olds and most 5 year olds to 'count on' in order to add one or two to a given number.

Hughes (1981) obtained similar results working with a sample of 60 Edinburgh children. Of these, 20 were in their first term in a nursery class (mean age 3.5 years), 20 were in their final term in a nursery class (mean 4.5 years) and 20 were in their first term in the infant school (mean age 5.2 years). Like the American sample none had had any formal teaching in arithmetic at school. The task given was essentially similar to that used by Starkey and Gelman except that the problem concerned bricks in a box. The task was given, for each pair of numbers, in each of five ways, four of which are described below:

a) *'Box open'*   The child could see some bricks in the box, watched a further set being added (removed) and could see the final total collection in the box.

b) *'Box closed'*   The child could see the bricks initially in the box, watched the experimenter add (or remove) some and heard a verbal description of how many were being put in (or taken out) but could not see the final collection.

c) *'Hypothetical box'*   The box was removed and the child given a verbal problem e.g. 'If there was one brick in the box and I put two more in, how many bricks would be in the box altogether?'

d) *Formal code*   The child was asked a formal 'sum' involving only abstract numbers e.g. 'What does two and one make?'

The results are given only for the combined sample of 3 to 5 year olds and are shown in the following table:

Number of children succeeding on each problem (N=60) (data from Hughes, 1981)

	Small numbers				Large numbers			
	1+1	2−1	1+2	3−2	1−1	5+1	6+2	8−1	7−2
a) Box open	56	57	50	57	56	35	27	25	28
b) Box closed	51	54	44	46	53	22	17	19	10
c) Hypothetical box	41	39	34	23	31	19	11	11	9
d) Formal code	22	7	8	3	4	6	3	4	2

Again there is evidence that many pre-school children can solve simple addition and subtraction problems even when no visual apparatus is present, although very few can solve 'abstract' problems involving only numbers. The proportions succeeding however fall as the size of the numbers increases, even when only counting is required, and as the degree of abstraction increases. Hughes noted the variation in attainment in his sample even for children of the same age group. In particular the half of the sample selected from a predominantly middle class school were, at 3½, already more than a year ahead in their development as compared to the group of children selected from predominantly working class school.

Although Hughes did not directly record the strategies used, he hypothesised that the ease of the small-number items was due to the child's ability to work with a mental image, whereas this was no longer feasible when the numbers got large. With larger numbers the child had to be able to work with the verbal number sequence. This would also help to explain why, for example, $7-2$ was more difficult than $6+2$, for in the former case the child had to count backwards along the number sequence.

Thus there is evidence from these two studies that many pre-school children are able to solve simple arithmetic problems in a concrete setting well before they are able, for instance, to conserve number. Nor is this phenomenon only observed in Western countries; for example, Hatano (1982) reports similar evidence from Japan.

3.3.2 Development of Strategies: 6 to 8 year olds

Even in younger children Starkey and Gelman had noted some relatively abstract strategies for solving addition and subtraction problems.

Carpenter and his colleagues (Carpenter and Moser 1979, 1982; Carpenter, Hiebert and Moser 1981) carried out a large-scale survey of 150 children aged 6 to 8 years in the United States, using a variety of verbal problems involving addition and subtraction. The first interviews occurred shortly after their entry to the first grade of elementary school and prior to any formal instruction in addition and subtraction.

At this stage around 60 per cent of the children could solve the addition problems when 'small' numbers (5 to 9) were involved, and 30 per cent with 'big' numbers (11 to 16). However when the children were allowed cubes to help them, the success rate for large numbers went up to around 45 per cent. Most of the errors were due to inaccuracy in counting rather than to any fundamental misunderstanding.

In subtraction, the difficulty depended on the problem type (see the later discussion in part 3.5 on page 227), but varied from around 60 per cent to around 25 per cent for small numbers, and from around 35 per cent to around 15 per cent for large numbers. Again the availability of cubes improved the success rate substantially for large numbers and marginally for smaller numbers, and again many errors were due to inaccurate counting.

Thus it would again appear that the majority of children even before they went to school at 6 years were reasonably competent at solving addition and subtraction problems, and a substantial minority could do so with numbers greater than ten. In particular Carpenter noted that hardly any children employed the wrong operation, an error frequently seen among older children.

Strategies used in one type of addition problem (A) at entry to elementary school (B) after a further six months (data from Carpenter and Moser, 1982)

| Condition | Interview | Percentage correct | Counting all | | Counting on | | Numerical | |
			With Models	Without Models	From First	From Larger	Recalled Fact	Derived Fact
Smaller numbers: Physical objects available	A	75	54	5	5	6	6	3
	B	91	33	1	4	19	30	1
Smaller numbers: No objects available	A	64	38	6	6	9	8	3
	B	91	14	1	13	16	36	0
Larger numbers: Physical objects available	A	50	51	1	3	3	1	1
	B	84	45	1	9	25	9	4
Larger numbers: No objects available	A	28	16	1	8	8	1	1
	B	66	13	0	21	33	6	3

Strategy (per cent responding)

Not surprisingly, most of the strategies used were based on counting. Thus in the addition problems with small numbers around 40 per cent of children with correct answers counted both collections separately whether they used cubes or fingers, and re-counted the combined collection from the beginning (the '*count all*' strategy). However around 7 per cent used the more sophisticated method of '*counting on*' from the larger number, and a further 7 per cent counted on from the first number given, which was in all cases smaller. But another 7 per cent used a '*recalled fact*' (e.g. 6 and 8 is 14) and 5 per cent used a '*derived fact*' involving a mental regrouping of the numbers e.g. '8 and 2 is 10, so 8 and 6 is 10 with 4 left over to make 14'.

Thus not only were pre-school children already able to solve many addition and subtraction problems, but in some cases they were able to use quite sophisticated strategies.

Carpenter and Moser monitored the development of the children's strategies over a six month period. (During this time they experienced some formal teaching of addition and subtraction.) A breakdown of the results is given in the table on page 192 for a particular type of addition problem (e.g. 'Sara has 6 sugar donuts. She also has 9 plain donuts. How many donuts does Sara have altogether?')

Thus even over a six-month period, substantial development can be seen in the type of strategies used. There is a gradual movement away from the use of concrete models (objects, fingers etc.) towards 'counting on' methods, and the use of recalled number facts. That this development might not have been entirely due to the teaching the children had experienced in school is suggested by results obtained by Groen and Resnick (1977). In their study ten 4½ year old children were trained individually to solve simple verbal addition problems by modelling each of the two collections with blocks, combining the two sets of blocks, and re-counting the total collection from the beginning (the 'count all' strategy). Half of the children, during the period of the study, spontaneously discovered that 'counting on' from the larger number was a more efficient method, and proceeded to use this strategy.

Steffe, Thompson and Richards (1982) and also Richards (1980) studied the ability of children to 'count on' by using a board onto which were stuck coloured discs; some of the discs were hidden behind a cover.

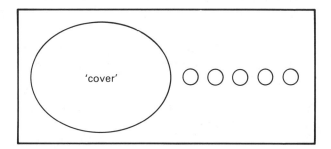

Fig. 3.8

The children, all American first graders (aged 6, in their first year of formal schooling), were told that there were, for example, seven discs under the cover and were asked how many there were altogether.

As a result five levels of performance were identified, depending on the flexibility of the counting process.

a. *'Perceptual' counters*   These children needed actual external items to count, whether objects, sounds, actions etc. They therefore could not progress at all with the task in which one group of the discs were covered. (5 out of 34 six year olds were included in this category.)

b. *'Figural' counters*   These children were able to 'picture' items mentally even though they could not perceive them directly through their senses. However they often failed to coordinate their pictures with the information given. For instance, in an example in which there were actually three discs covered and five exposed, this was the response of Glen (aged 6):

> Glen: Attempts to touch the covered discs again, but is stopped by the interviewer. He then looks at the cover, successively fixing his gaze on specific locations in a linear pattern across the cloth in rhythm with subvocally uttering 'one... ...six' and slightly nodding his head... He then finishes his counting activity, counting the visible checkers 'seven... ...eleven. Eleven.' (Steffe, Thompson and Richards, 1982)

(3 out of 34 six year olds were included in this category.)

c. *'Motor' counters*   These children were able to count motor actions, as well as objects, real or imagined. For example:

> Interviewer: 'There are seven checkers (or draughts) under this cover. How many checkers are there in all on the card?'
> Brenda: Begins by subvocally uttering 'one... ...seven' while pointing her pencil in the air over the cover each time indicating a different location, but clearly not referring to a specific place on the cloth. Pauses, then subvocally utters 'eight... twelve' while pointing her pencil at each of the visible checkers. 'Twelve.' (Steffe, Thompson and Richards, 1982)

(5 out of 34 six year olds were included in this category.)

d. *'Verbal' counters*   No definition is given of this category but presumably it includes children able to work with the verbal sequence of number names without any more concrete support, but otherwise similar to the previous group.

e. *'Abstract' counters*   These children were able to 'count on' from the total of the covered discs without repeating the whole sequence of number names. For example:

> Donna: 'Seven under here' (points to cover)?
> Interviewer: 'Yes.'
> Donna: 'Eight, nine, ten, eleven, twelve (points at the individual five checkers in synchrony with utterances). Twelve..' (Richards, 1980)

(21 out of the 34 six year olds were either 'verbal' or 'abstract' counters, but no further breakdown is given in the references.)

Fuson replicated Steffe's study using American 6 to 8 year olds from middle and upper-middle class backgrounds. She found, out of 27 children, the following distribution.

'Perceptual' counters: 0
'Figural' counters: 1

'Motor' counters: 1
'Verbal' counters: 5
'Abstract' counters: 16
(No categories are indicated for the remaining children.)

The 'verbal' counters recited all the number names for the 'covered' collection starting at one, whereas the 'abstract' counters started by repeating only the last number of the count sequence for the first collection. However it was noted that when the first collection was not covered, 5 of the 16 abstract counters reverted to the more primitive strategy of counting all the items, presumably for security. (This tendency to revert to more concrete strategies, when concrete apparatus is provided is also noted by Carpenter and Moser, 1982.) Fuson also notes that Davydov and Andronov (1981) have found similar results in Russia, and were successful in training some children to use a 'counting on' strategy.

Ginsburg gives an interesting example of a child who was a 'verbal counter' but not yet an 'abstract counter' in Steffe's terms. The problem concerned a birthday party to which twelve guests were invited, but only seven cups were available. James (5 years, 7 months) was asked how many more cups his mother should buy. His written response is given below:

James: (He wrote on the paper: 7 cups and 12) 'How do you spell children?'
I.   'C-h-i-l-d-r-en.'
J.   (He wrote, CHILDREN AD 2 cups) 'One two, three, four, five, six, seven – Oh! – eight, nine, ten, eleven, twelve.' (He erased AD 2 cups and wrote instead 8 9 10 11 2 13 14 15. Then he erased 13 14 15 and wrote in *cups* instead...)
J: 'Eight, nine, ten, eleven, twelve are missing cups.'
I: 'So now how many more cups should your mommy buy?'
J: 'She should buy eight, nine, ten, eleven, twelve cups.'
I: 'But how many cups is that?'
J: 'It's eight, nine, ten, eleven, twelve?'
I: 'How many children need cups?'
J: 'Eight, nine, ten, eleven, twelve.' (Ginsburg, 1977)

Fig. 3.9   James's solution (reproduced from Ginsburg, 1977)

What James seems to be unable to do is to detach part of a number sequence (eight to twelve) and count these number names themselves, arriving at an answer of five. One of Steffe's criteria for 'abstract counters' is that they can in this way use number names themselves as the items of a count, and can thus start new counts at any point in a sequence.

Steffe et al. (1982) found that the levels of flexibility in counting correlated well with children's ability and strategies in solving simple verbal addition and subtraction problems.

'Perceptual' counters were unable to produce any appropriate strategies; they seemed to lack the information processing capacity even to hold both the numbers in the problem in order to model the situation with concrete objects.

'Figural' counters could cope only with simple addition problems and then only if they were able to translate directly into a finger-counting model, which required that at least one of the numbers was very small. They became confused when trying to use a collection of objects.

'Motor' counters could generally use a 'count all' strategy using objects or fingers, both for addition and subtraction.

'Verbal/Abstract' counters. No detail is given, but by implication the latter group were able to use efficient 'counting on' and 'counting back' procedures.

3.3.3 Counting on, Class Inclusion, and Criteria for a Flexible Use of Abstract Numbers

From the preceding information it would seem that the ability to 'count on' marks an important step in the child's development of flexibility in handling numbers in a more abstract sense.

Fuson suggests that a major difficulty in using a 'counting on' strategy is in converting from a cardinal number representing the size of the initial collection to an ordinal number representing the final count word for that collection. For example if the child is told that twelve objects are in a box and five are added, then he must convert this cardinal use of twelve into an ordinal twelve in order to proceed 'thirteen, fourteen,..., seventeen'. This cardinal-to-ordinal shift is the inverse of the ordinal-to-cardinal shift which takes place after the child has counted the whole collection; he recites, for example, '...fifteen, sixteen, seventeen' and infers from the final ordinal, seventeen, that the cardinal number of the whole collection is also seventeen. (This cardinal-to-ordinal shift is referred to by Schaeffer et al. (1974) as the 'cardinality rule' and is discussed in Section 3.1.4 on pages 178 to 179).

Fuson notes that as many as one third of her sample of 6 to 8 year olds may obtain an answer one smaller than the correct answer by using the cardinal of the first collection (e.g. 12 in the previous example) to also represent the first element in the collection which is added to it.

In the task involving the covered discs, referred to previously, which was used by Steffe et al. (1982) and by Fuson there was little difficulty in counting on with the second collection since this collection, unlike the first, was physically present, and children could thus readily point at the numbers as the names were recited. However in most problems given verbally, there is the further difficulty of keeping track of the counting of the second collection. Thus if the problem involves adding 5 objects onto an initial collection of 12 objects, then the counting process for the second collection goes 'thirteen, fourteen…etc.' and the difficulty is knowing when to stop. Fuson showed that most children in fact used their fingers, first holding up five fingers and touching each in turn while saying 'thirteen, fourteen,… seventeen'. However a few were able to use aural or visual patterns to assist.

Critical to the whole 'counting on' strategy is the notion that a set can be simultaneously a set in its own right, and part of a larger set. For example James (page 195) was unable to detach the string 'eight, nine, ten, eleven, twelve' and see this as a separate set of five. Similarly in keeping track in an addition problem using fingers, the second collection is first represented as, say 'one, two, three, four, five' when putting up the correct number of fingers, and later as 'thirteen, fourteen, fifteen, sixteen, seventeen' when seen as integrated into the total collection.

This ability is what Piaget refers to as 'class inclusion' i.e. the simultaneous recognition of a total class and one of its subclasses. He gives a famous example of this in which when children of 5 and 6 are presented with poppies and bluebells, they judge that the bunch of poppies would be larger than the bunch of flowers (Piaget, 1952). (Other examples involve comparing collections of girls and children, brown beads and wooden beads, etc.) He also uses a situation more directly related to addition in which children need to simultaneously distinguish between a total collection and its subcollections. The child was told that on that day he was to receive 4 sweets for 'elevenses' and 4 for tea, whereas the following day he would be hungrier at tea time so he would receive 7 at tea time, but only one at 11 o'clock. The interviewer showed the child two collections of eight beans. Each collection was identically subdivided into two groups of four.

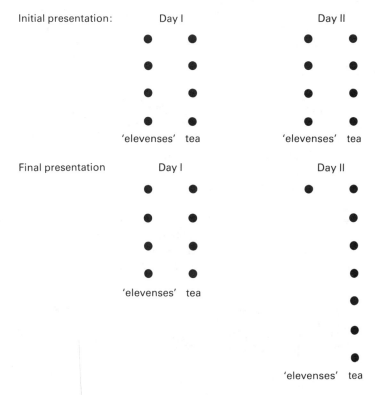

Fig. 3.10

When describing the allocation of sweets on the first day the interviewer used one collection of beans, and for the second day the interviewer re-grouped the second collection into a group of 7 and a single bean.

An example of a child who could not simultaneously recognise the total collection and the subcollection was An (6.11)

> 'Is there the same amount to eat there (I) and there (II)?' – *'No, there's one there (II) and there are 4 there (I)'* – 'How many were there before here (II)?' (The two sets of 4 in II were arranged again, then, before the child's eyes, 3 were taken from the first set and added to the 4 of the second.) 'Aren't they the same, these (II) and these (I)?' – *'No, now there's only 1 here (II) and there are 4 there (I)'* – 'Could we make them into 4 and 4 again (II)?' – *'Yes'* (doing so) – will you have the same amount to eat on both days (4 and 4 and 4 and 4)?' – *'Yes'* – 'And now' (changing again to 7 + 1)? – *'No, because there are less here (II).'*
> (Piaget, 1952)

However older children (the two quoted are aged 7 years 3 months and 7 years 6 months) have no difficulty in judging the total amounts for the two days to be equal, in spite of their different arrangements.

Piaget therefore suggests that in order to handle natural numbers flexibly in solving problems, children need the three operations of conservation (see part 3.2.1 on page 181), seriation (ordering) (see part 3.2.2 on page 185) and class inclusion, and that indeed these are achieved simultaneously. In fact there is some evidence to suggest that, even if Piagetian tasks are used as criteria for these three operations, they are by no means acquired simultaneously (e.g. Dodwell, 1962), but here as elsewhere, variations in the context of the task presented can cause substantial differences in the results. For instance Brainerd (1979), using his own form of the class inclusion task, reports that only a third of his 11 to 12 year old sample performed perfectly.

On the other hand a wide-scale investigation by Gonchar (1975) shows a high measure of agreement in these three areas, which gives strong support to Piaget's hypothesis.

Richards (1980) presents a similar view to Piaget in concluding that children can only operate efficiently and flexibly with numbers when they have both a grasp of conservation and are able to 'count on', like the 'abstract counters' identified by Steffe et al. (1982) (see page 194). Such children are able to detach numbers from objects and are beginning to operate with them as abstract entities.

An interesting illustration occurs in Ginsburg's study of a 10 year old girl, Stacy who was considered to be a very low attainer. This seemed to be confirmed when she seemed to be quite at a loss when attempting to solve a simple problem which she had actually set herself: 'Jimmy had 8 cats, he gave Brian 2 cats ... How many does Jimmy have?' She obtained various wrong answers for which she could only supply incoherent reasons. However it was later shown that Stacy could conserve and, at least in simple cases, count on:

> Interviewer: (after asking Stacy to get out some poker chips) 'You count 12, OK, How about two more?'
> S: (quickly) '14.'
> I: '14, OK. How are you doing that so fast? What are you doing in your head? Are you doing something, saying some numbers to yourself?'
> S: 'I say 13, 14, like that.'
> (Ginsburg, 1977)

The interviewer then asked Stacy problems in which she was shown two collections of objects, which she was allowed to count, and asked for the total while the objects themselves were hidden. She managed several of them up to 14 + 6, in some cases counting on from the larger number.

Thus although Stacy was indeed behind most children of her age group, she had the basic concepts needed both to conserve and to count on. Given time and encouragement, she was able to solve addition problems. Her initial failure had been due to a tendency to guess uncertainly on the basis of poorly remembered number bonds in the belief that this was the method expected, rather than building on what she did understand.

3.3.4 The Relationship between Understanding of Concepts and Arithmetic Performance

Numerous researchers have attempted to study the relationship between success on such tasks as Piagetian conservation and performance in formal tests of arithmtic (see Hiebert and Carpenter, 1982; also Brainerd, 1979, and Carpenter, 1980a, for summaries).

In general although they have found that the 'concept' tests are efficient at identifying children who are likely to make good progress in arithmetic, the overall correlations are not very high, especially if individual concept tests are considered rather than some composite score. This is not an unreasonable conclusion, for, as we have seen, some problems in arithmetic are soluble by 3 year olds, provided the numbers are reasonably small, and provided naive methods are allowed. It would seem likely that a necessary condition for the solution of arithmetic problems would be achievement of Schaeffer's stage three (see page 178) which depended on making the link between the counting process and the number of objects in a collection. Since the stage three children in Schaeffer's study had an average age of 4 years 2 months, it would be expected that most children involved in the studies reported above would have achieved this. It also seems likely that other factors, like willingness to engage in learning number facts, would also affect arithmetic performance marks. It is also the case that tests of 'understanding' based on a single 'pass-fail' criterion, like those of Piaget, are not particularly reliable.

Romberg and Collis (1980) give some indication that level of cognitive operation, as judged by several measures, is more closely related to the strategies used in solving arithmetic problems, but even here there is a tendency for able younger children to use safer, less sophisticated strategies than they would be capable of, given the choice.

Thus children who lack notions like conservation (although they might well be shown to possess some understanding if less stringent tasks were presented than those of Piaget) are not necessarily prevented from achieving progress in simple addition and subtraction problems. However their methods are likely to be less sophisticated than those which are available to children with greater understanding, and their grasp of number concepts cannot be regarded as very firm.

3.3.5 Implications for Teaching

It is clear that even very young children have a concrete notion of the effects of 'adding' and 'taking away' and realise the reciprocal nature of these operations.

When children learn to count, and in particular when they learn to relate the counting process to the size of a collection, they should be able to solve simple addition problems using their fingers or concrete objects as aids. (Subtraction problems are a little harder, in some cases significantly so – see section 3.5 on page 226).

Although the use of concrete models to solve problems may begin at age 3 or 4 for the average child, some will want to continue using these methods into the secondary school. In fact many of the studies referred to in this section emphasise the spread in children's attainments, right from the ages of 3 and 4 upwards.

It would seem wise, at entry to primary school, to build on the skills children already have, in addition and subtraction as well as in other areas. Gradually children evolve the more efficient strategies of 'counting on' or 'counting back' without needing to recount. It seems likely that the developoment of these strategies is related to the ability to perceive a collection of objects as simulataneously a part of another collection (Piaget's notion of 'class inclusion'). It is also related to the ability to coordinate the ordinal and cardinal aspects of number in a flexible way. The 'counting on' strategy itself is eventually replaced in most cases by the use of recalled facts and results derived from them.

Thus again we see a protracted period during which the child develops strategies for solving problems which progressively become more independent of concrete objects. This process of strategy development is sometimes ignored by primary texts, which often attempt to teach a single technique. This may be in advance of, or behind, the level of sophistication the child has reached for himself.

The evidence, such as it is, suggests that provided the child is presented with appropriate problems he will evolve these strategies for himself, and the case quoted above of Stacy (page 199) suggests that attempting to push the child on to more sophisticated strategies than he or she is able to confidently handle may prove counterproductive.

So far no mention has been made of the use of symbols for numbers, or for operations like addition. Children can clearly make considerable progress without the need of symbols, operating, first with real objects and later mentally. Indeed the one boy quoted who spontaneously used number symbols (James, page 195) found them of little assistance when he lacked the underlying concept necessary to solve the problem.

This would seem to counsel caution in pressing ahead too early with formal recording and thus confusing children who have only a tenuous grasp of number in any case. Of course if children are familiar with the number symbols (1, 2, 3,...) there is no need to refrain from using them, but there would seem to be every reason to wait until children have developed a confident oral understanding, as evidenced by conservation and efficient 'counting on' and 'counting back' strategies before encouraging formal recording of addition or subtraction operations. (This point is taken up again in section 4 on language and symbols.)

3.4 THE REPRESENTATION AND MEANING OF NUMBERS – THE PLACE VALUE SYSTEM FOR NATURAL NUMBERS

3.4.1 Early Encounters with Numbers Greater than Ten

The young child is first likely to come across numbers greater than ten in conversations or stories. At first he may not realise their meaning – 'thirty-seven Victoria Road' is probably heard, and maybe learnt, as a string of arbitrary words; later he may realise that 'a hundred' or 'twenty' mean 'a lot', without any notion of relative size.

However when he learns to extend his recitation of number names beyond nine, some of these words will become integrated into the sequence. Forty years ago, Brownwell (1941) recorded that American children entering kindergarten (at age 5) could generally count to about 20. More recently, also in America, Fuson (1980) recorded that the mean length of correct counting sequence for 5 to 5½ year olds was 44, and for 5½ to 6 year olds, 84. (However the children in these samples may not have been representative.) Ginsburg quotes the case of Paul, who at 5 years 1 month was able to count by ones as far as a million and by tens to a thousand.

Ginsburg notes that children quickly learn to count larger numbers by repeating word patterns, and indeed this activity sometimes gives them much pleasure. He quotes the example of Rebecca (4 years 11 months) who after hearing her twin sister count to 19 asked:

Rebecca: 'What's after nineteen?'
Mother: '20'
Rebecca: '20, 21, 22, 23, 24, 25, 26, 27, 28, 29. What's after?'
Mother: '30'
Rebecca: '31, 32, 33, 34, 35, 36, 37, 38, 39. Now 40.'
Mother: 'Good'
Rebecca: '41, 42, 43, 44, 45, 46, 47, 48, 49 (Pause) 50.'
Mother: 'Good'
Rebecca: '51, 52, 53, 54, 55, 56, 57, 58, 59' (Pause).
Mother: '60'
....
Rebecca: (on another occasion) '20, 21, 22, 23, 24, 25, 26, 27, 28'
Mother: 'You do an awful lot of counting.'
Rebecca: '15, 16, 17. What's after 21?'
Mother: '22'
Rebecca: '22, 23, 24, 25, 26, 27, 28, 29' (Pause)
Mother: '30'
Rebecca: '40, 50, 60, 70, 80, 90, tenny.' (Ginsburg, 1977)

In discussing Rebecca's guess at 'tenny', Ginsburg also quotes examples of children inventing 'tenty, eleventy, twelvety', 'forty-ten, forty-eleven...', and also 'five-teen'. The former two illustrate the fact that in verbal counting, there is nothing special about ten; it is only at 'thirteen, fourteen...' that any repetition occurs. Children seem to pick up these spoken patterns very quickly, even to the point as we have seen of making false generalisations, in the same way that they use 'I goed' or 'I buyed' in learning to use spoken language.

Hence many pre-school children are likely to acquire the verbal patterns of numbers expressed in 'tens-and-units' form. However as Schaeffer showed, the average child is at least five years old before he gains any appreciation that the order in which numbers come in the counting sequence is related to their relative size (see page 179). So although children can count to fifty, they may not realise, for instance, that forty-seven is more than twenty-one. And since they are unlikely to have ever consciously met a collection of forty-seven objects, they are likely to have little idea of the absolute magnitude of such a number.

All this verbal experience is likely to pre-date by quite a long period the recognition of numbers in symbolic form. For instance Ginsburg noted that Paul, the 5 year old who taught himself to count to a million, could not even recognise the written number 1. While Brainerd (1979) found that almost all 5 year old (pre-school) children in a large sample in Canada could recognise the written numerals from 1 to 5, Wang et al. (1971) found that only 42 per cent of a similar sample of American children could do so consistently, and only 22 per cent in the case of numbers from 6 to 10. It is certainly likely that children will see numbers around them, such as house numbers, bus numbers and so on. Higginson gives an example of a 3 year old who, given the stimulus of a digital clock, is learning to match written numbers with spoken numbers:

(Clock 7:57)	
Kate:	'Two sevens, seventy-seven'
(Clock 8:00)	
Kate:	'Oh! Eight, oh, oh
	Two ohs, and one eight...'
H (Father):	'What's it going to be next?'
Kate:	'Berry undecided'
(Clock 8:02)	
K:	'Is that two or five?'
......	
H:	(writing '33' in the notebook) 'What number is this?'
K:	'Three, three'
H:	(writing '44' in the notebook) 'What number is this?'
K:	'What?
	Tell me. I don't know the name of it.
	Forty-four?
	Give me another number.
	I want another number.
	Give me forty-five.' (H writes '45')
K:	'Forty-four
	That forty-four?'
H:	'No, that's forty-five (writes '46')
	What number is this?'
K:	'Forty-six
....	
H:	(writing '55') 'What number is this?'
K:	'Fivety-four
	Two fives.
	That's Aunt Kathy's favourite number. She likes three fives.'
H:	(Writing '66' in the notebook) 'What number is this?'
K:	'Ninety-nine
	That ninety-nine or sixty-six?'
....	
(Clock 8:08)	
K:	'Oh, eighty-eight'

(Higginson, 1980)

This illustrates a number of points, the main one being the picking up of the patterns in which '-ty' is attached to the first number (e.g. fivety-four'). However, it also shows the confusion of symbols (2 and 5, 6 and 9), and the failure to appreciate the meaning of the numbers.

The evidence presented above suggests that most children will have an intuitive acquaintance with the 'tens-and-units' system in spoken form, before they come to school. However they are very unlikely to recognise the significance of the representation of, say, forty-two (i.e. 4 tens and 2 units), nor indeed to have any idea of what 42 things would actually look like. They therefore need much experience in appreciating the size of numbers, including their relative size, in addition to more formal work on reading and writing numbers, before they can start to understand the place value significance of the numbers themselves.

3.4.2   The Idea of Grouping

The way we normally record numbers is known as a 'place value' system. For example, in 542 the position of the digit 5 in the third 'column' from the right informs us that the 5 refers to 5 hundreds, and not 5 units or 5 thousands. This type of system depends on a principle of successive grouping; units are grouped into tens, each set of tens then is grouped into a hundred, each set of ten hundreds into a thousand, and so on. Thus to find the size of a collection like this:

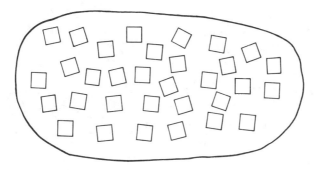

Fig. 11.a

we first group to obtain:

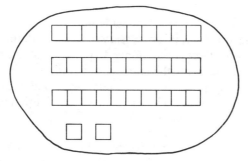

Fig. 11.b

Which allows us to write the number as 32. (The grouping may be done explicitly, using concrete apparatus, or implicitly by using the counting words.)

This idea of successive grouping is an important aspect of understanding our system of writing numbers.

Bednarz and Janvier (1979, 1982) report a study they conducted with 75 8 to 9 year olds which relates to the children's understanding of this idea of grouping without recourse to the symbolisation of it by numerals. Their concern was with strategies which showed understanding rather than with strategies that merely produced right answers.

They presented the children with a problem which involved the presence of concrete materials in the form of peppermints. These mints were loose, in rolls, and in bags of rolls, which the children could handle. They were told that:

> Mother had bought some loose mints some of which she had wrapped into rolls and bags of rolls ready for giving out at a party.

No mention was made of the number of mints in a roll or of the number of rolls in a bag. Mother started with:

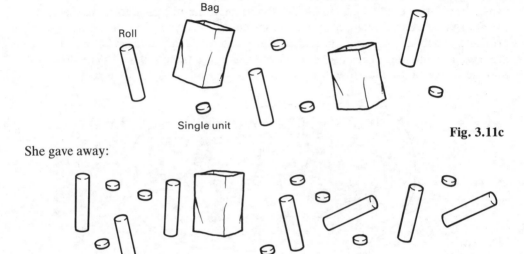

Fig. 3.11c

She gave away:

Fig. 3.11d

The task was to draw a picture of what she had left.

In order to solve this problem, it is necessary to know both the number of mints in a roll, and the number of rolls in a bag. Only 40 per cent of the 8 to 9 year olds gave evidence of realising the need for this information, either by inventing values for the two numbers, or by directly questioning the interviewer. Bednarz and Janvier identified two strategies (A and B), and in each case divided the children choosing that strategy into two groups. (See the table opposite.)

Group 1 were those on their way to understanding but had problems in the simultaneous handling of two groupings. Group 2 comprised those children whose strategies indicated a good understanding of numeration from the 'grouping' viewpoint.

The remaining 60 per cent of all the children

> did not even feel the need to inquire about the content of the rolls and bags. Either they considered the problem impossible or they carried out a subtraction taking away the smaller number of articles from the bigger one for each article or for the only articles where it 'works'. (Bednarz and Janvier, 1979)

Bednarz and Janvier point out that this sort of activity is not only useful for assessing the child's understanding of the necessity of knowing the basis of the groupings but also would serve as a useful teaching device in an oral and practical setting. They used

Strategy A They associated a number to each collection of mints.	*Strategy B* They proceeded directly through the operation "opening" bags of rolls.
G **R** **O** **U** **P** **1** Some of them did not associate the right number: rolls were confused with singles, or bags with rolls, in counting.	Some ran into trouble because either they operated only with the bags or with the rolls, or they played simultaneously with the bags and rolls.
G **R** **O** **U** **P** **2** 15% made the correct numerical association and (as it happened) did the correct computation.	15% successfully opened bags and rolls and borrowed the articles desired so as to make the operation possible.

(from data in Bednarz and Janvier, 1979)

another problem, again presented orally with specimen packets of mints available for handling. The child, although not told there were ten in the packet, could easily see this for himself.

Mother buys fifteen packets of mints. Does she have enough to present all these?

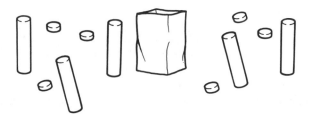

Fig. 3.12

Other groupings could easily be used to introduce notions of different bases, e.g. five mints to a packet, and five packets to a bag, for base 5 groupings. The important thing is that the child must recognise the need to know upon what basis the groupings are done – only then, with this understanding, is he anywhere near turning his attention to the symbolisation of these ideas in terms of abstract notation rather than the actual objects or pictures of them.

3.4.3 Reading and Writing Numbers

In the English language numbers are usually read and spoken, as with the written word, from *left to right* eg. 524 is five hundred and twenty-four. (There is the exception of the 'teens' where, for instance, 19 is said with the 'nine' first.)

However the size of an unfamiliar number with many digits may have to be evaluated before it can have meaning and the number read correctly. Many adults may be able to read large numbers like 19930 on sight, by mentally grouping the digits into threes and thus arriving at 'nineteen thousand, nine hundred and thirty'. However children take time to learn this trick, for instance Brown (1981c) quotes a thirteen year old who read 19930 from left to right as 'One thousand, ninety-nine, thirty'.

If the number is not recognised on sight , then it is necessary to evaluate it from *right to left*, i.e. in the opposite order to which it is read. Duffin (1976) points out the difficulty this creates for children. Brown quotes an instance of Frances, an above average 12 year old, working laboriously through this process in order to determine which of the numbers was larger, 20 100 or 20 010. (See section 4.3.2.2, page 355, for details.)

This problem also arises in work with different number bases, where the collections are grouped using numbers other than ten. For instance 2143 in base 5 can be read from *left to right* as 'two, one, four, three, base 5,' (it would be incorrect to read it as 'two thousand ....'). However in order to evaluate it, it is necessary to work from *right to left*, 2143_5 means 3 units (ones), 4 fives, 1 twenty-five, and 2 one hundred and twenty-fives $(5 \times 5 \times 5)$.

Ginsburg identifies three stages in developing an understanding of the theory of place value, where the written symbolisation of number is concerned.

i.   The first stage is where the child writes a number correctly with no idea as to why.
ii.   The second stage is where the child realises that other ways of writing a particular number are wrong – for example '31' is incorrect for 'thirteen'.
iii. Thirdly is the stage where the child is able to relate written notation of numbers to the theory of place value, e.g. Doug' a 7 year old, when asked why he had written a '1' followed by a '3' to indicate 'thirteen', replied that the '1' stands for ten and the 3 stands for three. Ten and three is thirteen.' (Ginsburg, 1977)

Ginsburg considers that not many children reach this third stage during their primary education. Only very limited evidence seems to be available on this question. It is true that in the APU Surveys, 69 per cent of 11 year olds (and 75 per cent of 15 year olds) were able to select the number in which the 7 represented '7 tens' from the list: 107, 71, 7, 710 (APU, 1981b, p.40), but this response does not necessarily indicate a firm understanding of place value.

Ginsburg also emphasises that a child's capacity for coping with number in its symbolic form is generally surpassed by his informal arithmetic, i.e. what he can do mentally on an oral level. The child tends to assimilate the symbolisation of number in terms of what he already knows such as the verbal names. Take the case of Rebecca, as described by Ginsburg; at the age of 7 she:

... was asked to write twenty-three. She did 203. Later she was asked to write thirty-five and wrote 305. She often made mistakes of this type.... (Ginsburg, 1977)

Her symbolisation of number reflected her spoken language: thirty-five was 305 (30 and 5).

Similarly Joe, aged 11, wrote the symbols shown in this table:

Spoken Number	Number written
Nineteen	19
Four hundred and seventy-two	472
Three	3
Six thousand and twenty-three	600,023
Seventy-one thousand, eight hundred and forty-five	710,00845
Fifty-six	56

(from data in Ginsburg, 1977)

Ginsburg points out that Joe was unable to write numbers with four or more digits without reverting to similar rules to those employed by Rebecca.

The reverse process where children are asked to say numbers from their symbolic form may cause similar difficulties. Dickson found 13 and 14 year old low attainers calling 3003, 'three hundred and three' and 4001, 'four hundred and one'.

It can be seen that of the ten numerals in our notational system, zero seems to cause the greatest difficulty. Leeb-Lundberg (1977) describes some of her problems as a teacher when dealing with zero. Its role as a placeholder, in the symbolic representation of number, is something not readily appreciated by children. She writes about her work on place value with 8 year olds. They considered numbers such as 100, 110, 101, and she describes a moment of enlightenment when one of her pupils cried 'I know. Zero is nothing – *of something.*'

Brown also found secondary children having difficulty with zero as a place holder, particularly with larger numbers. She quotes the example of Jack, aged 13:

… a mature and articulate middle-band boy who was thought by his teacher to be about average for his age group. But although he could read correctly the number 521 400, the number 8030 was read as 'eight hundred and thirty'. Similarly he rapidly calculated (mentally) the correct oral answers to 'add ten to 3597' and 'add one hundred to 19 930', but happily recorded them as 367 and 2030 respectively.

This example was not an isolated one. One of the test items used by Brown as part of the 'Concepts in Secondary Mathematics and Science (CSMS)' study was:

Write in figures: Four hundred thousand and seventy-three....

The success rate for this was:

1st yr	2nd	3rd	4th	secondary
42%	51%	57%	57%	

(Brown, 1981c)

The answers given contained most possible arrangements of varying numbers of zeros between the three digits 4, 7 and 3.

One area where secondary children show a definite weakness is with numbers over a thousand. Many seem to be unfamiliar with the place names of digits to the left of the thousands positions. Brown (1981c) points out that this may be due to a failure to recognise that spoken numbers, unlike written numbers, also make use of a thousand as a base as well as ten. Thus there is no new name for the columns between 'thousands' and 'millions', or between 'millions' and 'billions', as there is for each column up to thousands. (This is not the case in other languages such as Hindi.) A space, or comma, is used in our written notation to separate out each group of thousands, as in 21 730 000. This 'hidden' use of a thousand as an extra base seems to confuse children. Brown reports that many children interviewed gave the name 'millions' to numbers in the 'ten thousands' column. Evidence for this confusion was given by the CSMS item:

5214   The 2 stands for 2 *hundreds*.
521 400 The 2 stands for 2 ....

for which the success rate was only:

1st yr	2nd	3rd	4th	secondary
22%	32%	31%	43%	

(Brown, 1981b)

Much of what has been said here, concerning the difficulties in the development of place value, as experienced by children, is reinforced by virtue of a different kind of study which was conducted by the Russian, Luriya (1969). His work which was actually carried out during the Second World War, concerns the investigation of the effects of acalculia (also known as dyscalculia). Acalculia is the name given to the impairment of the concept of number and computation, resulting from a certain kind of localised brain disease (it affects the semantic aspect of speech). The relevance of Luriya's work rests on his belief that the breakdown of the concept of number and matters pertaining to number, due to this disease, reflects in reverse the original sequence of development of the concepts. So by studying the effects of acalculia it is possible to gain insight into the way these concepts came about in the first place. The most difficult aspects are the first to suffer. In all his patients Luriya found notions of place value being one of the initial aspects of number to break down.

All the patients cited by Luriya had, prior to their brain damage, been average or above in computational abilities. Although they could now cope with simple computation within the first ten numbers they had great difficulty in reading, writing and composing two-, three- or multi-digit numbers. Place value was affected. He gives the following examples: patients would read 22100 as '22' and '100'. 180 would be written from dictation as '100 and 80'. Some patients would evaluate 17 and 71 as identical in meaning. 489 would be considered larger than 701 because consideration was centred around single digits. Luriya also says that whenever the name of a multi-digit number did not coincide with its symbolic structure, especially when zeros were not specified in speech, then the numbers could not be written correctly. For instance, 'one thousand and three' was repeatedly written as '10003 and 'one thousand and twenty-eight' was either written as '128 or '10028'. Almost every patient made mistakes with these examples.

3.4.4   Ordering

The concept of place value can be looked at from the point of view of ordering numbers in their relative positions as they would appear on the number line. This also incorporates a notion of addition and subtraction in terms of obtaining larger or smaller numbers. Here we are going to consider the nature of some of the difficulties children experience with this ordering aspect of place value.

Bednarz and Janvier (1979, 1982) devised another task to investigate the understanding of place value in terms of the strategies employed by their 8 to 9 year olds in the following task. Each child played a 'game' with the investigator. He was shown:

4	2	3

and given an empty grid:

The task was to make within it any number larger than 423 and the first to do this was the 'winner'. It has to be done by taking turns to throw a die marked 0 to 5. When it was the child's turn he threw the die and wrote down on a separate sheet the number which came up. He then decided whether or not he wanted to use it in any of the three positions. If not he crossed it out.

By means of their performance in this game, Bednarz and Janvier found that:

i.   Fifteen per cent of the children did not even end up with a number greater than 423.

ii. Forty per cent used a positional strategy which meant they would only use a digit in a certain position if it was larger than the corresponding one already there, i.e. they would put only

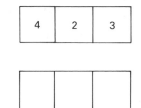

H	T	U
5	3, 4 or 5	4 or 5

They ended up with a number larger in all positions than 423.

iii. Thirty-five per cent used a mixture of strategies or unsystematic methods.
   e.g. a. Wait for a 5 to start in left hand position.
        b. Systematically refuse small digits e.g. 0 or 1, even though they have a 'winning 5' in the left hand position.

iv. Ten per cent of children displayed a good understanding of place value. This was operationally defined by actions such as:
   a. Selecting the first '5' being thrown and saying 'I'm the winner', after having put it in the left hand position.
   b. Accepting 0 and 1 and putting them in the middle and right hand positions when the left hand cell has already been filled with a '5'.
   c. Selecting '4' to go in the left hand position when conditions are favourable such as when a 3, 4 or 5 is already in the middle position, etc.

Again the benefit of this activity is not only its value in assessing the nature of the child's understanding but also its usefulness as teaching material.

It should be noted that children perform rather better on more routine items which require numbers to be ordered. Thus the APU found that 91 per cent of 11 year olds (and 93 per cent of 15 year olds) were able to select the largest of the set of numbers: 1998, 2012, 2004, 897 (APU, 1981b).

Ward, in his survey of English 10 year olds, found that the percentage of children successful was rather lower, 65 per cent, in the case of larger numbers. Children were asked to select the town with the largest population from:

Aberdeen	183 800
Bath	151 500
Fleetwood	28 800
Walsall	184 600
Winchester	31 000

(Ward, 1979)

3.4.5   Adding and Subtracting Mentally

Brown, in the report of the CSMS study, states that some children

...showed a superficial knowledge of place-names which could have easily led one to conclude that they understood the ideas of place value. It usually required a little more probing in order to expose the shakiness of their grasp. (Brown, 1981b)

She gives the example of an item which involved only the addition of one unit (or counting on one place)

This meter counts the people going into a football stand.

After *one* more person has gone in, it will read:

Almost one in three (32 per cent) of the 12 year olds in a large representative sample gave the wrong answer. In the case of a similar item used by Ward (1979), more than half (52 per cent) of ten year olds were incorrect. Some insight into the reasons for this was provided by the interviews:

Shakeel (aged 12, second year),... gave

0	6	3	1	0 0

since 'one to ninety-nine gives a hundred'. Raymond (aged 12, second year) first tried

0	6	3	9	9	1

and after the interviewer had explained that there were only the given number of 'holes', altered three to four thus:

0	6	4	9	9

He explained, 'It can't go there (points to 9 in "units") 'cos it makes 10, and it can't go there (points to 9 in "tens") as it makes 10, so...' (Brown, 1981b)

Brown also cites the examples of Maria (aged 13, third year) and Kim (aged 12, first year) who, when asked to *'add ten to 3597'* each gave the answer 35917. Maria explained why:

Maria: Three thousand, fifty-nine, seven. Do I add the one to the seven?'
Interviewer: 'Which do you think?'
Maria: 'The seven. Makes seventeen.'

(Brown, 1981c)

However Maria had earlier been able to identify correctly the digits occupying the 'tens' and 'hundreds' places in other numbers.

3.4.6 Decomposing

Another aspect of understanding place value involves the facility to reorganise or decompose a number. For instance:

523 is 52 tens and 3 ones
    or 5 hundreds, 1 ten and 13 ones
    or 3 hundreds, 22 tens and 3 ones, etc.

Three items from an American study by Flournoy, Brandt and McGregor (1963) indicate that their 106 above-average 13 year olds had some difficulty with this aspect of place value:
i.   Which means 25 hundreds and 4 tens?
      A.   25040   C.   2504
      B.   2540   D.   None of these
Less than 25 per cent of the pupils chose the correct answer of 2540

ii.   A meaning for 15320 is which one of these?
    A.   15320 tens   C.   1532 tens
    B.   15 hundreds 320 tens   D.   1532 tens 20 ones

Thirty-six per cent correctly chose C.

iii.   Which numeral shows another way to write this?

Thousands	Hundreds	Tens	Units
2	35	18	6

    A.   3486   C.   5686
    B.   5386   D.   None of these

Only 17 per cent answered this correctly. (C)

    A rather simpler item was given to 11 year olds by the APU: '7 hundreds 5 tens and 12 units make ......?' About 60 per cent gave the correct answer, with nearly 20 per cent writing 7512, whereas in a similar item 80 per cent of 15 year olds were successful (APU, 1981b, p.40).

    The problems of understanding the decomposition of number are particularly relevant to the decomposition method of subtraction. This is outlined in more detail in part 3.7 on 'Algorithms'.

3.4.7   Multiplying and Dividing by Powers of Ten

As was seen earlier the whole basis upon which place value is founded is the notion of grouping into tens of tens, etc. i.e. multiplication in terms of powers of ten,

$$\dots 10{,}000\text{'s} \qquad 1000\text{'s} \qquad 100\text{'s} \qquad 10\text{'s} \qquad 1\text{'s}$$

    Each column to the left is ten times its predecessor and going to the right each column is a tenth of its predecessor. So any real understanding of place value must include an appreciation of the fact that if any amount is multiplied by a power of ten, then the numerals representing this quantity will remain the same but in different positions, using zeros where necessary as place holders e.g. multiplying 32 by 100 gives 3200.

    The first APU Primary Survey (1980a) found that 71 per cent of the 11 year olds correctly wrote *the number which is ten times 100*.

    The Flournoy, Brandt and McGregor (1963) study involved items concerned with this notion of multiplication by powers of ten. Here are two of the items:

i.   The circled six ' ⑥ ' is how many times as large in value as the underlined six '6'?

               6⑥66

    Answers   A.   10   C.   1/10
            B.   1/100   D.   100

Thirty-five per cent of the above-average 13 years olds correctly answered (c).

ii.   In 1864 the 8 represents a value how many times as large as the 4?
    A.   2   C.   100
    B.   200   D.   20

Thirty per cent answered correctly with (B).

These areas of difficulty were also typical of 10, 11 and 12 year olds studied by the same authors.

Dickson has studied the performance of low attaining 13 and 14 year olds in the U.K. on this aspect of place value. These children were individually asked to compare like digits in multi-digit numbers.

For example, Julian, aged 13 years, was asked to compare the 2's in 23206.

Julian:	'The first two is bigger than the second.'
Interviewer:	'How much bigger?'
J:	'I'm not sure.'
I:	'If I say how many *times* bigger ...'
J:	'Nine thousand, eight hundred.'
I:	'How did you get that?'
J:	'Took two hundred from two thousand.'

Anthony, aged 13, compared the 6's in 56069.

Anthony:	'I think it's twenty times bigger.'
Interviewer:	'...Why?...'
A:	' 'Cos it's two columns away from the six.'
I:	And how do you get twenty then?'
A:	' 'Cos the nought's ten. And then there's another column which is another ten.'

The 'times' idea is very difficult for children to grasp. Further evidence of this is given in the section on 'Operations'. In the light of this, Dickson tried to study this aspect of place value from the 'division' angle – possibly a more meaningful operation for children.

Interviewer:	'How many ninety-fives are there in nine hundred and fifty?' (while saying this writing '95' and '950' for the pupil to see).
Lena:	(14 years) 'Oh! I'd say ... nine hundred ... I'd say ninety-five.'
I:	... Why? ...'
L:	'No I wouldn't say ninety-five. I'd say nine.'
I:	'... Why? ...'
L:	'Because nine into ...nine hundred goes ten. So it'd be ten.'
I:	'Ten.'
L:	'I think. And the five would go into fifty which would make... er ... eleven.'

Lena not seemingly grasping the relationship between 95 and 950 which underlies the place value concept, tackles this question in terms of performing a division sum – albeit incorrectly. She in fact was writing as she spoke.

$$95 \overline{)950} \begin{array}{c} 11 \end{array}$$

Using this kind of dialogue as a means of teaching is very useful. But to save the pupil getting confused in dubious computational procedures, the use of an electronic calculator could be invaluable. In a teaching situation Lena could well have done with a calculator for her $95 \overline{)950}$. She would then have had quicker and more accurate information and could have gone on to investigate the behaviour of other numbers.

3.4.8 Estimation and Approximation (with Whole Numbers)

With the recent expansion in the availability of calculators, the emphasis in teaching seems likely to shift away from pencil-and-paper computation procedures and towards those skills which seem to be necessary in using a calculator effectively (see for example Girling, 1977). One of these skills is the ability to estimate the approximate size of an expected result, in order to spot mistakes in operating the machine.

The ability to estimate must depend on a good grasp of basic place value principles. Ginsburg gives an example of a nine year old girl who appeared to generally have this grasp, although her performance of standard computational procedures was poor.

> ...the interviewer dictated some numbers for Jane to add... she wrote:
>
> 6
> 79
> 163
> 940
> 2342
> 15700
> ———
>
> She indicated that she would add by carrying from left to right.
>
> I: 'I'd like you to look at the numbers you have here and estimate – or guess, sort of an educated guess – as to what the sum should be when you add them all together.'
>
> Jane appeared to be thinking.
>
> I: 'What seems to be like a reasonable number to get?'
> J: '17... uh ... 18, and 18,392.'
> I: (quite surprised) 'That's a pretty good guess. I mean you were pretty precise all the way to the last unit. How did you estimate that?'
> J: 'I knew that it probably would be more than seventeen hundred.'
> I: 'Seventeen hundred? Seventeen thousand I think you mean.'
> J: 'Yes, seventeen thousand, and then I just decided, I said, and just *guessed* the rest.'
> I: 'How did you know that it would be more than seventeen thousand?'
> J: 'Well, that's two thousand (pointing to the 2,342) and that's fifteen thousand (15,700) which is seventeen thousand, and all of this (the rest) adds up to probably more than one thousand, and I know it would probably be eighteen thousand and I just guessed the rest...'
>
> (Ginsburg, 1977)

A similar question was given to 15 year olds in England and Wales as part of an APU 'practical test'. The children had first been asked to add 1056 + 672 on a calculator, which almost all did successfully. Then they were asked whether their answer was reasonable. Only two-thirds responded sensibly. A typical 'correct' answer was given by Dave, who obtained 1728 on his calculator:

> 'Yeh. Add together ... 1600. 72 and 56 are over 100. 2 and 6 equals 8.'
>
> When asked to work out an approximate answer to 2563 ÷ 97, Dave first mistook the division sign for a multiplication:
>
> 'about 19,000 to 19,500 ... 9 × 2000 = 18,000. Got the 563... that's got to be...' (The tester then pointed out the division sign) '... about 25 ... 97 goes into 256 about twice with a remainder – but not much under... (long pause)... much over... well...5.' (APU, 1980b)

Several points are worth noting here. The first is the tendency to 'round down' e.g. approximating to 97 using 90 rather than 'rounding up' to 100, which is a better approximation and also much easier to handle. Similarly 2563 is rounded down to 2000 rather than up to 3000.

A possible explanation seems to be that Dave is not using the techniques which are normally taught for approximating, which involve the use of one or two significant figure approximations.

(e.g. $\quad 1056 + 672 \simeq 1100 + 700 \qquad = 1800$
$\qquad 2563 \times \quad 97 \simeq 3000 \times 100 \qquad = 300\,000\,).$

Instead he seems to be attempting to carry out the usual pencil-and-paper methods in his head, but starting each time from the left rather than the right. This would explain the use of 9 rather than 90 when multiplying by 97, and of 256 rather than 2563 in the division. In fact, at least in the addition and multiplication cases, he shows the intention of also using the remaining digits to give the exact answer, so that in a sense he is here not really approximating but attempting to check that each digit is correct.

Nevertheless it is clear that the ability to assess the 'reasonableness' of an answer is dependent on a good grasp of place value principles.

3.4.9 Implications for teaching: Prerequisites and Early Work

Throughout the preceding discussion on place value, various ideas for teaching have been included. Here some more are presented but first certain prerequisites are listed which should perhaps be checked before work specifically on place value begins.

Ronshausen (1978) says that before place value is introduced to the primary school child he should be able to do the following:

1.   The child should know the cardinality of numbers from 0 to 9. That is to say he is able to select a set of say 5 objects from among different sized sets and associate this with the spoken word 'five'. Also when he is shown a collection of, say, seven objects, he can say there are seven. And when he hears, say 'nine', he can construct a set of nine objects.
2.   He can count from one through ten both in a rote manner and 'rationally'.
3.   He can read the symbols 0 through 9, so for instance if he sees '8' he can say 'eight'.
4.   He can associate these symbols with sets. So given a set of, say, four objects he can select '4' from the rest of the symbols to apply to it. Also given the symbol he can select the set, from a selection of sets, which has that many members.
5.   The child can write the symbols 0 through 9 when he hears their names or when he sees a set.

He is now ready to be introduced to the notion of 'ten', in a cardinal sense, orally and symbolically, as with 0–9 in the points above. He needs plenty of practical experience with grouping into tens as with bundling objects such as sticks etc.

A suitable introductory activity is illustrated in an example taken from a school visited as part of the Schools Council Project *Low Attainers in Mathematics 5–16,* and published in a companion volume to this.

In a class of 6- and 7-year olds, the teacher sat with a group of six children who were learning about tens and units. There were several cards on the table, similar in design. Edward, for example, had a card like the one in Figure 3.13a. He clipped tracing paper over the card and drew one cross on the tracing paper over each flower. Then he drew a line enclosing ten crosses. He counted this group again to check that there were ten, then counted those remaining. Next he recorded 5 in the *u* box and 1 in the *t* box (see Figure 3.13b), then handed it to the teacher and she checked it.

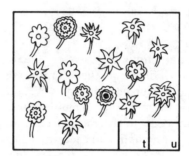

Fig. 3.13a Edward's card

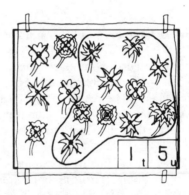

Fig. 3.13b Edward's completed
task

'How many tens?', she asked. 'How many units?' 'How many flowers have I drawn on this card?' Having answered the questions satisfactorily Edward chose his next card which was illustrated with twenty-one birds. (Denvir, Stolz and Brown, 1982)

The ILEA Abbey Wood Mathematics Centre (1975) issued guidelines for developing place value for tens and ones. They outlined the following sequence of activities:

1.   Bundling pencils into tens and ones. Putting beads into plastic bags – ten in each. Talking about 'tens' and 'singles' or 'ones'. Also getting into the habit of arranging the materials with 'ones' to the right of 'tens'.
2.   Joining rather than bundling into tens and ones, such as threading beads on a string, using unifix cubes joined into tens.
3.   Activities with ready made structural apparatus such as Dienes Base 10 blocks where the individual cubes are still distinguishable but inseparable.
4.   The next stage is to move onto tens and ones where the ten does not have the individual units marked but is, say, just a strip of card.

5. The child is now ready to use an object to represent a 'ten' which is only distinguish-able from the 'one' by its different colour and/or position. So if identical objects were arranged in a right to left fashion and separated clearly by columns they can represent 'ones' or 'tens' according to their position e.g.

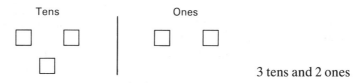

Tens   Ones

3 tens and 2 ones

6. The child is now prepared for using an abacus or similar model such as dried peas arranged in the 'troughs' on corrugated cardboard (with each trough representing one 'place', and with lines ruled between them).

The progression through these stages is necessary for the child to grasp the increasing abstraction from the actual act of grouping objects into tens and ones, to representing these by identical entities, such as peas, where the position is all important for determining whether the pea stands for a ten or a one.

With this foundation the child is then in a position to appreciate the significance of place value when the digits 0 to 9 are used instead of, say, peas or beads on an abacus. The Abbey Wood Centre emphasises

> that the abacus does not help to teach place-value for in essence it is as abstract as written numbers.

It must only follow the stages 1–5 outlined above.

Ronshausen (1978) outlines a programme of work which relates to the all important first stage of bundling. It is very similar to the sort of approach given in the series *Mathematics for Schools* (Howell, Walker and Fletcher, 1979). The important point she makes is that each child needs many of these experiences over many months to really begin to grasp the place value concept.

She tested a series of bundling activities with several hundred 6 to 7 year old children.

> ...the results showed that the achievement of children in all levels of ability improved significantly. Among ... children in the lowest ability level, improved achievement was even more noticeable than for children in higher ability levels (Ronshausen, 1978).

All that has been said so far is applicable to work with three digit and multi-digit num-bers. It should not be assumed that a child's activities with materials for two digit num-bers will automatically stand him in good stead for the 'hundreds' and 'thousands', and he can dispense with such materials. The need for concrete experience is emphasised throughout the literature – even well into the secondary level of education.

A particular example showing how one pupil was able to use concrete apparatus in order to grasp the idea of place value is provided by Magne. He describes a low attaining nine year old who had been introduced to the Dienes Base 10 blocks.

> Helen used these blocks with her exercises. To begin with it appeared nearly impossible for her to realize the meaning of the numbers. Nor did she seem able to see the significance of the material. But an improve-

ment was on its way. At first Helen succeeded in producing the correct set of blocks when the teacher had, for instance, instructed her to represent the number 207, or to tell the number 207 when the teacher laid the blocks in front of her. For a long time Helen could only work with the help of the blocks. But one day she achieved a dramatic triumph. Helen said to her teacher, 'I need no blocks. Because when I write 207, I have two flats and seven small cubes inside my head. And no longs at all. And so you see, miss, it is 207.'
(Magne, 1978)

The Abbey Wood outline to the development of place value points to the value of counting and mechanical counting devices; something like a car milometer is particularly useful. This can provide experience of large numbers and ordering especially in terms of when one more is added to a '9' digit. They recommend that, since pure counting becomes tedious, it should be put into some meaningful context such as collecting statistical data.

3.4.10   Implications for Teaching: More Advanced Activities

Once the child has grasped the notion of employing the digits 0 to 9 in each place, the concepts of place value can be further developed, at a relatively abstract level, by activities with digit cards.

Easterday (1964) suggests giving each child three blank cards. He is to write a different digit on each. With these Easterday suggests the following activities:

1.   Arrange the cards to name all possible numbers between, say, 100 and 1000 and list them.
2.   Arrange the cards to make the smallest possible number (or the largest) and arrange the list from (1) is ascending (descending) order.
3.   Suppose the cards are ⎡3⎤ ⎡4⎤ ⎡5⎤ . List the numbers which can be made from these cards which are less than 500 but greater than 345.

Easterday recommends that pupils should be allowed to manipulate the cards with a trial and error approach if necessary. Also similar activities may be carried out with more than three cards.

Easterday finds that work with digit cards is particularly suited to less verbal pupils who, in his experience at secondary level, seem to do well with this approach. He also points out that material of this nature readily lends itself to a larger group or class setting when digits are prepared on rectangular pieces cut from a transparency for use with an overhead projector. Children may then give oral or written respones.

Bednarz and Janvier (1979, 1982) offer another place value activity in the form of a game. The cards shown on page 219 are mixed up, face up on a table.

The teacher (or another child) says:

'I have this number' (402 is written down and read) 'and that one' (513 is also written down and read.) 'I'm thinking of a number that is bigger than 402 and smaller than 513, a number which is in between.' 'You must find out the number I have in mind but you must write it with the cards.' (Each time a guess is made, the interviewer makes clear that his number is bigger or smaller.)
   If the pupil cannot find the number, then it is given to him and he is asked to construct it with the cards.
(Bednarz and Janvier, 1979)

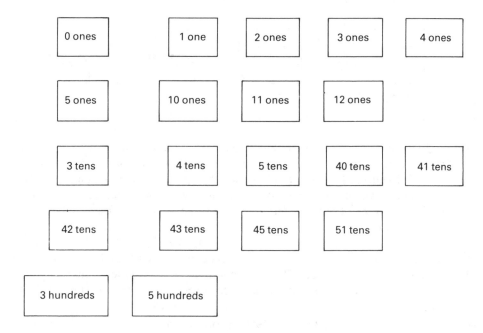

3.4.11 Implications for Teaching: the Use of Different Number Bases

As we have seen, in our own Hindu-Arabic system of representing numbers, units are grouped into tens, tens are grouped into 'tens-of tens' (hundreds), and so on. This is known as a 'base-ten' system; ten is probably used because of early finger-counting procedures. But mathematically speaking it would be possible to use any other number; for instance if seven were adopted as a base in a place value system the units would be grouped sucessively into 7's, sevens of 7's (forty-nines), sevens of 7's of 7's (three hundred and forty-threes), and so on. Thus the number 2534 (base 7) would represent the ordinary base ten number

$$(2 \times 343) + (5 \times 49) + (3 \times 7) + (4 \times 1) = 956$$

As long ago as the beginning of this century (e.g. Turnbull, 1903) educationists were recommending the use of number bases other than ten in order to help children to understand the idea of place value.

During the 1950s and 1960s, the use of different number bases became more popular at both primary and secondary levels, initially through the work of Dienes. Dienes (1960) suggested that in order to aid the abstraction of the place value idea, wooden blocks should be available in a number of different bases (Multibase Arithmetic Blocks or MAB). For instance, in base 3, the apparatus includes a number of blocks in each of the following shapes:

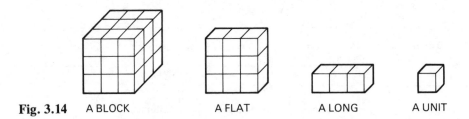

Fig. 3.14   A BLOCK   A FLAT   A LONG   A UNIT

In addition to this concrete embodiment of the place value system, he also suggested a more abstract representation using coloured counters in which, for instance, 3 whites would be worth 1 yellow, 3 yellows would be worth 1 green, 3 greens would be worth 1 red, 3 reds would be worth 1 blue, and so on. Glennerster (1980) reports a game, 'Trubloons' based on this type of representation in base three, which she found to be a good starting point for a first year class in a secondary school. The children collected 'treasure' according to their score on a pair of dice; thus a throw of 6 and 4 would be worth 10 whites using the system above, which could be traded for 3 yellows and 1 white, and then for 1 green and 1 white, and so on. The game continued for a number of turns, and the final totals were compared. Later subtraction was introduced, each child starting with 1 blue and subtracting the dice totals until someone was left with zero.

As a third representation, Dienes (1960) proposed an abacus-type model using washers arranged in 'columns', where this time each washer was equivalent to either one, three, nine, twenty-seven etc. (if working in base three) depending on the column in which it appeared. (This is similar to the 'pea and corrugated cardboard' representation quoted earlier.)

As a result of Dienes' work, the use of different number bases was incorporated into many 'modern mathematics' schemes in both primary and secondary schools.

However the evidence which has been gathered to evaluate the usefulness of the 'multibase' method has been somewhat equivocal. Of two studies conducted in England, the first (Brownell, 1964) gave evidence favouring an approach using base ten only, while the second (Biggs, 1967) supported a 'multibase' programme. However the first study tested only children in their first year in the junior school, and the second used a small and unrepresentative sample.

Results of American studies are no more clearcut. Callahan and Glennon summarise a number which compare 'base-ten-only' and 'multibase' methods and conclude:

> The hypothesis that the study of other base systems will enhance understanding of our own decimal system would seem to be a reasonable justification for its inclusion as a topic for study in the elementary grades. Evidence is not conclusive, however, that this is the only or best way of accomplishing this objective. (Callahan and Glennon, 1975)

The APU report of the first Primary Survey of 11 year olds notes that 'applications to numbers in bases other than ten ... were answered correctly by very few pupils.' However it is not possible to tell from the APU survey whether this poor result is due to the fact that few primary children now experience multibase work, or whether it is due to the fact that such work is frequently covered but less often understood. It may also be the case that even if children can technically cope with multibase work at this stage, they fail to see the point of it in terms of providing a deeper understanding of our own base ten system.

3.4.12 Summary

It can be seen that there are many facets to the understanding of the place value system. Evidence suggests that some of the ideas involved are not easy to grasp, perhaps not surprisingly in view of the time which elapsed before mankind invented such an efficient system. It would appear that misunderstandings occur with secondary as well as primary school children, and indeed, for some, their grasp is incomplete even at the end of the fourth year in the secondary school.

The teaching of place value would seem to be a long-term process, not simply limited to a few lessons, but needing a carefully designed progression over a long time period.

3.5 NUMBER OPERATIONS: UNDERSTANDING THEIR MEANING

3.5.1 Introduction

Here we are concerned in 3.5 and 3.6 with some of the literature reviewing the child's understanding of the meaning and structure of the four basic arithmetic operations: addition, subtraction, multiplication and division. An example of understanding the *meaning* of an operation is the child's grasp of both what is meant by, say, 3×2, and of the sort of situations to which such an expression is applicable. Appreciating the *structure* of number operations involves the understanding of mathematical relationships within and between operations such as, for example, that 9 fives are the same as 3 fives plus 6 fives.

The literature is organised roughly according to the stages of development in this area which are identified by Grossnickle (1959). These stages may be described briefly as

a. the meaning of the operation in concrete terms, from which emerges
b. computation and early structural properties of the operations.
c. understanding the structural properties of the operations.

In each part there follows an account of some instructional methods and ideas which may serve as useful guidelines for helping children understand the meaning and structure respectively, of the number operations.

Part 3.7 then concentrates on the procedures and strategies employed for performing arithmetic computations.

3.5.2 The 'Meaning' of Addition and Subtraction

The two concrete models used most frequently to illustrate the meaning of the addition and subtraction operations are those based on:

i.   *discrete objects*
     e.g.

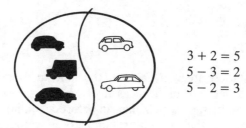

$3 + 2 = 5$
$5 - 3 = 2$
$5 - 2 = 3$

Fig. 3.15

ii. *continuous lengths*
     e.g.

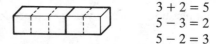

$3 + 2 = 5$
$5 - 3 = 2$
$5 - 2 = 3$

Fig. 3.16

The diagram above relates to problems involving the combining (for addition) or partitioning (for subtraction) of sets of objects or lengths.

In each of these two models 'comparison' problems (i.e. 'how many more than...' 'how much longer than...') can also be illustrated:

e.g.

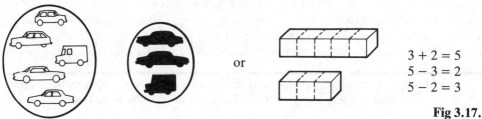

$3 + 2 = 5$
$5 - 3 = 2$
$5 - 2 = 3$

Fig 3.17.

Apparatus of the 'linking cube' variety (e.g. Unifix, Multilink) allows a direct translation between discrete (objects) and the continuous (length) models.

Surprisingly there seems to be little detailed research into children's ability to use these visual models in interpreting the meaning of addition and subtraction. There are only a number of global comparisons of, for instance, the 'Cuisenaire' (length) approach as against the 'traditional' (objects) approach, but a summary of these studies by Fennema (1972) shows that the results do not consistently favour any particular system.

The majority of the studies concerning the 'meaning' of addition and subtraction in fact centre on the ability of children to solve artificial 'word problems' (although in some of these studies concrete objects were provided).

Such word problems can be classified into different 'types' in the case of each operation. Possible classifications for addition and subtraction are proposed by Carpenter and Moser (1979), Vergnaud (1982), Nesher (1982), Brown (1981c), Lunzer et al. (1976) and Gibb (1954, 1956). Clearly it is not possible here to describe each system in detail, or to compare them; instead a flavour is given using a few examples in the case of each operation.

3.5.2.1   Addition Problems

Two examples of typical 'problems' are given below:
(A₁)   John has 3 large cars and 2 small cars. How many cars does he have altogether?
(A₂)   John has 3 cars. He buys 2 more. How many cars does he have now?
Vergnaud (1982) points out that both problems can be symbolised by $3 + 2 = 5$ or, pictorially, by

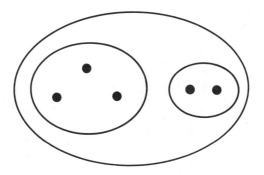

Fig. 3.18

However they have a different psychological structure; (A₁) describes essentially a symmetric composition of two sets (mathematically, a 'binary' operation) whereas in (A₂) a single set is transformed (in this case increased) to give a new set (a 'unary' operation).

Vergnaud therefore introduces the following diagrammatic representations to distinguish between them:

(A₁)

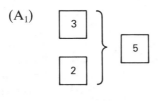

(A₂)

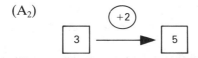

In a similar way other types of addition problems can be distinguished; five examples are given in the table which follows, together with suggested descriptive names and symbolisms. Each of these provides a concrete 'meaning' for 3 + 2 (although e.g. an alternative symbolisation for complementary subtraction would be ☐ − 2 = 3).

Examples of types of addition word problems

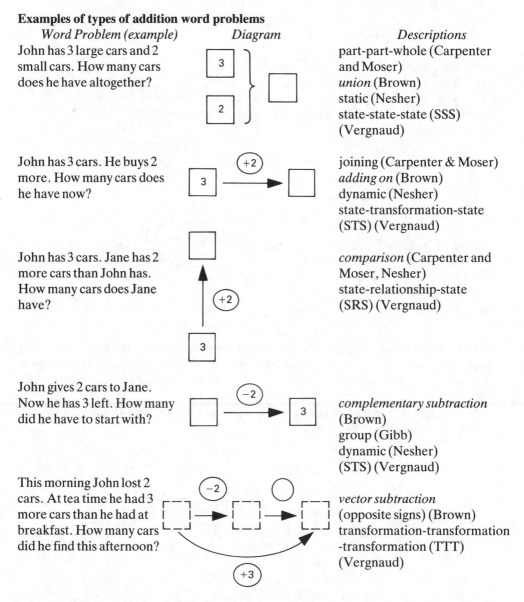

Word Problem (example)	Diagram	Descriptions
John has 3 large cars and 2 small cars. How many cars does he have altogether?		part-part-whole (Carpenter and Moser) *union* (Brown) static (Nesher) state-state-state (SSS) (Vergnaud)
John has 3 cars. He buys 2 more. How many cars does he have now?		*joining* (Carpenter & Moser) adding on (Brown) dynamic (Nesher) state-transformation-state (STS) (Vergnaud)
John has 3 cars. Jane has 2 more cars than John has. How many cars does Jane have?		*comparison* (Carpenter and Moser, Nesher) state-relationship-state (SRS) (Vergnaud)
John gives 2 cars to Jane. Now he has 3 left. How many did he have to start with?		*complementary subtraction* (Brown) group (Gibb) dynamic (Nesher) (STS) (Vergnaud)
This morning John lost 2 cars. At tea time he had 3 more cars than he had at breakfast. How many cars did he find this afternoon?		*vector subtraction* (opposite signs) (Brown) transformation-transformation -transformation (TTT) (Vergnaud)

(NB   The italic descriptions are those for ease of reference in the following text.)

Studies of the comparative difficulty of some of these problem-types have been carried out in various countries (France, Israel, the United States, Britain) by a number of authors (e.g. Carpenter and Moser, 1982; Nesher, 1982; Matthews, 1980; Brown, 1981a; Vergnaud and Durand, 1976).

Putting together the results of these studies suggests a hypothetical order of difficulty as shown below.

Hypothetical order of difficulty in addition problems

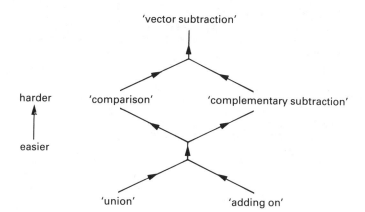

Fig. 3.19

It must be stressed that this is a hypothetical order since no study included all five types, nor are the separate results directly comparable due to variations in the ages of the children and the conditions under which the problems were set.

To give some idea of the differences in difficulty, Brown (1981a) found that 97 per cent of a sample of 11 to 12 year olds could successfully solve 'union' problems whereas Vergnaud and Durand (1976) report success rates of 25 per cent for 'vector subtraction' problems with the same age group. Similarly Matthews (1980) reports that 70 per cent of her sample of 6 to 8 year olds could solve 'union' problems in comparison with only 35 per cent for 'comparison' problems.

These results therefore indicate that 'addition' is a process which can be applied to solve a variety of types of problems, some rather straightforward and others quite difficult, even when the numbers involved are small.

An example of two 11 year olds who are finding difficulty in solving an addition problem is given by Brown. The question, which is not easily categorisable as any of the above types, was as in the following figure (adapted from Brown, 1981a)

The signpost shows that it is 18 kilometres west to Grange and 23 kilometres east to Barton. How many miles is it from Grange to Barton?

Fig. 3.20

Tracey (aged 11) asked:	'Does it mean that is, er, 18 kilometres to Grange and 23 kilometres to, er Barton, does that mean that it's from the same place?'
Interviewer:	'That's right; from this signpost here. It's 18 miles that way to Grange and 23 miles that way to Barton.'
T:	'Take 18 away from 23...5.'

Hilary (aged 11) was only able to do it when the situation was made more concrete.

Hilary:	'Oh no, I'm no good at these...(pause)...you times those two together, don't you?...No, you can't...(long pause).'
Interviewer:	'Imagine standing there and you're looking up at the signpost, OK? Now that way it's 18 kilometres to Grange and that way it's 23 to Barton; we want to know the distance between the two.'
H:	'23.'
I:	'23?'
H:	... (long pause) 'I'm not very good at doing kilometres...'
I:	'Let's try something else. We're sitting here, right? Say someone said it was three paces to the window and it was five paces to the window that way...'
H:	'You'd add them.'
I:	'...How far from one window to another?'
H:	...(long pause)...'eight.'
I:	'Yes, what are you doing?'
H:	'Adding them?'

(Brown, 1981a.)

Brown also asked 58 11 to 12 year olds to write 'stories' to match the expressions 9 + 3, and 84 + 28, in order to provide more direct evidence of the 'meanings' attached to the addition operation. She found that about one-third gave a 'union' model, e.g. Glenn (aged 12):

3 men on a belding siet tow of the men lade 84 breks and one men lade 28 How would you work it out.

A further third gave a straightforward 'adding on' example, e.g. Brian (aged 11):

John had 28p to go and spend and his mum gave him a nother 84p how much has he got now?

The remainder gave a comparison story e.g. Juliet (aged 11):

Karen has 9 eggs and Susan has 3 more.

Thus children can attach a variety of 'meanings' to an addition sum.

3.5.2.2 Subtraction Problems

In the same way as for addition, a variety of types of subtraction problems have been distinguished and described. Some examples, each providing a 'meaning' for the expression 5 − 3, are presented in the next table. (Again, this categorisation is not complete; some of the authors referred to on page 222 have identified additional types. Similarly there are in some cases alternative symbolisations like 3 + □ = 5.)

Examples of various types of subtraction word problems

Word problem (example)	*Diagram*	*Descriptions*

John has 5 cars. He loses 3.
How many does he have left?

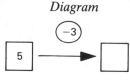

separating (Carpenter and Moser
take away (Brown)
left (Gibb)
pure subtraction (Lunzer et al.)
dynamic (Nesher)
state-transformation-state
(STS) (Vergnaud)

Jane has 5 cars and John
has 3.

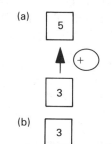

comparison (Carpenter
and Moser, Brown, Nesher)

(a) How many more cars does
   Jane have than John?

(b) How many fewer (less)
   cars does John have
   than Jane?

(c) What is the difference
   between the number of
   cars Jane has and the
   number John has?

(a) comparative – more ⎫
(b) comparative – less ⎬ (Gibb)
(c) difference (Lunzer et al.) ⎭
state-relationship-state
(SRS) (Vergnaud)

John has 5 cars. 3 are large.
How many are small?

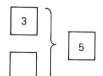

part-part-whole with missing
addend (Carpenter and Moser)
complementary addition
(i) union (Brown)
static (Nesher)
state-state-state (SSS)
(Vergnaud)

John wants 5 cars. He already
has 3. How many more does
he need?

joining with missing addend
(Carpenter and Moser)
complementary addition
(ii) adding on (Brown)
more needed (Gibb)
complementary addition
(Lunzer et al.)
comparison (Nesher)
STS (Vergnaud)

John had some cars. He
bought 3 more. Now he
has 5. How many did he
have to start with?

complementary addition
(iii) 'adding on' (Brown)
how many had (Gibb)
complementary addition
(Lunzer et al.)
STS (Vergnaud)

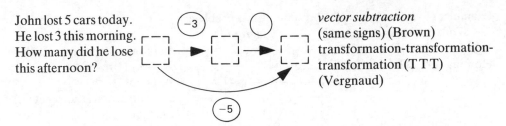

John lost 5 cars today.
He lost 3 this morning.
How many did he lose
this afternoon?

vector subtraction
(same signs) (Brown)
transformation-transformation-
transformation (T T T)
(Vergnaud)

NB   Both the 'Fletcher' series *Mathematics for Schools* and the Nuffield Maths series refer in their teacher's handbooks to the 'take away' and the 'comparison'/difference models given here (Howell et al., 1979; Williams and Moore, 1979.)

Again, putting together the results of studies by Gibb (1954), Nesher (1982), Vergnaud and Durand (1976), Lunzer et al. (1976), Brown (1981c), Schell and Burns (1962), Matthews (1980) and Carpenter and Moser (1979), a hypothetical 'order of difficulty' can be arrived at as shown in the following figure:

Hypothetical order of difficulty in subtraction problems

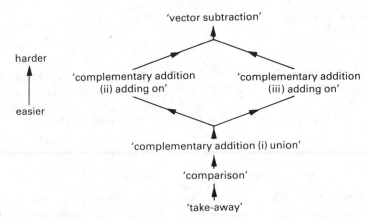

Fig. 3.21

Again, it should be stressed that this order is hypothetical since no study compared all these types on a single sample.

But, to give some idea of the range of difficulty, Vergnaud and Durand (1976) found that half the 7 year olds and all the 9 year olds in their French sample were successful with 'take-away' whereas there was a lag of about a year for 'complementary addition (ii) adding on' and of about 4 years for 'vector subtraction', with only just half of the 11 year olds able to solve problems of this type.

It is not surprising that the 'take away' form is found easiest, since this is the 'meaning' which is most commonly taught when an expression like 5 – 3 is introduced. Indeed Schell and Burns (1962) found that the order of difficulty in their study of 7 to 8 year olds was (a) 'take away', (b) 'comparison' and (c) 'complementary addition (ii) adding on' which exactly matched the frequencies with which problems of these types were presented in American textbooks then in common use. In these, 66 per cent of the problems were of the 'take away' type, 20 per cent were of the 'comparison' type and the 'complementary addition' type was the least frequent.

Of course this does not necessarily mean that the textbooks were responsible for the children's results; the authors may simply have decided to include fewer of the examples that they felt to be more difficult. 'Take away' is certainly an easy form to associate with a physical action.

All the results quoted above concerning the ease with which children solve various subtraction problems must be interpreted with caution, since there is considerable evidence that in many of these cases children do not view the problem as having any connection with the operation of subtraction. This is particularly true of those types of problem which arise from addition problems i.e. the 'complementary addition' forms of subtraction problems.

As reported on page 192, Carpenter and Moser (1979) found that most of the 'complementary additions' forms of subtraction problems they used were actually solved by additive strategies. Similarly Gibb (1954) found that three-quarters of a sample of 7 to 8 year olds solved complementary addition problems by using some form of addition, and APU (1980a) record that three-fifths of 11 year olds use a form of 'mental adding on' to solve these, and other, subtraction problems. A 'comparison' problem used by the APU in their practical testing of 11 year olds was:

> George has 65 stamps in his album and Susan has 102 in hers. How many more stamps has Susan than George? (APU, 1980)

The APU note that in this case

> ...the most usual counting on process was from 65 to 70, then in tens to 100, adding a further 2 to arrive at 102 having counted on a total of 37 which was the answer to the problem. (APU, 1980a)

However given the related question:

$$102 - 65$$

the APU found that four-fifths worked it out as a written 'subtraction'. Comments by testers include the following example:

> Pupils did not seem to relate the (different) answers achieved; Pupil treated problem and algorithm as two separate entities; Pupil did not recognise the problem and the algorithm as the same. (APU, 1980a)

Similarly Brown (1981c) found that when a subtraction problem of a 'complementary addition' type which involved numbers over a hundred was given to a child in the 11 to 13 age range during an interview, additive strategies were used on 10 of the 41 occasions. On half of these the child was not able to recognise that the problem could be represented by a subtraction expression. (See section 4.3.3.1 page 361 for an example of an interview with such a child who did not 'know the sign for adding on'.) In a wide-scale test, the complementary addition item in Fig. 3.22 was correctly identified as a 'subtraction' by 79 per cent of 11 year olds (although nearly a quarter of these ticked $87 - 261$ rather than $261 - 87$).

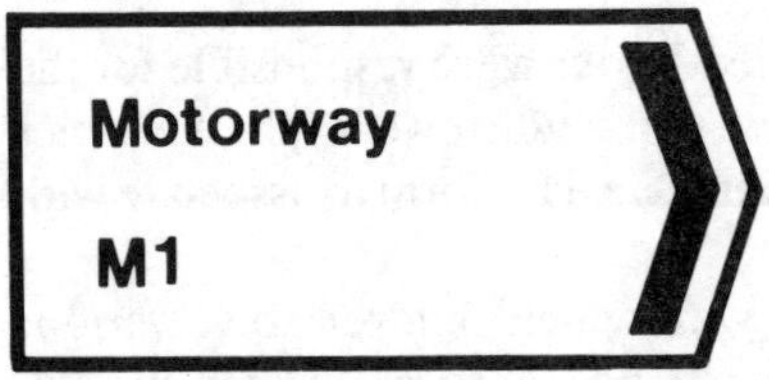

6.

The Green family have to drive 261 miles to get from London to Leeds. After driving 87 miles they stop for lunch.

How do you work out how far they still have to drive?

....

87×3	$261 + 87$
$87 \div 261$	$261 - 87$
261×87	$261 \div 87$
$87 - 261$	$87 + 174$

Fig. 3.22   (reproduced from Brown, 1978)

Brown (1981c) also found that when children were asked to write 'a story' to match the expression $84 - 28$, the answers were predominantly of the 'take away' form. In the case of 61 11 to 12 year olds, the results were as in the following table.

Types of subtraction model given as stories to match $84 - 28$
(data from Brown, 1981c)

Type	Frequency (n = 61)
'take-away'	39
'complementary addition (iii) adding on'	2
'complementary addition (ii) adding on'	1
'complementary addition (i) union'	1
wrong/omitted	18

McIntosh gives some examples of stories produced to illustrate $72 - 29$ by British children aged 7 to 11 years, all of whom were able to complete the calculation successfully. (Ages, where known, are given in brackets.)

A crocodile had 72 teeth and it was eating something when 29 teeth fell out. How many were left? (9/10)

When I went to catch a bus I had 72 minutes in which to catch it. I was at the Coop and it took 29 minutes to get to the bus so how much time had I to spare? (10/11)

In Summer a boy went to the sea and found 72 pebbles, 29 were black, how many were red? (11)

John is aged 29, Grandad is 72, find mother's age by taking 29 from 72, answer 43 (10)

There were 72 sweets in a jar. 29 people guessed how many sweets there were and they won, how many sweets did they get each?

In the year 2000 there is a new scheme going out to change all ships into Nuclear power. There are 72 ships 29 were using nuclear power, how many had to be changed? (10)

72 balls in a shop at ten pounds each. 29 footballs how much would all the footballs cost together (10)

Paul and Mark went to town. Paul got on the 72 bus and Mark got on the 29 bus and they got off at the 43 bus stop (9)

One day in school I was told to do the sum 72 - 29. I did, it was right, I got a tick. (8)

One day there was 72 boy and 29 girls and that was 43 girls and boy. (10/11)

Tim has 72 marbles and Brian has 29 marbles, how many more marbles has Tim than Brian? (10)

Daisy has 72 pieces in her puzzle and Larry has 29 pieces in his puzzle. So how many less has Larry? (10)

You take 72 rulers and 29 pen and then you do the taking away.

72 girls and 29 boys: each boy pushed a girl into the water. There were 43 left.

Geoffrey had 72 sweets and Ann had 29 sweets, if they shared their sweets they would have 43 sweets. (8)

You can use this sum 72 – 29 for maths and you can use this sum for adding but not for times.

In real live you could find a sum in a Fletcher book in a subset with 72 cakes and cross out 29 and see how many are left. (8)

(McIntosh, 1978)

These examples illustrate that children can interpret a subtraction expression like 72 – 29 in a variety of ways. The results also suggest that in several cases children were able to do the calculation *without* being able to attach any valid meaning to the expression.

3.5.3   The 'Meaning' of Multiplication and Division

A number of authors have pointed out that understanding of the meaning of the operations of multiplication and division is considerably more difficult than that of addition and subtraction. Nesher and Katriel (1977) demonstrate that if operations are defined in terms of mathematical logic ('the predicate calculus') then addition and subtraction are identical whereas multiplication and division require much more complex definitions.

Luriya, studying brain-damaged adults in the USSR, found that in many cases his patients could cope satisfactorily with addition or subtraction problems, but when given those involving multiplication or division containing expressions like 'three times as much', they tended to simply add three.

One of the tasks given was to divide a group of 12 blocks into two equal parts. The action was generally performed successfully, but it was written down as $12 - 6 = 6$ and not as $12 \div 2 = 6$, Luriya concluded:

A complex abstract association which expresses a relationship is everywhere replaced by the concrete association of adding or subtracting real units (Luriya, 1969).

Since in other areas it tended to be more complex concepts which first deteriorated in brain-damaged patients, Luriya felt that his results gave evidence of the extra complexity involved in multiplication and division as compared with addition and subtraction.

Similarly Dienes (1959) gives an example of a child comparing the two blocks:

  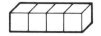

Fig. 3.23

He notes that the child will consider the single block first, and mentally build *three more* to make the longer one. He is unlikely to see the longer one as being *four times as long* as the single block, without considerable experience with such constructive activities.

Similarly Hart (1981) found that over 30 per cent of secondary children consistently chose to add rather than to multiply when solving ratio problems.

When Brown (1981a) asked children to 'make up stories' to 'match' various arithmetic expressions, the results for a large sample of 11 year olds were as shown in the following table:

**Percentage of 11 year old children successful in providing stories
(from data in Brown, 1981a)**

Expression	Percentage success (n = 497)
84 − 28	77
9 ÷ 3	60
84 ÷ 28	42
9 × 3	45
84 × 28	31

This order of difficulty was the same as that obtained when children were asked to select the expression which best matched a given story, as is shown in the table below. (The full test, with five word problems of a variety of types in the case of each operation, was given to a sub-sample of 81 11 to 12 year old children.)

Percentages of 11 to 12 year old children who successfully identified the operation which best matched word problems (data from Brown, 1981a)

Operation	Mean percentage successful (n = 81)
+	88
−	67
÷	63
×	53

In both sets of results, addition and subtraction are shown to be easier to identify than multiplication and division.

Brown (1981c) attributes the extra difficulty of multiplication and division to the structure of the operation. Thus addition and subtraction are associated with a situation in which two sets of *similar* objects are combined or disassociated:

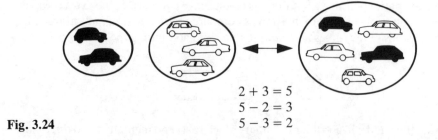

$$2 + 3 = 5$$
$$5 - 2 = 3$$
$$5 - 3 = 2$$

Fig. 3.24

However for multiplication and division, not only are the objects in the two sets generally of *different* types, but in each case each *element* of one of the sets has to be associated with an equivalent *subset* of the other set:

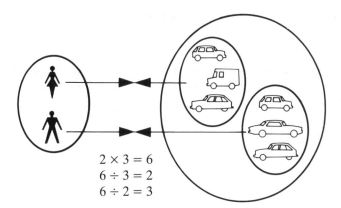

$$2 \times 3 = 6$$
$$6 \div 3 = 2$$
$$6 \div 2 = 3$$

Fig. 3.25

Setting up such a situation clearly causes children problems such as those demonstrated by some of the examples collected by McIntosh from 9 to 11 year olds to illustrate 6 × 3:

> Tim had 6 books × Mary had 3 books = 6 × 3.
> In the farmyard were 6 chickens and 3 pigs. My father said six times three is 18. (McIntosh, 1979)

Similar problems occur in the case of division e.g. (for 84 ÷ 28):

> If a one boy has 84 sweets and another boy has 28 and they share, how many will they have? (11 year old)
> (Brown, 1981c)

In this example, the word 'share' is associated with division, but in the wrong sense. Brown points out that the words most commonly used to interpret the operation symbols provide powerful clues to the type of situation which match it i.e.

> + means 'add' or 'and'
> − means 'take away'
> ÷ means 'share'
> × means 'times'

However whereas 'adding', 'taking away' and 'sharing' are concrete actions easily visualised, 'times' has no such obvious active reference. This may explain why children appear to have particular difficulty with attaching any meaning to the operation of multiplication, as evidenced in the previous tables. (See section 4.3.3.1 on page 362 for further references to the role of 'cue words' in matching problems to operations.)
McIntosh concludes:

> The idea of multiplication is not easy; addition is much simpler. Many children confused the two in stories, even when they performed the computation correctly. Several teachers reported that younger juniors were quite unable to produce situations requiring multiplication. On the other hand, many stories were really about division, showing that many children appreciated the use of multiplication to solve division problems... (McIntosh, 1979)

Similarly, Brown and Küchemann suggest that:

...the results show that teachers of first year secondary school children should not, except in the case of very bright children, take their understanding of multiplication and division for granted (Brown and Küchemann, 1977).

As in the case of addition and subtraction, various different types of 'word problem' associated with multiplication and division have been identified. These are described in parts 3.5.3.1 and 3.5.3.2.

3.5.3.1   Multiplication Problems

Some examples of different types of multiplication word problems associated with the expression 3 × 4 are given in the next table. Diagrams are taken from Brown (1981c), and descriptions from Brown (1981c), and Vergnaud et al. (1979). (Again, the types given here probably do not form a complete classification.)

Examples of types of multiplication word problems

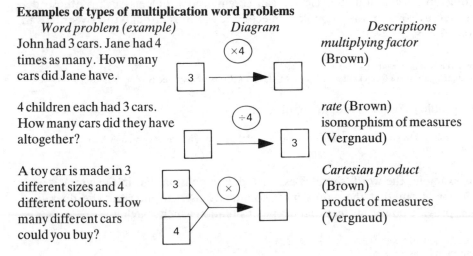

Word problem (example)	Diagram	Descriptions
John had 3 cars. Jane had 4 times as many. How many cars did Jane have.		*multiplying factor* (Brown)
4 children each had 3 cars. How many cars did they have altogether?		*rate* (Brown) isomorphism of measures (Vergnaud)
A toy car is made in 3 different sizes and 4 different colours. How many different cars could you buy?		*Cartesian product* (Brown) product of measures (Vergnaud)

NB   The Nuffield Maths series suggests the terms 'repeated addition' for the aspect referred to here as 'rate', and 'partnering' to correspond to 'Cartesian product' (Williams and Moore, 1979).

As in the case of addition, there are alternative concrete interpretations using either discrete ('objects') or continuous ('length') analogues. Thus the 'multiplying factor' and 'rate' forms could be shown as:

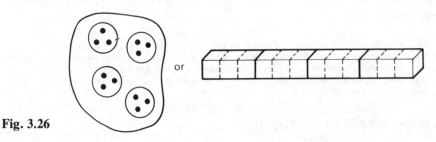

Fig. 3.26

However the 'Cartesian product' model might best be illustrated as

Fig. 3.27

Brown (1981a) found that the 'rate' model was very much easier to recognise as a multiplication problem than the 'Cartesian product' model (the 'multiplying factor' model was not tested). For example 73 per cent of a sample of 11 year olds were able to select an expression (4 × 8 or 8 × 4) which matched a 'rate' problem as against only 46 per cent who succeeded with a 'Cartesian product' model of comparable complexity. It seems likely that the 'rate' model is more easily identified as a 'repeated addition' e.g. ('4 lots of 3', or '4 times 3' or '3, 4 times') whereas with the Cartesian product model there is more difficulty in seeing how to coordinate the two numbers. For example, given a problem relating to the number of different types of crisp packet, if these come in four flavours and three sizes,

Interviewer:	(reads question, then long pause)… 'I'll explain this one… We have four different flavours, what flavour might we have?'
Joseph:	'Cheese and onion.'
I:	'So you'd have cheese and onion, three different sizes, so in one box we might have small cheese and onion, in another box we might have middle size cheese and onion packets… how many boxes would you need altogether?'
J:	'Three.'
I:	'Three for a small, medium…'
J:	'And big…'
I:	'All right, that's all right for cheese and onion, now there are four different flavours, each of them comes in three sizes. So how would you work out how many boxes there would be altogether?…(pause)…You said that for cheese and onion there'd be three boxes, that's right. Now there are four different flavours, cheese and onion's only one flavour, there are three other flavours, there's four altogether…'
J:	'Oh, there's four different flavours, one for each flavour.'
I:	'Right, but in each flavour we've got three different sizes.'
J:	'Oh, small, large and medium…'
I:	'Right, so for each… there are three boxes for each of the four flavours. How many boxes would you need altogether?'
J:	'Four.'
I:	'That would hold the four different flavours. Now for each flavour you've got three different sizes.'
J:	'Oh!… I know what you mean.'
I:	'Tell me how many boxes you'd need altogether?'
J:	'Oh, three! 'Cos there's three different sizes…'
I:	'That's for each..one flavour, but you'd have three for cheese and onion, then you might have three for plain…'
J:	'Oh! I know what you mean…you mean four three's.'
I:	'That's right.'
J:	'I get that now, you ought to've explained that a bit more(!).' (J then selects 3 × 4 as the correct expression, and hesitates over 3 ÷ 4 also.)

The above results were confirmed by the classification of the types of stories given by 66 11 to 12 year olds to match the expression 9 × 3:

Types of multiplication model given to match 9 × 3 (adapted from data in Brown, 1981c)

Multiplying factor	5
Rate (repeated addition)	23
Cartesian product	0
Wrong (addition problem given)	11
Other wrong	18
Omissions	9
Total	66

McIntosh gives examples of 'rate' and 'multiplying factor' models given by 9 to 11 year olds to match the computation 6 × 3 = 18:

> One day a little boy named Andrew went to a park. He saw 6 trees and every tree had three birds in them. So Andrew saw 18 birds.

> Sandy had six sweets and Danny had 3 times as many how many sweets has Danny.
>
> (McIntosh, 1979)

Both Brown (1981c) and McIntosh note that although multiplication is an easier process to carry out than division, children appeared to often be able to most easily interpret the 'meaning' of multiplication in terms of what was basically a division situation. For example, several stories for 6 × 3 = 18 were of the form:

> Mummy and daddy had 18 sweets so me and my brother and sister had 6 sweets each (McIntosh, 1979).

This suggests that, at least psychologically speaking, division may be prior to multiplication. (See also the results in the tables on page 232, which suggests that children find multiplication problems more difficult to both recognise and to construct than division problems.)

3.5.3.2   Division Problems

The two major types of division problems, recognised in most primary school texts, are 'sharing' ('partition') and 'grouping' ('measurement'). These will be the only types of division discussed, although again they do not exhaust the field; for example many types of 'complementary multiplication' would give rise to division problems.

Examples of types of division word problem

Word problem (example)	*Diagram*	*Description*
John had 12 cars. He wanted to arrange them in 4 equal rows. How many should he put in each row?	÷4 12 → □	'sharing' (Brown)
John had 12 cars. He wanted to put them in rows of 4. How many rows could he make?	÷ 12 → 4	'grouping' (Brown)

NB   The 'Fletcher' series *Mathematics for Schools* suggests the terms 'partitive' and 'measure' respectively for the two aspects of division (Howell et al., 1979); the Nuffield Maths series uses 'sharing' and 'repeated subtraction' as the corresponding descriptions (Williams and Moore, 1979.)

Results obtained by Hill (1952) and Brown (1981a, c) suggest that there is very little difference in difficulty between these two models. Since the structure of the operations appear to be very similar in the two cases, with one as the complement to the other, this is perhaps not surprising.

However Gunderson (1955) and Zweng (1964) both report that younger children (8 year olds in both cases) found it easier to solve 'grouping' problems than 'sharing' problems, perhaps because the former lend themselves to a more direct concrete strategy of repeated subtraction.

Brown (1981c) found that in spite of the fact that 11 year olds appeared to find the two models equally easy to recognise as division, with 70–75 per cent of success in each case, when asked to supply a problem to match the expression $9 \div 3$, ten times as many gave a 'sharing' situation as gave a 'grouping' example. This is presumably due to the fact that children learn from an early age to associate the sign '$\div$' with the word 'share'.

3.5.4   Summary: Implications for teaching

The two previous sections of text have highlighted the importance of children being able to recognise which of the four common number operations is appropriate to solve a particular problem. This skill relates to the child's ability to construct a 'meaning' for the operation. It seems that in each case the easiest meanings to grasp are those which can be directly translated into an action e.g. 'add', 'take away', 'share'. (This corresponds to Piaget's definition of operations as 'internalised actions'.)

Only at a later stage are children able to interpret the operations in terms of a static structure of relationships (e.g. 'comparison', 'multiplying factor', etc.). Similarly the appreciation of complementary forms (e.g. subtraction as the operation needed to solve an addition problem with the total known and one of the smaller numbers unknown) also comes later. Even at 11 years old, a minority of children experience difficulty in e.g. the construction of a multiplication problem, or the combination of numbers which themselves represent transformations or relationships (e.g. 'vector addition', 'vector subtraction' forms).

Thus, as in other mathematical areas, the understanding of number operations develops slowly over a number of years; the concept of each operation gradually extends to cover a wider and more abstract range of situations. This is true even when the numbers involved are small whole numbers, as was the case in most of the studies referred to in this section. (In part 3.8.3 the need for further extensions of the number operations concepts to include the use of fractions and decimals is considered.)

In the past it was thought that children, having safely 'grasped' the meaning of, say, multiplication at the age of 7, could then simply go on to master more and more complex methods of calculation. However the results given previously suggest that explicit attention should be paid to ensuring that children gradually become familiar with the many different models of multiplication and that this needs to extend well into the secondary school period. One way of doing this is by discussion of how many essentially different 'stories' a child can produce to match an expression such as 5 × 3.

It is important to emphasise the progression from a child's ability to solve practical problems involving a particular operation by an informal method (e.g. using addition to solve a subtraction problem, or repeated subtraction to solve a division problem) and the further step of being able to recognise and symbolise the problem in terms of one of the expressions:

$$a+b \qquad a-b \qquad a\times b \qquad a\div b$$
$$ \qquad b-a \qquad \qquad b\div a$$

in which a and b are numbers occurring in the problem.

The informal method suffices in many practical situations; however when large or complex numbers are involved, and especially when a calculator has to be employed, then the latter ability becomes virtually essential. Again, many children have not achieved this ability on entry to secondary school.

Brown (1981a) draws attention to the likely shift from methods of calculation towards recognition of which operation to use, given the ready availability of electronic calculators. She also points out that using calculators can fulfil a positive role in focusing a child's attention on the structure of the problem in order to identify which operation buttons should be pushed and in what order.

Others e.g. Vergnaud (1982), Botsmanova (1972a,b) have highlighted the role of diagrams in assisting children to identify the essential structure of a word problem and thus to select the appropriate operation. Botsmanova demonstrated that children can be trained to use abstract schematic diagrams effectively in solving problems, although unfortunately the papers do not give details of the form of the diagrams. Vergnaud, whose diagrams have already been used extensively in this section, also reports success from the explicit teaching of the use of such diagrams.

There were however no other references found in which research was aimed at the improvement of children's understanding of the meaning of the number operations.

3.6   NUMBER OPERATIONS: UNDERSTANDING THE STRUCTURAL PROPERTIES

3.6.1   The Development of Awareness of the Structure of Number Operations

The previous text was concerned with the child's understanding of number operations in terms of the ability to relate these to real-world situations. However another aspect of such understanding is the ability to grasp and use more abstract properties of the number operations themselves i.e. an awareness of their internal structure and of how they relate to each other.

Grossnickle (1959) considers that early awareness of structure only begins to arise when the child can derive number bonds and relationships on the basis of what he has previously discovered using concrete means. Thus to appreciate structural properties it is necessary for the child to be able to view numbers as entities which exist and can be manipulated in their own right, independent of any particular context. For instance, in observing Russian 6 year olds solving verbal problems, Menchinskaya notes that for some children:

the arithmetical operations merge with the tangible operations;

whereas

stronger first-grade pupils performed external operations with *numbers* (for example, '3 added to 5 is 8.') (Menchinskaya, 1969).

Similarly reference has already been made to Carpenter and Moser's sample of American 6 year olds, some of whom were able to use known number bonds of this type in solving problems, although they had previously had no formal mathematical teaching in school (Carpenter and Moser, 1982). Hence at least some children appear to be able to use number in an abstract way on entry to school.

Nor is this knowledge always purely rote-learned; Carpenter notes that 15 out of the 94 children in his sample in fact were able to derive further number bonds from the ones they already knew. For example, to solve a verbal addition problem involving the sum 4+7, one child gave:

'Seven and 3 is 10, so I put 1 more on there and got 11.' (Carpenter, 1980).

Carpenter termed strategies of this type '*decomposition*' strategies, since they involve breaking up one of the numbers involved. He noted that often they involved obtaining two numbers which added to 10, as in the above case. Also such strategies frequently made use of 'doubles' as in the example of another child solving a problem involving the sum 6+8:

'6 + 6 = 12 and 2 more is 14.' (Carpenter, 1980).

A further type of reasoning was employed which involved some sort of '*compensation*' strategy, where a change in one of the terms is offset by a compensating change in the other. For example, a second child solving the problem involving 6 + 8 reasoned:

'I took one from the 8 and gave it to the 6. 7 + 7 = 14.' (Carpenter, 1980).

Similar strategies not always correct, were noted in subtraction problems e.g.

9–6: 'Nine take away 4 is 5, take away 2 is 3.'
11–3: 'Three from 10 is 7. It would have to be 1 more because it's 11.'
14–8: 'Seven and 7 is 14; 8 is 1 more than 7. So the answer is 6.'
13–9: 'Thirteen and 10 would give 3; so 13 and 9 give 2.'
(Carpenter, 1980)

The number of children who at some stage were able to derive a solution from a known number bond in this way increased from 15 out of 94 at the entry to the first grade of elementary school at 6 years of age to 79 out of 96 midway through Grade 2, 16 months later.

This suggests that over the age range 6 to 8 years there is a very considerable growth in the ability of many of these children to reason using abstract properties of number rather than concrete aids.

Carpenter suggests that the use of such strategies can be encouraged by teaching them explicitly. Hatano (1982) reports that explicit teaching of the decomposition strategy using 10-bonds (e.g. $8 + 5 = (8 + 2) + 3 = 10 + 3 = 13$) is used in Japan, and quotes a study of Yoshimura who observed that a small sample of 6 year olds had successfully adopted this strategy although a year previously they had all used a 'counting on' strategy.

Similar reasoning strategies based on the properties of number can be observed in multiplication problems. For example, Weaver (1955) describes the method used by one teacher to discover the skills, thinking patterns and levels of understanding of the basic facts of the multiplication process for each child in her class. This was her normal preparation for structuring her teaching which was modified according to the needs of each pupil. In this case she was teaching 10 year olds who during their previous school year had worked with multiplications involving the numbers 2, 3 and 4. Weaver states that on the basis of past experience this teacher was aware that each child would vary in his mastery and understanding both in relation to other children in the class and between various multiplication combinations.

She prepared six cards each with either a vertically or horizontally presented multiplication item as in Fig. 3.28. Some of these cards involved 'previously taught' facts and others, 'untaught' facts. It was felt that the untaught multiplication items would help to give more insight into the child's understanding of multiplication, whereas automatic replies might be given to the familiar bonds. Each child was presented with one card at a time.

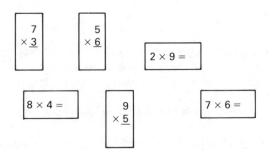

Fig. 3.28

When immediate responses were not forthcoming the child was encouraged to think aloud – an activity which this particular teacher nurtured as a regular feature of her classroom style. The following table is reproduced from Weaver's article and summarises the responses of five children to three of the combinations.

Sample multiplication responses (reproduced from Weaver, 1955, p.42)

Pupil	Previously 'Taught' Facts		'Untaught' Fact
	7 ×3	8 × 4 =	9 ×5
Sally	Automatic Response Correct Product	Hesitated, then recited full 'table': $1 \times 4 = 4,$ $2 \times 4 = 8,$ $3 \times 4 = 12,$ etc. to $8 \times 4 = 32.$	Hesitated, saying 'I don't know that.' Then put down 5 rows of 9 dots and counted by ones to reach 45.
David	Automatic Response Correct Product	Said, 'Let's see: 6 fours are 24, 7 fours are 28, 8 fours are 32. That's it.'	Looked up a bit, then said, 'Now I know how to work that one! 5 tens are 50, so take away 5 and that's 45.'
Linda	Hesitated, then said: '7 and 7 are 14—15, 16—17, 18—19, 20—21.'	Hesitated, then said: 'Oh, I know! It's the same as 4 eights, and I remember that's 32.'	Hesitated, saying 'We never had that one before.' Began to add 9 and 9; then suddenly she stopped and counted by fives to 45.
Carole	Automatic Response Correct Product	With seemingly no hesitation, said: '4 fours are 16; and 10, that's 26; and 6 more—30, 32.'	Very little hesitation, then said: 'Well, 4 nines are 36; and 10, that's 46; so take away 1, that's 45.'
Jerry	Automatic Response Correct Product	Automatic response, but incorrect product of 36. When asked to 'prove' it, looked puzzled and said, 'I can't, but I'm sure that's what it is—36.'	Looked very confused saying: 'I don't remember *that* one. Did we have it before?' Was unable to attack it sensibly, asking: 'Is it near 14?'

This table illustrates the point that although these children may be obtaining the correct answers they are not comparable in their mastery of the facts and their understanding of the fundamental nature of the multiplication process. For example for the $\begin{smallmatrix}9\\\times5\end{smallmatrix}$ item Sally conceptualises it from a concrete basis in terms of rows of dots, whereas David is able to approach it abstractly by means of the distributive principle (i.e. $9 \times 5 + 1 \times 5 = 10 \times 5$). Linda makes effective use of her understanding of the commutative nature of multiplication (i.e. $9 \times 5 = 5 \times 9$), while Carole uses her knowledge of number bonds in conjunction with her understanding of the base ten number system. Weaver describes Jerry's performance in the following manner:

> Multiplication as a mathematical process has very little meaning for Jerry, if any. He could respond automatically to 'previously taught' combinations – sometimes correctly, sometimes incorrectly – but always mechanically to a meaningless stimulus. He has no understanding to enable him to cope with the new situation of 'untaught facts'. (Weaver, 1955)

The case of Jerry illustrates a point made by Hendrickson:

> Premature use of symbols before the concepts referred to have been developed through activity with real objects results in rote learning and overloads on short-term memory processes. (Hendrickson, 1979)

3.6.2   The Structural Properties of Number Operations

In the previous section it has been shown that children gradually, and at different rates, develop an awareness of some of the properties of number operations, presumably as a result of generalising from previous concrete experience of these operations.

Williams (1964) notes that most structural properties are related to the notions of *equality* and *invariance* as fundamental logical concepts. Thus a young child may combine a set of 3 cars and a set of 5 cars and make the empirical discovery that there happen to be 8 cars. If he then puts together 3 vans and 5 vans he may not be at all surprised to find that there are 9 vans, due to inaccurate counting. However at a later stage he becomes aware that the fact that 3 cars and 5 cars make 8 cars is a necessary consequence of the way numbers behave i.e. 3 and 5 will always be equal to 8, and is quite independent of the nature of the objects themselves.

Williams gives a list of structural properties of number operations which relate to the concepts of equality and invariance and which derive from the analogous properties of sets of concrete objects:

1. *Identity*   A number is unaltered if zero is added or subtracted. It also remains the same if multiplied or divided by one.
2. *Commutativity*   Addition and multiplication (but not subtraction or division) can be performed in the reverse order e.g. $2 + 3 = 3 + 2$, $2 \times 3 = 3 \times 2$.
3. *Associativity*   If more than one operation of addition (or multiplication) is performed, the numbers can be combined in pairs in any order e.g. $(2 + 3) + 4 = 2 + (3 + 4)$ or $(2 \times 3) \times 4 = 2 \times (3 \times 4)$.
4. *Distributivity*   Multiplication can be 'decomposed' into the sum or difference of other multiplications.
    E.g. 10 fives are equal to 6 fives added to 4 fives $[10 \times 5 = (6 \times 5) + (4 \times 5)]$;
    10 fives are equal to the difference between 12 fives and 2 fives $[10 \times 5 = (12 \times 5) - (2 \times 5)]$.
    Similarly for division,
    E.g. a fifth of 30 is equal to a fifth of 20 added to a fifth of 10 $[30 \div 5 = (20 \div 5) + (10 \div 5)]$
    and so on.

5. *Invariance* Two equal numbers remain equal if identical operations are performed on each.
   e.g. if $3 + 4 = 6 + 1$, then $(3 + 4) \times 5 = (6 + 1) \times 5$ etc.
6. *Invariance of difference* If two numbers are unequal, the difference between them is preserved if the same number is added to, or subtracted from, both of them. This property, which does not hold if the numbers are multiplied or divided, is the basis of the 'equal addition' method of subtraction. For example   73 becomes 7'3;
$$\frac{-28}{} \qquad \frac{-3\,8}{}$$
   since ten units have been added to both top and bottom lines the answer will be unchanged.
7. *Invariance of ratio* The ratio of one number to another is preserved if both numbers are multiplied or divided by the same number e.g. $2 \div 10 = 1 \div 5 = 20 \div 100$
   $\left(\dfrac{2}{10} = \dfrac{1}{5} = \dfrac{20}{100} \right)$ etc. This property, which is the principle behind equivalent
   fractions, does not hold if the numbers are added to or subtracted from.

Williams also cites three further properties which are specifically concerned with the relationships between operations. ('Distributivity', listed above, could also be included in this set.)

8. *Multiplication/division as repeated addition/subtraction* The multiplier can be regarded as a count of the number of times the number is added
   e.g. 4×7 is equivalent to $7 + 7 + 7 + 7$
   $$\text{or } 4 + 4 + 4 + 4 + 4 + 4 + 4$$
   Similarly for division, $8 \div 4$ is equivalent to the number of times 4 can be subtracted from 8.
9. *Inverse operations* Subtraction is the inverse operation of addition (e.g. $3 + 5 = 8$ is equivalent to $8 - 5 = 3$ and $8 - 3 = 5$) and similarly multiplication and division are inverse to each other.
10. *Non-associativity of subtraction and division* Subtraction cannot be performed in any order
    e.g. $12 - (6 - 2) \neq (12 - 6) - 2$
    Nevertheless there is a related expression incorporating addition which is true
    e.g. $12 - (6 - 2) = (12 - 6) + 2$
    Similarly for division,
    e.g. $12 \div (4 \div 2) = (12 \div 4) \times 2$.

Of course the child does not consciously learn these properties in an abstract form. Rather, he comes to accept some expressions equal, generalising from whatever concrete 'meaning' he has attached to the operation. For example, if he associates $5 + 3$ with the combining of the elements of two sets, then it may be obvious to him that $5 + 3$ will give the same answer as $3 + 5$, and so on. Some of the above properties, for example, invariances of difference and ratio, and non-associativity of subtraction/division, are clearly much more complex than the others and are not likely to be perceived by most children until much later.

It is not always easy to assess whether a child is aware of such structural properties of number except by observing whether he or she makes use of them in solving problems. For example children who work out the answer to $8 + 1$ by counting on one from 8, but, given $1 + 8$, count on 8 starting from 1, clearly have not really grasped the associativity of addition.

However some children appear to develop awareness of some of these properties at an early age; for example, the 6 year old child from Carpenter's sample quoted on page 239, who in answer to a problem involving $4 + 7$ replied:

'Seven and 3 is 10, so I put 1 more on there and got 11' (Carpenter, 1980).

As Carpenter notes, this solution used not only the commutative property to change $4 + 7$ to $7 + 4$, but also the associative property to go from $7 + (3 + 1)$ to $(7 + 3) + 1$.

Of course, given large numbers, the child may not have been able to use such relationships. For example Collis (1975) shows that 70 per cent of 7 year olds can correctly fill in the missing numbers in items of the type:

$$4 + 3 = 3 + \square$$
$$8 + 4 - 4 = \triangle$$

but virtually none can answer correctly:

$$429 + 379 = 379 + \square$$
$$4283 + 517 - 517 = \triangle$$

However by the age of 9 years, 70 per cent can give correct answers in the case of each of the latter items. It seems likely that the 6 year old is able to reason by analogy with sets of objects he can visualise using neither symbolism nor consciously abstracted properties, whereas in order to solve the large-number problems the child is required to attend to the structure of the symbolic statement, ignoring the numbers themselves.

Commutativity Ginsburg feels that it is important that children should be encouraged to perceive the properties underlying an operation, in addition to practising computations, for otherwise the child becomes rigid and inflexible in the application of learned rules. He cites the examples of Barbara. Barbara, aged 6, who was learning simple addition with the aid of blocks, was given the following problems to do using the blocks.

5	7	3	2	4	3	5	3
+7	+5	+2	+3	+3	+4	+3	+5

She treated each item in the same way, combining 5 blocks with 7 blocks for the first item and then counting the total number. She then combined 7 blocks with 5 blocks counted these and so on. The point stressed by Ginsburg is that although she is using a sound strategy and gaining correct answers it must be recognised that her understanding is very limited. She is failing to perceive the regularity brought about by virtue of the commutativity of addition, or, even if she does see some pattern, she is failing to act upon it.

Another example given by Ginsburg is that of Joe, aged 11 years,

> Joe was presented with three addition problems, all involving the same numbers, and solved them as follows:
>
> | 379 | 427 | 16 |
> |---:|---:|---:|
> | 16 | 379 | 427 |
> | +427 | + 16 | +379 |
> | 922 | 812 | 822 |
>
> I: 'If you have the same numbers, are your answers going to be the same?'
> J: 'No. Because they are switched around and one number is harder than another sometimes.'
> I: 'What happens to the answer?'
> J: 'It might get bigger or smaller.'
>
> (Ginsburg, 1977)

Thus it seems that Joe has been so concerned in the details of routine computation that he has not had the opportunity or encouragement to abstract important properties of number.

However, Diana, a 9 year old:

> ...believed quite firmly that 20 + 14 is the same as 14 + 20. 'Twenty and 14 is the same question... that's the same thing... All you do is put that number backwards.'
> I: 'Do you think it always works with other numbers?'
> D: 'Yeah. Except for times.'
>
> (Ginsburg, 1977)

Thus Diana was able to be quite certain about the commutativity of addition, but did not yet accept the commutativity of multiplication.

Willington (1967) tested a small sample of British children on the commutativity of multiplication in a concrete framework by arranging 4 flowers in each of 6 vases and then asking the child how many flowers would be in each vase if there were only 4 vases.

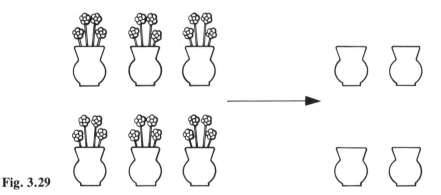

Fig. 3.29

This was part of a wider task, derived from Piaget (1952), dealing with multiplicative relations. (For example, children were also asked, given 4 vases with 6 flowers in each, how many flowers would there be if they had to be equally divided between 12 vases, and so on.) Willington found that the problems were correctly solved by most 9 and 10 year olds, and an occasional 8 year old, but that the majority of these children used multiplication to find the total number of flowers and division to find the number in each vase and only a minority used the properties of commutativity and ratio, for example, reasoning that 3 times as many vases would give one third as many flowers in each.

Thus it seems possible that a functional use of the commutativity of multiplication does not generally arise until the later primary/early secondary school stage (10+). However it may be that children are able to complete written examples of the type $3 \times 4 = 4 \times \square$ rather earlier than this.

Brown (1981a) suggests that one of the problems is that children not only come to accept the commutativity of addition and multiplication, but assume that it is also a property of subtraction and division. For instance when given a verbal division problem and asked to select the 'sum' which best matched the problem, more 12 year olds actually chose $26 \div 286$ than chose the correct answer of $286 \div 26$. Almost all the children interviewed who had selected $26 \div 286$ did so at random, feeling that the two expressions were interchangeable (see responses quoted in section 4.3.4 on pages 366 and 367). Brown concluded that at most 30 per cent of 12 year olds recognised that division was not commutative, whereas the figure for subtraction was probably about double this.

Associativity   Willington (1967) also devised a concrete test for associativity of addition [e.g. $(2 + 3) + 4 = 2 + (3 + 4)$]. He put:

    2 marbles into each of 2 red bags
    3 marbles into each of 2 yellow bags
    4 marbles into each of 2 green bags
    5 marbles into each of 2 blue bags.

He arranged one bag of each colour on each side of a table. He then took an empty white bag and emptied into it the contents of the red and blue bags on the left side, and asked the child whether there were now more marbles on the left or right side.

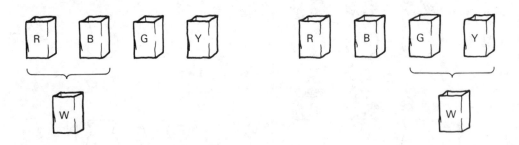

Fig. 3.30

He also used other variations e.g. putting the contents of the red and blue bags into a white bag on the left hand side, and contents of the green and yellow bags into a white bag on the right side, removing 2 marbles from a yellow bag on one side and adding, or removing, 2 marbles from a different bag on the other side, and so on. For younger children he reduced the number of bags to two or three. He also repeated the whole experiment with each colour of bag being filled from a different tube of unknown capacity.

Some 7 year olds and most 8 year olds were able to solve the task provided they knew the number of marbles in each bag, but they operated by adding up the totals on each side and not by logic. Most 10 year olds, and a few 9 year olds, were able to argue logically from equivalences without knowing the numbers in the bag. (However it should be noted that the samples were very small.)

Distributivity   Both Brown, P.G. (1969) and Willington (1967) devised practical tests of distributivity of multiplication over addition [e.g. $5 \times 4 = (2 \times 4) + (3 \times 4)$].

Brown used sets of dot-problems and a peg board and asked for example, which pattern of dots in section B contains the same number of dots as are in *all* of section A.

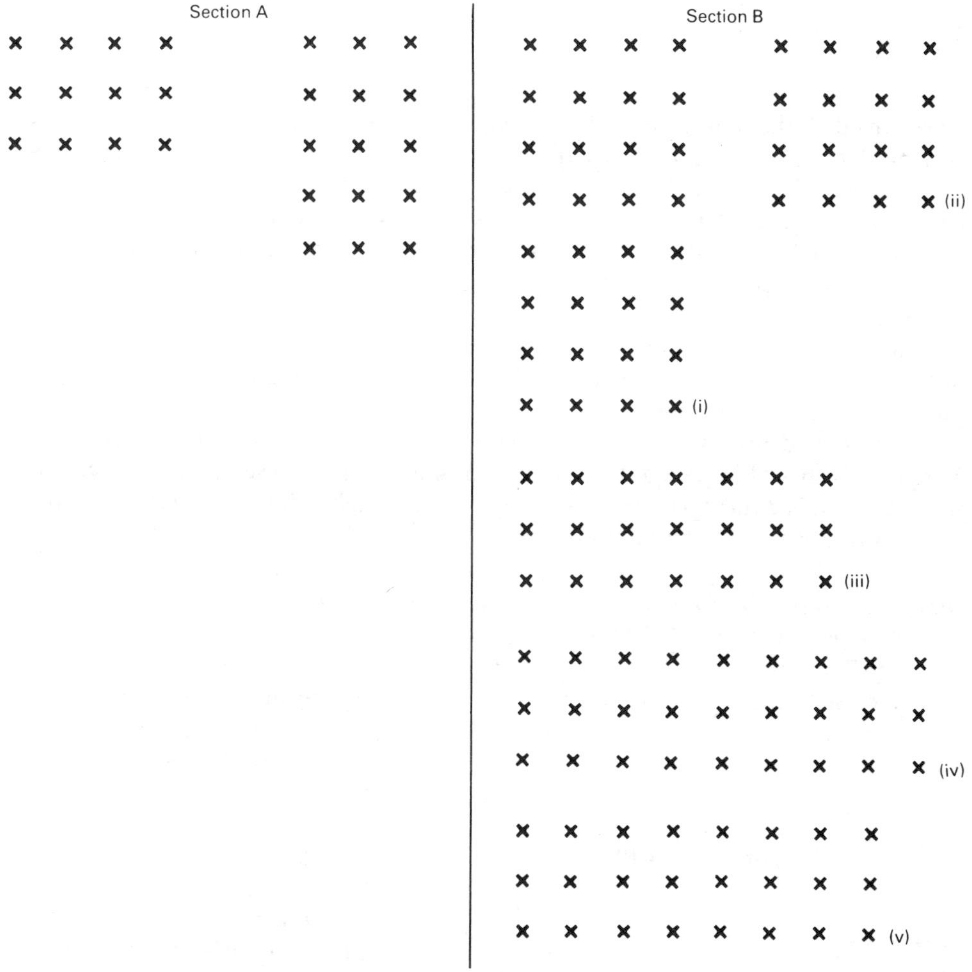

Section A

```
x  x  x  x        x  x  x
x  x  x  x        x  x  x
x  x  x  x        x  x  x
                  x  x  x
                  x  x  x
```

Section B

```
x  x  x  x          x  x  x  x
x  x  x  x          x  x  x  x
x  x  x  x          x  x  x  x
x  x  x  x          x  x  x  x (ii)
x  x  x  x
x  x  x  x
x  x  x  x
x  x  x  x (i)

x  x  x  x  x  x  x
x  x  x  x  x  x  x
x  x  x  x  x  x  x (iii)

x  x  x  x  x  x  x  x  x
x  x  x  x  x  x  x  x  x
x  x  x  x  x  x  x  x  x (iv)

x  x  x  x  x  x  x  x
x  x  x  x  x  x  x  x
x  x  x  x  x  x  x  x (v)
```

Fig. 3.31

Willington used a more concrete setting:

 picture showing
5 boys each with *3* buttons
on his blazer.

 picture showing
3 girls each with
3 buttons on her
blouse

 picture
2 girls each with
3 buttons on her
cardigan

Fig. 3.32

He removed all the buttons worn by the boys (they were on press-studs) and all those worn by *all* the girls (both blouses and cardigans) and asked which pile contained more. In some cases there were the same number of girls, but those wearing cardigans had fewer buttons each, and in other cases there were the same number of buttons but different numbers of boys and girls. The numbers were much larger in some cases (e.g. 12 boys with 7 buttons each, 10 girls wearing blouses with 6 buttons and 2 with cardigans with 6 buttons).

Both Brown and Willington found about half the 10 year olds and a few 8 and 9 year olds able to argue correctly using distributive ideas (i.e. without first totalling the numbers using multiplication).

A few isolated results are quoted in the next table from surveys by Collis (1975), Ward (1979) and APU (1980a,b; 1981a,b), but since most of these relate to pencil-and-paper items, it is difficult to know whether they accurately reflect children's awareness of the corresponding structural properties.

Survey results relating to understanding of properties of number
(see list on page 242 for definitions of properties)

Property	Source	Result
(1.) Identity	APU, 1980a, p.37 APU, 1980b, p.22	90 per cent of 15 year olds and 60 per cent of 11 year olds can deduce that if $\square \times 9 = 9$ then $\square$ must equal 1.
(2.) Commutativity	APU, 1980a, p.37	60–90 per cent of 11 year olds showed that they understood the commutativity of addition and multiplication by the way they answered the problems (items not released).

	Property	Source	Result
(3.)	Associativity	APU, 1980b, p.22	40 per cent of 15 year olds answered correctly on question involving associativity (item not released).
(4.)	Distributivity	APU, 1980a, p.37	About half of 11 year olds answered correctly question testing an implicit knowledge of the distributive law (item not released).
(6/7.)	Invariance of difference/ratio	Collis, 1975, p.134	When children were asked to work out whether expressions of the following types were equivalent or not:

$$479 \times 231 \text{ and } 456 \times 231$$

$$475 + 234 \text{ and } 471 + 232$$

the percentages of questions answered successfully were: 58 per cent for 9 year olds (i.e. only just above the chance level) 73 per cent for 10 year olds 93 per cent for 11 year olds

	Property	Source	Result
(9.)	Inverse operations	Ward, 1979, p.85	About 30 per cent of 11 year olds could solve $\Box \div 4 = 14$
		APU, 1981b, p.27	51 per cent of those 15 year old children given a practical test involving the calculation of 95.78−7.9 on a calculator were able to suggest using addition as a method of checking their answer.

(Further information on children's understanding of the properties of number operations, in particular the 'identity' property, is given in the context of the errors made in computation in section 3.7.3 on pages 257 and 261.)

The results quoted now suggest that although some children as young as 6 years old are able to use some structural properties of number in certain circumstances, these properties may not be generalised to hold for all whole numbers until at least the age of 9. It is clear that in the case of other properties, many secondary children are not able to apply them flexibly in solving problems.

Grossnickle (1959) felt that once a stage was reached at which children were reasonably familiar with and able to apply the basic structural properties of number operations, then they should be introduced to a third stage of development involving the investigation and discovery of number patterns (e.g. the digit-patterns arising in the multiplication tables, etc.) The aim at this stage is to provide enrichment, interest and reinforcement.

3.6.3   Implications for Teaching

It should again be stressed that in none of the above research is there any concern that the child should learn explicitly the properties of number as expressed in either words or symbols. Rather, the focus is on basic numeracy i.e. the ability to operate flexibly with numbers in solving real problems, and especially in operating efficiently either mentally or using an electronic calculator. (Written forms of calculation are considered in section 3.7.) For example, given a problem requiring the difference to be found between 587 and 569, one efficient method would be to realise that this would be the same as the difference between 87 and 69, which is the same as that between 88 and 70 i.e. 18. Although this uses the principle referred to above as 'invariance of difference', the child would not generally be expected to be familiar with either the name or any formal description of this property.

Such knowledge is not readily teachable in the same way as particular techniques or procedures are; it depends first on the child both perceiving and internalising the various structural patterns and then applying them spontaneously in appropriate circumstances.

It seems likely that much of this facility with structural properties is based on a firm grasp of the meaning of the operations in terms of concrete referents, both discrete (sets) and continuous (length/number line). For instance Grossnickle (1959) suggests using strips of geometric shapes which could be folded into strips e.g.

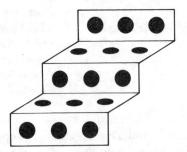

Fig. 3.33

as a discrete model of the multiplication tables (in addition to other models like jumping on along a number line). This helps to demonstrate the commutative property (5 × 3 = 3 × 5) if the sheet of paper is turned the other way up, and the distributive property (2 threes and 3 threes make 5 threes). It can also be used to highlight the inverse relationship of division to multiplication ('how many rows of 3 circles make 12 circles?'), and so on.

Collis (1975) also stresses the need for concrete referents with older children. He identified a group of children in the 8 to 13 age range who were unable to consistently give correct answers to problems of the following type in which children were asked to tick all of the right-hand expressions which would be numerically equal to that on the left:

$$
94 - 47 = \quad
\begin{array}{l}
95 - 46 \\
93 - 46 \\
93 - 48 \\
95 - 48
\end{array}
$$

$$
567 + 486 = \quad
\begin{array}{l}
568 + 487 \\
568 + 485 \\
566 + 487 \\
566 + 485
\end{array}
$$

(Collis, 1975)

He provided a programme of instruction based on simple examples and using concrete materials. For example students were asked to demonstrate addition equalities and inequalities by moving along cardboard squares

e.g.

Fig. 3.34

Similarly in the case of subtraction, although this was more difficult to demonstrate

e.g. $\quad 8 - 3 \quad = \quad 7 - 2$

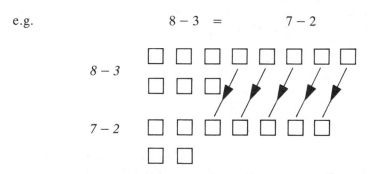

Fig. 3.35

Collis records very significant improvement in the children's grasp of the structural properties of addition and subtraction as a result of this instruction.

Weaver (1955) stressed the importance of providing experiences at the right level for each individual child. Weaver describes some of the prescriptive measures used by the teacher referred to earlier on page 240 who decided to assess children's understanding of the structural properties of multiplication. Weaver followed this up by introducing a class activity based on the problem of determining how many wheels would be needed in order to build various numbers of toy vans, and then divided the class into three groups with separate activities. The children who had little structural understanding carried out concrete activities designed to build up their understanding of multiplication tables in order to notice, for example, that with four wheels per vehicle it was necessary to add on a set of four wheels each time an extra van was added. The second group were asked to derive previously unknown products from known ones e.g. if 28 vans need 112 wheels how many would 29 vans need? The third group, with a firm grasp of structure, carried out more abstract activities based on patterns in multiplication tables.

Denvir, Stolz and Brown, in the companion volume to this, stress the importance of the role of discussion and questioning, with a whole class, a small group or simply a teacher and a single student, in establishing relationships of this type. For example:

> Before the child does a sum: 'How big do you think the answer will be?'
>
> Or after he or she has done it: 'If that number had been 27 instead of 26, would the answer have been bigger or smaller?'
>
> 'Can you think of a story to go with this sum?'
>
> Or, 'Is that a sensible answer? How can you convince me that it is?'
>
> (Denvir et al., 1982)

3.7 NUMBER OPERATIONS: COMPUTATIONAL PROCEDURES

3.7.1 Introduction

'*Algorithm*' is a word used to denote a mechanical, step-by-step procedure. In this section the term will be used very broadly to describe not only the standard written computational procedures (formal algorithms)

$$
\begin{array}{r}
\text{e.g.} \qquad 74 \\
5 \\
+\,\underline{193} \\
\underline{272} \\
{\scriptstyle 11}
\end{array}
$$

but also to refer to mental procedures (informal algorithms) e.g. for the same calculation, 193 plus 5 is 198, plus 4 is 202, plus 70 is 272 (performed in the head).

As mentioned by Carpenter and Moser (1979), it has traditionally been the case that children are taught to master computational skills before beginning to apply them to problem situations. There has been, and still is, much emphasis placed on children becoming skilled in the standard written procedures of computation regardless of whether or not they understand the basis of such techniques.

However, with the advent of the pocket calculator and a growing body of research into the ad hoc methods employed by children to perform their calculations there is a concern that a preoccupation with standard algorithmic approaches to computation is both unnecessary and to some degree positively harmful.

In this section the intention is to first examine the nature of standard algorithms and the hazards inherent in teaching such procedures. There will then be a review of some of the literature concerned with the difficulties children have with these methods in terms of the errors they make and the nature of their understanding of these processes. This will be followed by some suggestions as to how written computational methods may be adapted to facilitate developing an awareness of place value. The latter part of this text will concentrate on non-standard, especially mental procedures – informal algorithms – and how they reflect a natural development in performing operations on numbers and how an understanding of what is going on is an integral part of their usage.

3.7.2   Problems Associated with the Use of Standard Algorithms

Plunkett asserts that

> ...it is a fair bet that almost all children in this country are taught standard written algorithms for the four rules of number ... which are laid out something like this:

$$
\begin{array}{cccc}
\begin{array}{r} 57 \\ +38 \\ \hline 95 \end{array}
&
\begin{array}{r} {}^{4}\!/^{1} \\ \cancel{5}\cancel{7} \\ -38 \\ \hline 19 \end{array}
&
\begin{array}{r} 57 \\ \times\ 8 \\ \hline 456 \end{array}
&
\begin{array}{r} 19 \\ 3\overline{)5\,7} \end{array}
\end{array}
$$

Furthermore:

> The vast majority of children are taught these methods as the primary way of dealing with whole numbers (and decimal) calculations: and for the majority ... these are the *only* methods they are taught.
> (Plunkett, 1979)

Thus Plunkett emphasises the fact that most children are unaware that there is any choice of method, and in particular of written method, open to them.

As Plunkett notes, these standard algorithms are universally taught because they are probably the most contracted and efficient way of dealing with written computations. The method is learnt and then applied to any numbers, large, small, whole or decimal.

Plunkett considers that possibly the most significant characteristic of such computational procedures is their 'analytic' quality. They demand that numbers are broken up into hundreds, tens and units, and so on, and the digits dealt with separately, thus denying attention to the global wholeness of numbers. This is totally alien to how number concepts develop in children and to intuitive computational approaches (as will be seen later).

This analytic approach to computation involves a series of rules to be learnt and even if they

> ...can be remembered they are learnt largely without reasons and are not related to other number knowledge. They are far from aiding the understanding of numbers: rather they encourage a belief that mathematics is essentially arbitrary. (Plunkett, 1979)

Similarly Ginsburg states:

> Often children think of mathematics as an isolated game with peculiar sets of rules and no evident relation to reality. (Ginsburg, 1977).

The following example from Ginsburg helps illustrate this point: Seslie, 9 years old,

was asked to add 123 + 52 + 4. She wrote the numbers down, misaligning them:

123   She got 219 by doing
52   3 + 2 + 4 = 9; then 12 (from the 123)
4   + 5 + 4 = 21. Then 21 combined
219   with 9 is 219.

The interviewer attempted to correct her. He did the problem properly.

S: 'You did it different. I put the 4 in the middle.'
I: 'Why did I put the 4 all the way over there?'
S: 'Probably because you wanted to. You wanted to add this together (3, 2, 4) and this (2, 5) and this (1) go over here.'

Seslie seemed to think that the interviewer's procedure was arbitrary – that one can do addition any way one wants to. He asked her to do the problem again. This time she did it properly, adding 3 + 2 + 4 to get 9 and then 12 + 5 to get 17, with the overall answer 179.

But which answer did she think was correct? 219! She had no good reason for this. She just preferred it. For Seslie, one calculational method is about as good (non-sensical?) as another. To decide among them seems a matter of caprice. (Ginsburg, 1977)

Although the use of squared paper may prevent children from counting digits twice over, as Seslie did here, it cannot act as a substitute for a proper understanding of place value concepts.

Lowry (1965) points out that often the thinking behind teaching standard algorithms as step by step procedures, lies in the expectation that the process will become better rationalised by the child in the future. This may be the case for some children, especially the more able, but the less able child will continue to operate without understanding. (See Section 3.7.4, page 262, for further evidence relating to this point.)

The standard algorithms are often introduced at a stage of development of the child when he does not have an adequate grasp of the underlying concepts (as listed by Williams, 1963) e.g. place value. Since they are 'short cut' devices the nature of the procedures is often concealed e.g. the long multiplication algorithm depends on distributivity (see previous section on operations 3.6.2, page 242) and the more complex methods depend on component operations, for example, the long division algorithm depends on subtraction. It is also difficult to relate the steps to concrete situations.

Williams (1963) says that the 'average' child is likely to lose track of what is going on in these short cut calculative devices. He also maintains that these methods encourage, what he calls, 'cognitive passivity' i.e. the child is unlikely to consider why and how a set procedure works if it is the only option open to him. Even if he has a limited understanding of the procedures they may become so mechanised that he loses sight of this initial comprehension and fails to develop any further understanding. In turn this tendency not to interpret the steps involved in standard computational procedures may lead to inappropriate, inefficient and muddled use of such techniques. It is also likely to lead to failure, boredom, anxiety and adverse attitudes towards mathematics.

One of the major dangers of becoming preoccupied with standard written algorithms is that concept and algorithm may become synonymous, e.g.

to teach division you teach a method rather than an idea (Plunkett, 1979).

3.7.3 Errors in the Performance of Standard Algorithms

Much of the literature concerning the errors made by children in their use of standard computational procedures emanates from the U.S.A., but the findings are generally in accord with those from British studies.

Cox studied the systematic errors made by 8 to 14 year olds in the four vertical algorithms for $+, -, \times, \div$. She used two samples, one drawn from normal schools and one drawn from special schools. These latter pupils included educable mentally retarded children, and others finding it difficult to make progress in normal schools. She discovered that the most frequent error in the entire study occurred in subtraction and involved subtracting the smaller digit from the larger one, regardless of whether the digits were on the top or bottom line.

e.g.
$$\begin{array}{r} 43 \\ -17 \\ \hline 34 \end{array}$$

This error occurred at all skill levels requiring regrouping (borrowing) and with a very high frequency: 57 – 88% for handicapped populations and 63 – 83% for normal pupils. (Cox, 1975).

For the addition algorithm Cox found 51 different systematic errors being made. Twenty-three of these were to do with what she describes as 'renaming' (carrying) e.g.

$$\begin{array}{r} 26 \\ +\ 7 \\ \hline 213 \end{array} \qquad \begin{array}{r} 56 \\ +28 \\ \hline 74 \end{array}$$

Similarly for the vertical subtraction algorithm, 28 of the 52 identified systematic errors were to do with regrouping ('borrowing') e.g.

$$\begin{array}{r} \overset{2}{\cancel{3}}\ \overset{10}{\cancel{4}}\ \overset{14}{\cancel{4}}\ \overset{16}{\cancel{6}} \\ -\ \ \ 5\ 8\ 9 \\ \hline 2\ 5\ 6\ 7 \end{array} \qquad \begin{array}{r} 5\ 3 \\ -\ 1\ 4 \\ \hline 4\ 1 \end{array} \qquad \begin{array}{r} \overset{7}{\cancel{8}}\ \overset{13}{\cancel{3}} \\ -\ 3\ 2 \\ \hline 4\ 1\ 1 \end{array}$$

Examples of the errors made with the vertical algorithm for multiplication from Cox's study illustrate just how confused many children appear to be with the various steps involved in the procedure.

$$\begin{array}{r} 3\ 1\ 3 \\ \times\ \ \ \ \ 3 \\ \hline 3\ 1\ 9 \end{array} \qquad \begin{array}{r} 8\ 0\ 5 \\ \times\ \ \ \ \ 4 \\ \hline 3\ 2\ 6\ 0 \end{array}$$ (Adds renamed digit to the multiplier)

$$\begin{array}{r} 5\ 3\ 9 \\ \times\ \ \ \ \ 3 \\ \hline 7 \\ 5 \\ 1\ 8 \\ \hline 1\ 8\ 5\ 7 \end{array}$$

(i.e. $9 \times 3 = 27$
$2 + 3 = 5$; $5 \times 3 = 15$)
$1 + 5 = 6$; $6 \times 3 = 18$)

$$\begin{array}{r} 5\ 3 \\ \times\ 2\ 0 \\ \hline 1\ 0\ 3 \end{array}$$ $\quad \begin{array}{l}(3 \times 0 = 3 \\ \ 5 \times 2 = 10)\end{array}$

Similarly with division:

$$\begin{array}{r} 11 \\ 5\overline{)85} \end{array} \qquad \begin{array}{r} 20 \\ 3\overline{)702} \end{array} \qquad \begin{array}{r} 19 \\ 4\overline{)436} \end{array}$$

Cox points out that in her study

> many of the systematic errors related to a poor understanding of the algorithmic process. (Cox, 1975).

It is noteworthy that in a follow-up study one year later she found almost a quarter of the children making the same systematic errors.

Roberts in his study of the failure strategies of 148 9 year olds in arithmetic found very much the same kinds of mistakes being made as Cox. More than a third of all his errors were caused by an incorrect application of an algorithm. He emphasises the frequency with which pupils mixed operations. They usually started with the correct operation and then seemingly lost track of what they were doing and resorted to a different operation, usually addition, e.g.

$$\begin{array}{r} 314 \\ \times\ \ 6 \\ \hline 364 \end{array} \qquad \begin{array}{r} 112 \\ -\ 36 \\ \hline 136 \end{array}$$

A further 20 per cent of the errors were due to children applying the wrong operation from the beginning. Although the incidence of these types of errors fell in relative frequency as the ability level of the children increased, Roberts found that there was little difference across the ability range in the number of errors due to misremembered number bonds. (However for the lower ability children not all the 'number bond' errors may have been counted if they appeared in association with other errors.)

Roberts makes a very pertinent statement with respect to the importance of the role of place value in computational procedures:

> Although much emphasis is apparently placed on teaching the fundamentals of the decimal system of numeration and of positional numeration systems in other bases … it seems clear that we have not been uniformly successful in making this knowledge relevant to the processes children use in computation. (Roberts, 1968)

He suggests that this is in part due to the child's perception of a computation task; he feels he is expected to perform in a rote manner given a stimulus which is essentially non-meaningful to him.

Engelhardt investigated the computational errors of 198 9 year olds across the whole range of achievement on a computation test. The distribution of types of error is shown in the table on page 257.

It can be seen that the type of computational error which most dramatically distinguished between highly competent performance, i.e. the highest quartile of attainment, and the other three quartiles was to do with executing the correct procedure. There were virtually no errors involving incorrect procedure (i.e. 'defective algorithms') in the

Frequency of errors by type (reproduced from Engelhardt, 1977)

Error Type	(Low) 1	2	3	(High) 4	Total No.(%)
Basic Fact	304	245	218	103	870 (33)
Grouping	159	157	97	90	503 (22)
Inappropriate Inversion	211	174	24	63	472 (21)
Incorrect Operation	48	20	19	10	97 (4)
Defective Algorithm	155	164	99	4	422 (18)
Incomplete Algorithm	45	55	48	14	162 (7)
Identity	7	4	6	0	17 (1)
Zero	66	29	43	6	144 (6)

(The header above the quartile columns reads *Quartile*.)

(NB   Definitions of some of the error types are included in the following paragraphs.)

highest quartile. Pupils in the three lowest quartiles frequently made these procedural errors.

Another striking result in contrast to those of Cox and Roberts is the higher incidence of basic number fact (number 'bond') errors for each quartile in relation to the other types of errors, with the lowest quartile of attainment making nearly three times as many such mistakes as the highest quartile. This may be explained by the fact that Engelhardt's analysis appears to allow for all such errors to be counted. Engelhardt states that in his study these mistakes occurred most frequently within the context of multidigit computations.

$$\text{e.g.} \quad \begin{array}{r} 2\,7 \\ \times \quad 9 \\ \hline 2\,3\,6 \\ \end{array} \quad (9 \times 7 = 56)$$
$$5$$

A certain amount of speculation seems inherent in classifying errors under the heading of incorrect operation rather than as misremembered number fact; for example Englehardt cites

$$\begin{array}{r} 2 \\ \times 3 \\ \hline 5 \end{array}$$

as an 'incorrect operation'. Without talking to the child the origin of this and other errors must remain uncertain.

Engelhardt describes an identity error as being a mistake which indicates that a pupil performs a computation in a way suggesting some confusion with the operation identities 0 and 1. A zero error is where there is an indication of some difficulty with the concept of zero. He gives the following examples of these errors:

$$\begin{array}{r} 5 \\ \times 1 \\ \hline 1 \end{array} \qquad \begin{array}{r} 1 \\ -0 \\ \hline 0 \end{array} \qquad \begin{array}{r} 3 \\ \times 0 \\ \hline 3 \end{array} \qquad \begin{array}{r} 0 \\ -5 \\ \hline 5 \end{array}$$

It would appear that there are again considerable difficulties in Englehardt's classifi-

cation of an error such as $\dfrac{\begin{array}{r}1\\-0\end{array}}{0}$ on the basis of a written response without probing further

into the child's reasoning behind such an answer. The child may be failing because of a lack of understanding of zero as the identity in the operation of subtraction, or because of some inadequacy in his concept of zero, or because of a misremembered number fact or for all or none of these reasons.

Engelhardt acknowledges these shortcomings when he says

> Examining only written performance without the opportunity to investigate a given error further greatly increased the possibility of misjudging a pupil's erroneous approach (Engelhardt, 1977).

He recommends an interview approach to help overcome this problem. He also admits the limitation of his study in terms of not considering the correct responses, some of which he feels were derived from 'erroneous approaches'. Again it is only by the child attempting to verbalise his thinking that one can gain a better insight into the sources of error.

Difficulties with zero within the context of the arithmetic operations are also discussed by Oesterle (1959) in his summary of some of the literature of the preceding two decades or so which is concerned with the teaching of the zero number bonds. It appears that the consensus of opinion supports the view that at the primary level to operate in any meaningful way with zero is somewhat nonsensical. One may have occasion to add 3 pence and 5 pence but not to add 0 pence and 3 pence. The empty set in concrete situations is not attended to; there is no necessity to incorporate it into one's thinking. However when the need to operate with two or multi-digit numbers arises then it is necessary

to operate with zero as in $\begin{array}{r}67\\+20\end{array}$

Oesterle points out the dangers of regarding zero as being synonymous with 'nothing', the implication being that if it is nothing then it need not be considered. Rather zero should be looked upon as being 'not any of something'. He states,

> The significance of zero as a place holder and as a vital part of our number system appears, as many writers have stated, when we begin addition and multiplication of two-digit numbers. Until this time there seems to be no real reason for introducing the zero facts in any of the basic operations. (Oesterle, 1959)

Once the need for facts involving zero has arisen

> ...the student should be given specific practice with these processes both in isolation and as integral parts of real problems. Generalisations from numerous contacts with zero should be derived from the student's personal experience with these facts as they occur in real problems. (Oesterle, 1959)

Oesterle quotes Wilson (1951) when he says

> ...specific attention to zero in the teaching programme rapidly eliminates the zero errors...
> (Wilson, 1951).

It may be that teaching does not dwell sufficiently on the zero combinations during the addition, subtraction, multiplication and division of two digit numbers. The child is being confronted with zero within a context of the symbolic, abstract notation of arithmetic demanding an understanding of place value to make it meaningful. In part 3.4 it was seen that place value is a complex idea which children find difficult to grasp. With the added problem of coping with zero – a concept with little concrete foundation in terms of a child's earlier counting and measuring experiences – it is little wonder that there are problems in this area.

Even relatively high achievers sometimes have difficulty remembering the standard algorithms. For example Brown quotes the case of Anthony, an above average 12 year old, who is able to solve a problem involving subtracting 78 from 204 in his head, but then writes $\begin{array}{r} 78 \\ -204 \\ \hline \end{array}$ and proceeds to become confused, admitting

...that messes me up, all that (Brown, 1981a).

Similarly a bright 13 year old who demonstrated a firm grasp of the meaning and structure of number operations confessed to an inability to recall the long division algorithm:

Claire: 'I'd do one of those things... (writes $23\overline{)391}$)... but I don't think I know how to work it out like that. I've had it explained to me but... I'd have to do it the long way.' (Brown, 1981a)

As Duffin (1976) points out, the division algorithms may present particular difficulties since they work from left to right, in contrast to those for addition, subtraction and multiplication. Certainly Dickson found some low attaining 13 to 14 year olds who attempted to work division from right to left, e.g. (for $216 \div 9$)

(a) $9\lfloor216$ '6's in 9 is 1 remainder 3; add 3 to 1 gives 4;
       321   4's in 9 is 2 remainder 1; add 1 to 2 gives 3;
                   3's in 9 equals 3'.

(b)   2 1 6   '9's into 6 is 0
        9   9's into 1 is 0
    0 0 0   9's into 2 is 0'.

Note the vertical format in setting out as with the other operations. This pupil realised this was not a sensible answer, but she could not remember how to do it.

The Assessment of Performance Unit's studies in Britain do not distinguish the exact nature of the errors made in their computational items. Early reports of the results (APU, 1980a,b; 1981a) indicate that about 90 per cent of the 11 year olds were successful in gaining the correct answers to vertically presented addition and subtraction items where carrying was not involved. For items requiring carrying in addition, 10 per cent fewer children gave correct answers. When addition was presented horizontally, involving 3 numbers with 2, 3 and 4 digits, then 65 per cent completed it successfully. When borrowing was required for subtraction, be it horizontally or vertically presented, then the success rate was reduced to between 50 per cent and 70 per cent. (A typical item was 9417–283, which was solved correctly by 67 per cent of 11 year olds and 83 per cent of 15 year olds).

For multiplication about 57 per cent of 11 year olds and 83 per cent of 15 year olds gained the correct answer to 76 × 7, whereas 29 per cent of 11 year olds succeeded with

$$314$$
$$\times \underline{201}$$

For 84 ÷ 4 the success rate for 11 year olds was about 70 per cent but fell to 40 per cent when this was presented as $\dfrac{84}{4}$

For 816 ÷ 8 about 50 per cent of 11 year olds were correct, many of the remainder experiencing problems with the zero in the answer.

The first primary report (APU, 1980a) suggests that the poor results on computation are due to a lack of place value concepts, which is also demonstrated by the items directly testing place value ideas.

Ward (1979) in his survey of teachers of 10 year olds from 40 schools asked for examples of common mistakes which occurred in mathematics lessons. He found that by far the commonest mistakes which were mentioned were to do with subtraction and notation. Forty per cent of the 87 teachers cited errors in subtraction mainly to do with taking the smaller from the larger regardless of position, problems with zero, whether on top or bottom, and regrouping (borrowing).

Thirty-nine per cent of teachers mentioned notation problems based on a lack of understanding of place value, e.g. 7 + 12 set out as

$$7$$
$$+ \underline{12}$$

and regrouping ('carrying' or 'borrowing'), e.g.

$$15$$
$$+\underline{19}$$
$$\underset{4}{61}$$

In some of the above studies there was mention that some computational errors were not readily classifiable as it was not possible to spot any recognisable procedure. This emphasises the need to ask the child to explain what he has done for as Ginsburg says

> Errors are seldom capricious or random.... Children's faulty rules have sensible origins. Usually they are a distortion or misinterpretation of sound procedures. (Ginsburg, 1977).

Brown, J.S., and Burton (Brown and Burton, 1978; Burton, 1981) have followed up Ginsburg's hypothesis by investigating the errors arising in the subtraction algorithm using a total of 2500 American children. They found that in most cases the errors were indeed systematic, and they analysed them by using the analogy of 'bugs' in a computer program. Thus they reasoned that the child was following some well-defined procedure which could be programmed on a computer, and which was in general identical to a 'correct' procedure except for one or more faulty steps in the program. Examples of seven of the commonest 'bugs' are given below, as summarised by Resnick.

1. Smaller-From-Larger.   The student subtracts the smaller digit in a column from the larger digit regardless of which one is on top.

$$\begin{array}{r} 326 \\ - \underline{117} \\ 211 \end{array} \qquad \begin{array}{r} 542 \\ - \underline{389} \\ 247 \end{array}$$

2. Borrow-From-Zero.   When borrowing from a column whose top digit is 0, the student writes 9 but does not continue borrowing from the column to the left of the 0.

$$\begin{array}{r} 6\overset{6}{\cancel{0}}2 \\ - \underline{437} \\ 265 \end{array} \qquad \begin{array}{r} 8\overset{8}{\cancel{0}}2 \\ - \underline{396} \\ 506 \end{array}$$

3. Borrow-Across-Zero.   When the student needs to borrow from a column whose top digit is 0, he skips that column and borrows from the next one. (This bug requires a special "rule" for subtracting from 0: either $0 - N = N$ or $0 - N = 0$.)

$$\begin{array}{r} 602 \\ -\ 327 \\ \hline 225 \end{array} \qquad \begin{array}{r} 804 \\ -456 \\ \hline 308 \end{array}$$

4. Stop-Borrow-At-Zero.   The student fails to decrement 0, although he adds 10 correctly to the top digit of the active column. (This bug must be combined with either $0 - N = N$ or $0 - N = 0$.)

$$\begin{array}{r} 703 \\ -\ 678 \\ \hline 175 \end{array} \qquad \begin{array}{r} 604 \\ -387 \\ \hline 307 \end{array}$$

5. Don't-Decrement-Zero.   When borrowing from a column in which the top digit is 0, the student rewrites the 0 as 10 but does not change the 10 to 9 when incrementing the active column.

$$\begin{array}{r} 902 \\ -\ 368 \\ \hline 344 \end{array} \qquad \begin{array}{r} 205 \\ 9 \\ \hline 1106 \end{array}$$

6. Zero-Instead-Of-Borrow.   The student writes 0 as the answer in any column in which the bottom digit is larger than the top.

$$\begin{array}{r} 326 \\ -\ 117 \\ \hline 210 \end{array} \qquad \begin{array}{r} 542 \\ -384 \\ \hline 200 \end{array}$$

7. Borrow-From-Bottom-Instead-Of-Zero.   If the top digit in the column being borrowed from is 0, the student borrows from the bottom digit instead. (This bug must be combined with either $0 - N = N$ or $0 - N = 0$.)

$$\begin{array}{r} 702 \\ -\ 368 \\ \hline 454 \end{array} \qquad \begin{array}{r} 508 \\ 489 \\ \hline 109 \end{array}$$

(reproduced from Resnick, 1982, p.140; adapted from Brown, J.S., and Burton, 1978)

Brown and Burton found that they could diagnose the vast majority of student performances in terms of one or more of the 88 'bugs' which they have so far identified. Brown, J.S., and Van Lehn have gone further and produced a theory of why such 'bugs' occur.

> ...When a student has unsuccessfully applied a procedure to a given problem, he or she will attempt a repair. Suppose he or she is missing a fragment (subprocedure) of some correct procedural skill, either because he or she never learned the subprocedure or maybe forgot it. Because the missing fragment must have had a purpose, attempting to follow the impoverished procedure rigorously will often lead to an impasse. That is a situation in which some current step of the procedure dictates a primitive action that cannot be carried out, usually because one of its preconditions or input/output constraints has been violated. For example, an attempt to decrement a zero will lead to an impasse. When a constraint gets violated the student, unlike a typical computer program is not apt to just quit. Instead he or she will often be inventive, invoking problem-solving skills in an attempt to repair the impasse and continuing to execute the procedure, albeit in a potentially erroneous way. We believe that many bugs can best be explained as patches derived from repairing a procedure that has encountered an impasse while solving a particular problem. (Brown, J.S., and Van Lehn, 1982)

The 'repair rule' invented by the child clearly depends on the extent to which he understands the basis of the algorithm; erroneous 'repair rules' are generally in conflict with the structural properties of number or number operations. For example, the 'smaller-from-larger' bug is equivalent to the false assumption that subtraction is commutative (e.g. that '6 from 2' is equivalent to '2 from 6'). Similarly the 'Borrow-Across-Zero' bug ('I can't take one off the 0 on top of the next column on the left, so instead I'll take one off the top number in the next column on the left of that') is a contradiction of place value principles.

3.7.4   The Relationship Between 'Understanding' and Algorithmic Skills

By 'understanding' in this part is meant a grasp of the meaning and structure of number and number operations, as described in parts 3.4 to 3.6.

It is clear that some children can, and do, perform computational procedures correctly by rote with little basis of numerical understanding. For example Brown, M. (1981b) quotes Shakeel, a 12 year old whose concept of place value was so shaky that he thought that the number one more than 6399 was 63100, yet who correctly carried out a long multiplication involving decimals. Brown found overall that there was quite a high overall correlation between tests of understanding of number operations and tests of computation, but

> ...those with hardly any understanding at all of the operations were not in general successful with the calculations. (Brown, M., 1981a).

Dickson investigated to what extent 13 to 14 year old children in the lower half of the ability range understood the basis of the commonly used algorithms. In fact she found evidence of very little understanding of these standard procedures, even when correctly applied.

e.g. Karen, a low attaining 13 year old, was presented with $78 + 932 + 6 + 4001$.

She wrote
$$
\begin{array}{r}
78 \\
932 \\
6 \\
\underline{4001} \\
\underline{5017} \\
{\scriptstyle 111}
\end{array}
$$

Interviewer:	'You added the first column and it came to seventeen and you said' "Seven down and carry one." What do you mean by "carry"?'
Karen:	'I said seven, carry one because you have to add the next column. You've got another column to add so if you've got one left over you add it to the column.'
Interviewer:	'One what left over?'
Karen:	'This one unit and you add it up to the column and you do the same all the way along.'
Interviewer:	'Then you added the seven, three and the one and got eleven and you put one there and ...'
Karen:	'Carry one again.'
Interviewer:	'And that's one what?'
Karen:	'Just say carry one.'
Interviewer:	'And nine and one makes ...'
Karen:	'Ten, nought down, carry one.'
Interviewer:	'Ten what is it in that column?'
Karen:	'What do you mean, ten what? It's ten; just say ten and carry the one. When you come to your last column if you've got like no more numbers to carry and you've got a two figure sum and you like say you had a sum equal to fifteen – instead of putting five down, carry one, you've got no more columns to add up so you just put your whole figure down.'

Wendy from the same class was given $651 - 74$ which she did like this:

$$
\begin{array}{r}
{\scriptstyle 5}\ {\scriptstyle 14}\ {\scriptstyle 1} \\
\cancel{6}\,\cancel{5}\,1 \\
-\ \ 7\,4 \\
\hline
5\,7\,7
\end{array}
$$

Wendy explained how she did this and an excerpt from the interview follows. She has 'borrowed' one from the tens and subtracted 4 from 11. She is now explaining the next step.

Wendy: '...you take seven away from four, which you can't do so you borrow another ten from six hundred. That leaves five: you put one on there: that's seven away from fourteen.'

Interviewer: 'You borrow ten from six hundred and that leaves five!'

Wendy: 'You borrow ten from six hundred and that leaves five hundred and ninety.'

Interviewer: 'Where's the ninety?'

Wendy: 'I don't know how to do these ones.'

Interviewer: 'You started off with six hundred and fifty-one on the top. You've crossed these numbers out and now you have five fourteen and eleven. Do you still have the same sized number as six hundred and fifty-one on the top?'

Wendy: 'No, it's smaller; it's five hundred and forty-one now because you've been borrowing from the six hundred and fifty.'

Interviewer: 'Does it matter that you've got a different number?'

Wendy: 'Dunno – confusing.'

Ginsburg (1977) says:

> What is interesting is the rare case of a child who *does* understand the algorithms (Ginsburg, 1977).

In some cases this failure to understand the basis of an algorithm may be due to lack of understanding of number ideas in general. However, as Resnick concludes from her research:

> ...children's difficulty with place value in addition and subtraction does not necessarily derive from an *absence* of semantic knowledge about the base system. Rather, it results from an inadequate *linking* of the semantics of the base system with the syntax of the written algorithms. (Resnick, 1982)

Brown, M. cites the example of Richard, a 14 year old with above average knowledge of the meaning and structure of number, who had tried unsuccessfully to make such a link for himself in the case of 'equal additions' ('paying back to the bottom') algorithm for subtraction. He wrote:

$$\begin{array}{r} 2\ 3\ 1\ '2 \\ -\ 5\ ^5\!\!\not{4}\ 7 \\ \hline 5 \end{array}$$

MB: 'Explain why you changed that 4 into 5.'

R: 'Cos if you take one away you got to put one back to the bottom.'

MB: 'Why do you put it back to the bottom when you took it from the top?'

R: 'Oh – well I always thought it was funny but it gets them right. I did try and think but I couldn't work it out.' (Brown, M., 1982)

In fact of the six British secondary children interviewed by Brown who used the 'equal additions' algorithm, none could work out why this method worked. The more usual 'decomposition' method was found easier to explain, but it was interesting that 6 out of 7 children who used it successfully were able to explain why it worked, whereas none of the 3 who made errors in the procedure were able to do so. Perhaps not surprisingly of the 6 children able to explain the algorithm, 5 were judged on other grounds to be 'above average' in their grasp of the meaning and structure of number, whereas none of the children unable to explain the algorithm were judged to be 'above average'.

This corresponds to Engelhardt's findings, quoted previously (page 256) that brighter children make very many fewer procedural errors than weaker children.

The explanation for the evidence given may be that the learning of algorithms imposes a considerable memory-load, so that most children are likely to 'forget' some of the rules sooner or later. For example Suydam and Dessart (1976) quote studies by Osburn and Foltz (1931), Kurtz (1973) and Grenier (1975) which show a significant loss in computational skills in children, ranging from 6 to 13 years, over the summer holiday period. (Interestingly Grenier found that, in contrast, in the case of her 13 year olds there was a gain in concept and application scores over the same period, in spite of a lack of formal education during this time.)

If a step, or steps, are lost, then as Brown, J.S., and Van Lehn (1982) point out, the child has to use his own knowledge and inventiveness to effect a 'repair', and thus the validity of this repair is likely to depend on the child's grasp of the meaning and properties of number operations.

This explanation would also account for the finding of Brownell and Moser (1949) that meaningful (as opposed to mechanical) teaching of the subtraction algorithm gave better results in the long term, although in the short term, the mechanical teaching was more efficient.

Skemp makes similar points in a more general context of mathematics learning. He contrasts rote learning ('instrumental understanding') with meaningful learning ('relational understanding') and concludes that the former may be easier to teach and gives quick results in the short term, but is harder to remember, as well as being lacking in intrinsic motivation. He uses the analogy of a person finding his way about an unfamiliar town:

> A person with a set of fixed plans can find his way from a certain set of starting points to a certain set of goals. The characteristic of a plan is that it tells him what to do at each choice point: turn right out of the door, go straight on past the church, and so on. But if at any stage he makes a mistake, he will be lost; and he will stay lost if he is not able to retrace his steps and get back on the right path.

> In contrast, a person with a mental map of the town has something from which he can produce, when needed, an almost infinite number of plans by which he can guide his steps from any starting point to any finishing point, provided only that both can be imagined on his mental map. And if he does take a wrong turn, he will still know where he is, and thereby be able to correct his mistake without getting lost; even perhaps to learn from it. (Skemp, 1976)

3.7.5 Some Suggestions for Teaching Written Algorithms

Much of the literature concerned with the teaching of computational skills in a written form again emanates from the United States. Suydam and Dessart (1976) have produced a booklet in which they offer suggestions for algorithm teaching which are based on ideas from research in this area. They also give a comprehensive account of American research studies up to that date.

It should be clear from the previous text that a good understanding of the basis of the written algorithms will assist retention in the long term, even if it makes little difference in the short term. Suydam and Dessart state:

A cardinal rule has been evolved through experience and affirmed through research: *Develop mathematical ideas and skills from a concrete physical basis*…. Generally, researchers have concluded that understanding is best facilitated by the use of concrete materials, followed by semi-concrete materials (such as pictures) and finally by the abstract presentation with symbols. It is as important to use concrete materials in introducing algorithms as it is in introducing basic facts. (Suydam and Dessart, 1976)

They also emphasise that the progression 'concrete → semi-concrete → abstract' holds as much for new ideas presented at the secondary level as it does at the primary level.

Resnick stresses the importance of making explicit links between the concrete representation and the written algorithm. She suggests a method using Dienes' Base 10 blocks which is based on the procedure shown in the following figure.

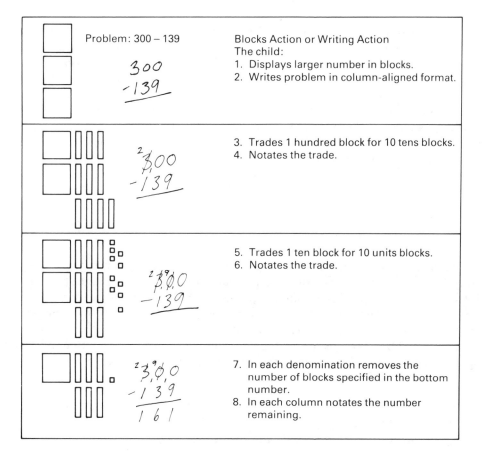

Fig. 3.36   Method suggested for making explicit links between concrete and written form of algorithms (from Resnick, 1982)

Resnick reports considerable success in using this method with three low attaining children all with different 'bugs' in their subtraction algorithms. Not only were they soon able to cope without the blocks, but they also demonstrated the ability to explain the algorithm; in each case the child in fact used a correct idea not explicitly taught.

Resnick stresses the importance of linking the concrete to the written abstract algorithm in terms of:

a.   the place value 'coding' i.e. can the child map between a written number and its concrete representation in base 10 blocks?
b.   the result i.e. does the child expect the same result from manipulating blocks and manipulating the corresponding numbers? (She quotes some children in the 7 to 9 year age range who were quite unperturbed when different results were obtained in the two cases.)
c.   the operation i.e. can the child translate each step of the written algorithm into a concrete action, and vice versa?

Apart from this developmental approach to algorithmic teaching one must also consider the prerequisites in terms of the conceptual basis required for meaningful (written) computational work. Thus teaching a child to do long multiplication is going to be a meaningless pursuit for him if he has little understanding of large numbers, of the nature of the multiplication operation – its meaning and structure – and of the symbolic notation for these.

With an appropriate background of experience with the nature of number operations and our place value symbolic system, Lowry then considers the child to be set for developing computational skills which

> should take place in such a way that a child first uses procedures which are rather long and even tedious compared with methods used by adults. However, these procedures are so designed that the child can see each step and can justify it at his own level of understanding of our numeral system and principles governing the operations on numbers. The purpose after this beginning is to work toward more and more efficient methods of computation. (Lowry, 1965)

He gives the example of a ten year old child who has an appreciation of the multiplication of whole numbers in terms of repeated addition e.g. 3×6 means three 6's, $6 + 6 + 6$. He knows some basic multiplication facts and has multiplied multiples of 10 by single digits e.g. 3×40. It is considered that the next stage is to find a method for coping with a product such as 3×23. The child can cope with it as $23 + 23 + 23$, but the aim is to move away from addition towards 'understanding multiplication as an operation in its own right'. Lowry suggests steps towards this goal in the following sequence:

$$
\begin{array}{lll}
\begin{array}{r} 23 \\ 23 \\ +23 \end{array} \quad \rightarrow &
\begin{array}{l} \text{Add } 20 + 3 \\ 20 + 3 \\ \underline{20 + 3} \end{array} &
3 \times 23 \;\rightarrow\; 3 \times (20 + 3)
\end{array}
$$

which from this $\nearrow$ is 3 lots of 3 and 3 lots of 20 or $(3 \times 20) + (3 \times 3)$

$$
\rightarrow \quad
\begin{array}{r}
20 + 3 \\
\times \qquad 3 \\
\hline
\end{array}
$$

With further encouragement it may be possible to shorten this to
$$
\begin{array}{r}
23 \\
\times\; 3 \\
\hline
9 \\
\underline{60} \\
69
\end{array}
$$

which Lowry claims pupils can arrive at by themselves and most certainly will if given a slight hint. He recommends that pupils work with such longer methods for as long as they need to. However, this is said with one major reservation. Children who have already met the standard algorithms will not take kindly to having to perform a lot of calculations by these longer more cumbersome methods. Although Lowry sees it as being necessary to show such children these non-standard algorithms to help justify their shorter methods and encourage them to devise other methods of their own, they cannot be expected to adopt them to the extent that is recommended for the 'uninitiated'.

Suydam and Dessart offer some alternative methods of calculation for each of the operations $+ - \times \div$ which involve knowledge, skills and understandings which they say should be acquired by children.

For addition, using an algorithm such as

$$
\begin{array}{ll}
\begin{aligned}
20 & + 6 \\
+ \underline{40 + 7} \\
60 & + 13 = 73
\end{aligned}
&
\begin{aligned}
26 \\
+ \underline{47} \\
13 \\
\underline{60} \\
73
\end{aligned}
\end{array}
$$

For subtraction:

$$
\begin{array}{lll}
50 + 4 & = & 40 + 14 \\
- (20 + 5) & & - \underline{(20 + 5)} \\
& & 20 + 9 = 29
\end{array}
$$

For multiplication:

$$
\begin{array}{ll}
\begin{aligned}
50 & + 2 \\
\times \underline{(30 + 8)} \\
400 & + 16 \\
\underline{1500 + 60} \\
1900 & + 76 = 1976
\end{aligned}
&
\begin{aligned}
52 \\
\times \underline{38} \\
16 \\
400 \\
60 \\
\underline{1500} \\
1976
\end{aligned}
\end{array}
$$

For division:

$$
\begin{array}{ll}
35 \overline{)463} \\
\quad \underline{350} & 10 \\
\quad 113 \\
\quad \underline{105} & \underline{\ 3} \\
\qquad 8 & 13\,\text{r.}8
\end{array}
$$

(Suydam and Dessart, 1976)

Other alternative 'low-stress' algorithms are suggested by Ashlock (1976). These generally work on a basis of reducing the amount of information a child has to hold in his head at once. They include:

Addition – the 'tens' method (suggested by Fulkerson, 1963):

e.g.　468　　　(Adding down each column, each time the total exceeds ten a stroke
　　　79̸5̸　　　is written across that digit and only the last digit of the total re-
　　　8̸0̸9̸　　　tained to add to the next digit. (e.g. $8 + 5 = 13$, cross through the 5
　　　9̸8̸6　　　and start $3 + 9 = 12$ etc.) Then count up the crossed digits to obtain
　　3058　　　the 'carry' figure. This obviates the need for mental addition with num-
　　　　　　　bers over ten.)

Subtraction – Hutching's low-stress method (from Hutchings, 1975)

e.g.　　4352　　　4352　　　4352　　　(The top number is first re-written in its 're-
　　　−1826　　　4̲2̲　　　3̲3̲4̲2̲　　　grouped' form to make the final subtraction
　　　　　　　−1826　　−1826　　　straightforward.)
　　　　　　　　　　　　　2526

Division – the doubling method (also known as the 'Russian peasant method')

e.g.　290 ÷ 8　→　$8 \times 1 = 8$　　　　　　290
　　　　　　　　　$8 \times 2 = 16$　　−256 → 32
　　　　　　　　　$8 \times 4 = 32$　　　34
　　　　　　　　　$8 \times 8 = 64$　　　32 → 4
　　　　　　　　　$8 \times 16 = 128$　　2　　36
　　　　　　　　　$8 \times 32 = 256$

$$290 \div 8 = 36 \, \text{r.} 2$$

(The divisor is doubled until the next 'double' would be too large, and subtracted. Other doubles are subtracted from the remainder until the number left is smaller than the divisor.)

(Some other alternative algorithms based on informal procedures are given in 3.7.7 on page 272)

Barclay (1980) reports successful use, as a tool for teaching, of the computer program BUGGY, devised by Brown, J.S., and Burton (1978) to simulate the bugs children have in their computation procedures. Children were shown on a computer terminal several examples of faulty calculations containing a systematic error, and asked to 'spot the bug'. The program itself allowed children to propose hypotheses and to ask for further examples. They were also asked to invent plausible bugs. Although the use of terminals is particularly suitable for this activity, there is no reason why it could not be adapted for blackboard or worksheet, as in the book by Ashlock (1976).

Brown, S.I. suggests that work with algorithms must stress the need for children to ask questions and explore different methods of computation. He says

...when the pupil feels confident with what he is doing, he is frequently less threatened by examining its alternatives and in addition he has a strong perspective from which to view alternatives (Brown, S.I., 1974).

He also feels that

> The point is that the search for worthwhile algorithms and comparison among them according to any number of interesting criteria is itself a much more significant activity than merely acquiring a strategy for working efficiently with a particular algorithm, a task the hand calculators will be taking over for us soon enough. (Brown, S.I., 1974)

3.7.6 Informal Algorithms

Ginsburg points out that children's informal arithmetic is powerful. What they can understand and do on an intuitive level outstrips their comprehension and achievement at a symbolic, written level of computation. He says

> ...often children who botch up standard written calculations nevertheless understand the relevant concepts on an intuitive level (Ginsburg, 1977).

For example

> Caroline aged six was asked, 'If you bought twenty-four flowers and six of them were tulips and the rest were daffodils, how many daffodils would you have?'
>
> Caroline responded correctly, 'eighteen'.
>
> Now can you write down what you have done in your head?
>
> Caroline wrote: 6   Then   6
>   24 -24
>   22
>
> She then said, 'But it ought to be eighteen oughtn't it?' (Ginsburg, 1977)

Ginsburg emphasises the strength of counting methods employed by children. He says that they provide the basis for informal procedures in a safe and secure way. He claims that almost all 6 year olds who succeed at such calculations as $4 + 3, 7 + 12, 23 + 13$ do so by means of some counting strategy.

> They count on their fingers, they count tallies, they count starting at one, or they count on from the larger number. (Ginsburg, 1977)

The ability of children to evolve strategies prior to formal instruction was investigated by Carpenter and Moser in a study already referred to on pages 191 and 222. They summarise the main findings of their study by saying:

> The tremendous variability between and within children in the solution processes used suggest that before receiving formal instruction, young children do not transform problems into a single type and apply a single strategy. The results indicate that children have available a rich repertoire of strategies and that they make use of many of these to solve various problem types. It is still not clear what triggers the use of a particular strategy; but it seems plausible that children solve each problem type directly, rather than collapsing them and applying a single strategy consistently. (Carpenter and Moser, 1979)

They also suggest that:

> Perhaps by introducing operations based on verbal problems and integrating verbal problems throughout the mathematics curriculum, rather than using them as an application of already learned algorithms, chil-

dren will develop their natural ability to analyze problem structure and will develop a broader conception of basic operations. (Carpenter and Moser, 1979)

Ginsburg says that

Despite all the effort expended in teaching algorithms, children often do not use them.... They assimilate school mathematics into their own mental framework: the result is invented procedures, methods that are partly based on codified written arithmetic and partly on the child's distinctive approach. Often children find their invented methods more comfortable than the algorithms taught in school. (Ginsburg, 1977)

Ginsburg cites Lori, aged 9 years, who was asked to add 83 and 5 and 294.

Well 294 is almost in the three hundreds. Then 5 more is 299 ... What's the other number?

The interviewer told her. She then said, 'It'd be 382.' (Ginsburg, 1977)

Ginsburg sees this as being

a clever rearrangement of the order of numbers, Lori took advantage of certain properties that allowed her to do easy mental calculation (Ginsburg, 1977).

Brown, M. also found during her interviews as part of the CSMS project that many secondary children, from the whole range of attainment, would employ informal procedures, which were often addition/subtraction based, to solve practical problems.

e.g. For the problem – 'A gardener has 391 daffodils. These are to be planted in 23 flowerbeds. Each flowerbed is to have the same number of daffodils. How do you work out how many daffodils are to be planted in each flowerbed?

Stephen, aged 12, gave the answer:

Interviewer: 'What would you do with those two numbers to work it out?'

Stephen: '23 ... there ... take away 23, 23, 23, ...'

Interviewer: 'Keep on taking away 23?'

Stephen: 'Yes'.

Interviewer: 'Can you think of any of those that would fit?'
(showing $391 - 23$ $23 \div 391$
                    $23 - 391$ 391×23
                    $391 + 23$ $23 + 23$
                    23×17 $391 \div 23$)

Stephen: 'That one.' ($23 \div 391$)

Interviewer: 'Any others?'

Stephen: 'And that one' ($391 \div 23$)

(Brown, 1981a)

Brown, M., also reports that Tim, aged 13, considered to be 'below average', had his own invented methods for coping with the vertical forms of addition and subtraction.

For   38
         +27

Tim: 'When I first did it I thought I'd try doing it in my mind – I said 3 add 2 – 30 add 20 is 50, so I went to add 7 and 8 together. Then I said 8, 9, 10 take 2 off the 7 is 5 and that makes another one, that's 6. You're supposed to carry but I don't always do that.' (Brown, M., 1981c)

For   51
         −28

Tim: '2 from 5 is 30, then I said 8 take 1 leaves you 7, then I said 7 from 30 leaves you 23.' (Brown, M.,
1981c)

A similar but more formally expressed method used by an 8 year old is reported by
Davis, as quoted by Brown, S.I.

A few years ago, in the elementary school in Weston Connecticut, a third grade boy named Kye invented a
new algorithm for subtracting. His teacher had been solving the problem

$$\begin{array}{r} 64 \\ -28 \end{array}$$

and had said: 'We can't subtract eight from four, so we have to regroup the sixty as...' At this point Kye
interrupted, took the chalk, and did this.

Kye said: Kye wrote:
'Oh yes, you can! Four minus   64
eight is negative four... -28
                                        $\overline{\;-\;4}$

and sixty minus twenty is forty...   64
                                        -28
                                        $\overline{\;-\;4}$
                                               40

...and forty and negative four   64
are thirty-six -28
                                        $\overline{\;-\;4}$
                                        $\underline{40}$
                                               36

so the answer is thirty-six...

(Davis, 1973, as quoted by Brown, S.I., 1981)

In Figure 3.37 is reproduced a 'slow' 14 year old's method of working 24×53. It is
reproduced from Fitzgerald and illustrates well the blend of inventiveness and
written symbolism giving rise to a sound addition-based strategy for performing this
computation.

3.7.7 The Nature of Mental Algorithms: Implications for Teaching

Many invented procedures for computation rely heavily on a mental approach to devis-
ing strategies. Some of the characteristics of mental algorithms which have been outlined by Plunkett are particularly worthy of consideration from the point of view of
appreciating the advantages they hold in terms of the developmental nature of the
child's understanding of computation. He lists the following points concerning mental
algorithms:

1. They are *flexible* and can be adapted to suit the numbers involved. Do you have different methods for
    83–79, 83–51, 83–7?

2. They are *active* methods in the sense that the user makes a definite, if not always very conscious, choice of
    method and is in control of his own calculations.

3. They are usually *holistic,* in that they work with complete numbers rather than separate tens and units
    digits,
    e.g. $4 \times 35 \;= 2 \times 70 = \underline{140}$
    $\quad 4 \times 28: \quad 4 \times 30 = 120, \, -8, \, \underline{112}$

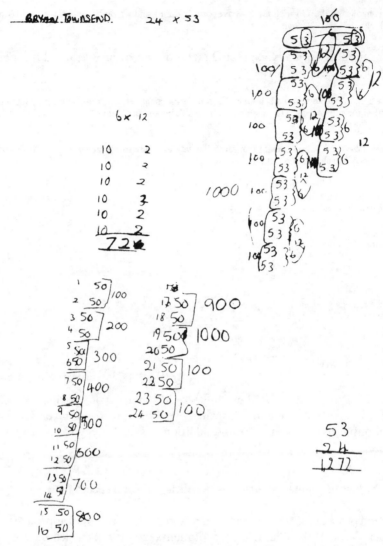

Fig. 3.37   A slow 14 year old's method of working out 24 × 53 (reproduced from Fitzgerald, 1976).

4. They are frequently *constructive*, working from one part of the question towards the answer, e.g. 37 + 28: 37, 47, 57, 67, <u>65</u>

5. They require *understanding* all along. A child who gets his mental calculations right almost certainly understands what he is doing. Equally their use develops understanding.

6. They are *iconic*. Either they relate to an icon such as the number-line or a number-square, or they depend upon serial enunciation as in 32 + 21: 32, 42, 52, <u>53</u>. In either case some overall picture of the numbers is being used.

7. Often they give an *early approximation* to the correct answer. This is usually because a left-most digit is calculated first, but in the context of complete numbers,
    e.g. 145 + 37: 175, <u>182</u>
          34 × 4: 120, <u>136</u>

                                                      (Plunkett, 1979)

There are, of course, certain drawbacks associated with mental algorithms especially from the teacher's standpoint. For the teacher to ascertain the nature of a child's mental computational techniques he has to ask each child individually to reflect upon and vocalise his methods. As Plunkett points out, because of the fleeting nature of these procedures they are sometimes difficult to capture. He also says:

> They are not designed for recording. So written down they tend to sprawl.... But they *can* of course be recorded where this is desirable. (Further, mental algorithms) ... are limited in the sense that they cannot be applied to the most difficult calculations such as 269 × 23. Nevertheless they are suitable for a greater range of problems than a casual observation of school number work might lead one to suppose.
> (Plunkett, 1979)

Fostering informal computational procedures helps to develop the child's appreciation of the meaning and structure of number operations. Plunkett proposes an approach to teaching elementary arithmetic based on this assertion which involves three stages:

Stage 1   This stage is concerned with developing an understanding of numbers and of the place value system. It concentrates first on number bonds up to about 10 + 10, 10 × 10 etc. and second on calculations which can be performed mentally in one step, e.g. 135 + 100   85 − 20   5 × 30   90 ÷ 3

Plunkett emphasises the value of number lines in helping the place value idea to develop. He feels that time should be taken to help children appreciate facts like 47 + 10 = 57 and 13 × 10 = 130 mentally before progressing to say 47 + 18 or 13 × 12. With a sound basis such as this they are then able to develop their own techniques for coping with, say, 47 + 34 (47, 57, 67, 77, 81), 17 × 4 (20 × 4 = 80, 76, 72, 68) which are performed mentally and can be recorded.

Stage 2   With a good foundation in Stage 1 work the child is now set to make sensible use of calculators for more difficult problems such as:
    592 − 276   931 × 768   8391 ÷ 57.

Stage 3   This stage of Plunkett's approach centres on the development of 'casual written methods'. He considers that primarily there is a need for this in terms of money calculations where it may be necessary (when a calculator is not at hand) to have some non-standard written method available. Such techniques can be adapted from mental algorithms. e.g. totalling £54.75 and £32.80:

    starting from   54.75
    the left   32.80
                       86ꞌ55

or subtracting by an 'adding on' method e.g. subtracting £32.80 from £54.75:

                    32.80
                    42.80   10
                    52.80   20
                    54.80   22
                    54.75   21.95

The important thing is a process which is *intelligible* (to the user), rather than one which is standardised or quickest (Plunkett, 1979).

Brown, M., reinforces Plunkett's view when it is stated that:

> It may well be the case that a combination of reliable mental methods and the ability to use a calculator are sufficient for all practical purposes .... Perhaps the present (standard) methods should be abandoned in favour of others, maybe less efficient, but more related to children's own informal methods, and hence easier to remember. (Brown, M., 1981a)

And furthermore as Ginsburg says

> Even children with severe learning problems may have unsuspected informal strengths. The gap may originate because instruction does not devote sufficient attention to integrating formal written procedures with children's already existing and relatively powerful informal knowledge. (Ginsburg, 1977)

3.8   FRACTIONS, DECIMALS AND PERCENTAGES: MEANING, STRUCTURE AND OPERATIONS

3.8.1   Developing an Understanding of the Meaning of Fractions, Decimals and Percentages

3.8.1.1   Introduction: the Complex Nature of the Meaning of Fractions, Decimals and Percentages

One of the difficulties of operating with fractions, decimals and percentages is that they have a multiplicity of meanings. Thus any particular number, say ⅗ (or 0.6 or 60%), can be interpreted concretely in many ways, all of which occur in everyday life applications. This is in contrast to the whole numbers, which are used mainly either for counting discrete objects, or counting repetitions of measuring units as in working out lengths, and so on.

For example, the fraction ⅗ can be interpreted variously as:

(a) a sub-area of a defined 'whole' region
      e.g.

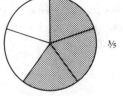

⅗   'divide the whole into 5 equal
                  parts and take 3 of them.'

Fig. 3.38

(b) A comparison between a subset of a set of discrete objects and the whole set
      e.g.

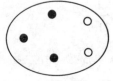

'³/₅ of the dots are black'

Fig. 3.39

(c) a point on a number line which lies at an intermediate point between two whole numbers

e.g.

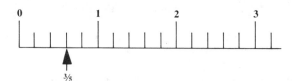

Fig. 3.40   ⅗

(d) The result of a division operation (in this case, 3 ÷ 5)

e.g.

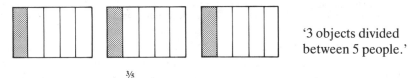

'3 objects divided between 5 people.'

Fig. 3.41   ⅗

(e) a way of comparing the sizes of two sets of objects or two measurements

e.g. (i)

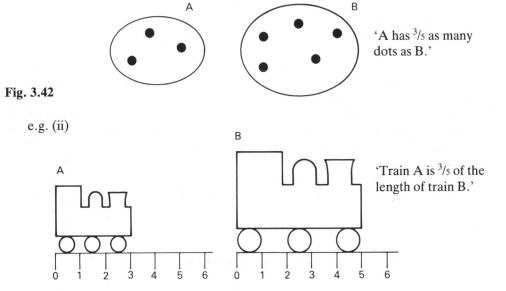

'A has ³/₅ as many dots as B.'

Fig. 3.42

e.g. (ii)

'Train A is ³/₅ of the length of train B.'

Fig. 3.43

(In a similar way, any decimal or percentage, say $0.37 = 37\% = {}^{37}\!/_{100}$, can be interpreted in each of these ways.)

The list of meanings above is not exhaustive; in particular none involve the notion of equivalence (e.g. ³/₆ = ½) and all of the more abstract meanings are omitted. Nor is it unique; although it draws on the classification systems of other authors, for example Kieren (1976), Novillis (1976) and Lesh et al. (1980), it differs to some extent from each of these. Nevertheless it gives some idea of the variety of meanings and usages of fractions, many of which may not be linked in children's minds. Hartung comments:

> The fraction concept is complex and cannot be grasped all at once. It must be acquired through a long process of sequential development (Hartung, 1958).

Evidence relating to possible sequences of development, which suggest in turn sequences of instruction, is now reviewed. What is important is that:

> Because each interpretation of rational numbers *(i.e. fractions/decimal fractions)* relates to particular cognitive structures, ignoring a conglomerate picture or failing to identify particular necessary structures in developing instruction can lead to a lack of understanding on the part of the child. (Kieren, 1976)

(Kieren uses the mathematical term 'rational numbers'; *rational numbers* are numbers which can be expressed as a ratio of two whole numbers (e.g. ⅔), and thus include all fractions, percentages, and all decimals representable as fractions i.e. finite and recurring decimals.)

In the text which follows, several of the more common meanings are reviewed in more detail. In parts 3.8.2 and 3.8.3, which relate to the structure of, and operations on rational numbers, it will be seen that not all the models lend themselves equally well to illustrating all aspects.

It should be noted throughout this chapter that much of the research which is referred to is concerned with rational numbers in the abstract, rather than as used in applications. For example 1.5 appears as an abstract number rather than in the context of, say, 1.5 kg or 1.5 m. (There is however some discussion of the use of rational numbers in section 2.1.8.1, and of decimals and money in section 2.5.2.4.) This emphasis probably reflects how number work is taught in many mathematics classrooms, but good practice would involve much more stress on applications than is the case here, especially at an early stage. Also, many children may be able to cope quite well with fractions and decimals in an everyday context but would find abstract number work difficult; for example they may be able to work out that dividing up a 3m strip of wood into 5 equal pieces would require 60cm pieces, without any understanding of the result that $3 \div 5 = 0.6$.

3.8.1.2   Fractions as Sub-Areas of a Unit Region ('Parts of a Whole')

A child's earliest encounters with fractional concepts and terminology are likely to be of a spatial kind, and they are likely to be 3-dimensional rather than 2-dimensional in nature. Thus 'half-full' is likely to refer initially to a glass of milk which is neither full nor empty but somewhere in between, and 'half an apple' may be understood merely as 'a piece of apple'. Some children's ideas seem to be still at this imprecise, qualitative level even at secondary school. For example, Hart (1980), as part of the CSMS study, gave the following question to a representative sample of about 550 English 12 and 13 year olds.

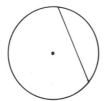

Has this circle been split into 2 halves?....................
Why do you think this?...............................

Fig. 3.44

While 89 per cent answered correctly that the circle had not been split in halves because the pieces were unequal, 6 per cent thought that the circle had been split into halves because 'there were two pieces'.

Piaget, Inhelder and Szeminska (1960) performed a number of experiments relating to the 'area' ('region') aspect of fractions. Children were given shapes of flat clay or paper and asked to divide them up 'fairly' between 2, 3 or more dolls, either by cutting, folding or drawing. They noted that the particular children in their sample were able to cope successfully with this task at the following ages, on average:

4–4½ years for halves of small or regular shapes
6–7 years for thirds
7–9 years for sixths, by trial and error.
10 years for sixths, by using a precise plan (e.g. dividing into halves, then each half into thirds).

(They noted however that some children were as much as two years ahead of this.)

One of the problems in this experiment was that of finding a method of dissecting a shape into a given number of congruent pieces; for example rectangles were divided into three more easily than circles. Thus children may well be able to recognise 'thirds' at a lower age than they are able to construct them.

Piaget, Inhelder and Szeminska give seven criteria for operational understanding of the spatial 'part-whole' aspect of a fraction:

i.   a 'whole' region is seen as divisible (2 year olds often refused to cut into the shape at all);
ii.   the 'whole'can be split into any required number of parts;
iii. the parts must exhaust the 'whole' (some children asked to divide a 'pie' between 3 dolls cut off 3 pieces and ignored the remainder);
iv   the number of parts do not necessarily match the number of cuts (thus if you cut a pie twice you may produce four pieces, and not two);
v.   the parts have to be equal in size;
vi.   parts can be seen as wholes in their own right (thus a sixth of the whole can be obtained by dividing each half into thirds);
vii. the whole is conserved, even when it is cut up into pieces.

There is considerable evidence that children find this spatial 'parts of a whole' notion the easiest aspect of fractions to grasp. Thus Hart (1980) in the study of 12 to 13 year olds previously referred to, found that 93 per cent could correctly shade in two-thirds of the shape in Fig. 3.45:

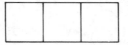

Fig. 3.45

A slightly more difficult task required children to perform the inverse operation and record as a fraction the part of Figure 3.46 which is shaded.

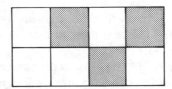

Fig. 3.46

Here the success rate for 12 to 13 year olds although reduced to 79 per cent, was still relatively high. A very few children recorded the fraction as ⅗ instead of ⅜, relating the shaded to the unshaded area rather than to the whole region.

The APU (1980b) report very similar results for a corresponding item involving a diagram of a circle divided into 5 sections, some of which were shaded. In this case 70 per cent of 11 year olds and 80 per cent of 15 year olds were able to supply the fractions represented by both shaded and unshaded regions.

It should be noted that in all these cases the areas were not only equal in size but also congruent in shape; it may be for instance, that children would be less happy to accept that both the shaded areas in this figure represent quarters of the square.

Fig. 3.47

Payne (1976) reports a number of studies on the initial teaching of fractions directed from the University of Michigan. He cites studies by Galloway (1975) and Williams, H.B. (1975) to demonstrate that the basic 'part-whole' view of a fraction can be successfully taught to pupils of 8 years old and above. The teaching programmes involved the folding and cutting of sheets of paper and drinking straws, leading on to the use of diagrams of regions. Symbolism was introduced only after a great deal of oral work. In both these studies, and in a further one by Muangnapoe (1975), it appeared that the 'area' aspect of fractions was found to be much easier than either the 'set' aspect or the 'number line' aspect. (Both these are described on pages 274 and 275 and are discussed in detail below.) Even so in the three studies, each involving pupils in the 8 to 12 age group, various difficulties were encountered repeatedly:

i.   the realisation of the need for equal-sized areas;

ii. the transition from the diagram to the words (e.g. three-quarters) and symbols (e.g. ¾);
iii. the understanding of fractions greater than one unit;
iv. the identification of a unit from a diagram showing more than one unit.

As an example, of points, (iii) and (iv), Payne quotes from Williams results for an item relating to the diagram shown in Figure 3.48. When asked what fraction was represented by the diagram, many students wrote ⁷⁄₁₀ rather than ⁷⁄₅.

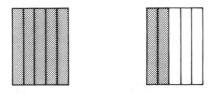

Fig. 3.48

It is clear that, for this reason, the representation of fractions as sub-areas of a unit area does not lend itself very well to the representation of improper fractions such as ⁷⁄₅; in fact the acceptance of the definition of a fraction as meaning 'part of a whole' is inconsistent with the very existence of such improper fractions.

Other problems arise over the use of the 'area' model, in particular for addition. These will be discussed in 3.8.3 on page 306.

Nevertheless there is considerable evidence that it is the easiest aspect to learn; indeed many textbooks use it exclusively, and it seems likely that, as a result, children find it difficult to go beyond this model and to apply fractions in situations embodying the other aspects.

3.8.1.3 *Fractions as Subsets of a Set of Discrete Objects*

This representation is essentially similar to that of fractions as sub-areas of a unit region; indeed if the sub-areas are separated it becomes almost indistinguishable:

area representation   set representation

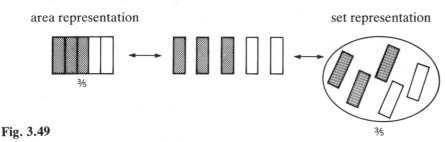

Fig. 3.49

However, it is probably true that the 'whole' is more easily perceived as a single unit in the case of the area representation.

Novillis (1976), in her study of the hierarchical development of various aspects of a fraction among American 10 to 12 year olds, found that the two aspects were roughly of the same difficulty; there was no clear indication that one arose prior to the other.

However this is in contrast to the results of the Michigan group quoted by Payne (1976). In this case the 'sets' approach was not only found to be significantly more difficult than that of areas or number lines, but a decision was taken to remove it from the initial teaching programme because it appeared to cause confusion among children in their understanding of the other models.

The reason for this discrepancy between two research studies is not clear, although the Michigan group were using younger children who might have found it more difficult to handle two aspects simultaneously.

The first APU (1980a) primary report describes the results of a practical task on fractions. As part of this, pupils were presented with 4 square tiles, 3 yellow and 1 red (see Figure 3.50) and asked 'what fraction of these squares are red?'

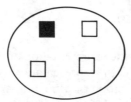

Fig. 3.50

Sixty-four per cent of 11 year old children were successful, but many of the remainder gave 'one third' instead of 'one quarter' failing to take the complete set as the 'whole' unit. This success rate is very similar to that on a question using the area model which was set in a written test, and suggests that, at any rate with the older children, Novillis' findings that set and area models are of similar difficulty may be correct.

It is interesting to compare results on the APU 'sets' item with that of a comparable verbal item from the CSMS study reported by Hart:

> 5 eggs in a box of 12 are found to be cracked. What fraction of the box of eggs is cracked?..............
> What fraction of the box of eggs is not cracked?.............. (Hart, 1980)

In this question 70 per cent and 66 per cent respectively of 12 year old children obtained correct answers to the first and second parts of the question, again giving similar results to those of APU (allowing for the fact that the children tested by CSMS were a year older).

However the results in both 'sets' and 'area' models do suggest that about one-third of children may have no clear concept of a fraction, even in a very concrete sense, at entry to secondary school.

The 'sets' model discussed here has some of the same drawbacks as the area model when it comes to illustrating improper fractions e.g. ⅞ would have to be represented:

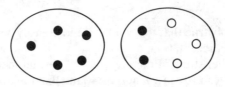

Fig. 3.51

which could again easily be interpreted as showing ⁷/₁₀, due to a confusion about the nature of a 'unit'.

However the sets model leads on naturally to the idea of ratio and percentage in more abstract numerical situations (e.g. ¾ of 20, 75% of 50). This application involves a fraction seen as an *operator* acting on another number, rather than an entity standing on its own.

The 'area' model on the other hand leads more normally to the idea of a fraction as a part of a standard unit used in *measurement,* via the number line aspect to be discussed in 3.8.1.4.

Thus it would seem necessary to provide plentiful concrete experiences of both 'area' and 'set' representations at an introductory level in order to lay the foundations for both *measurement* and *operator* aspects, since fractions (together with decimal fractions and percentages) are commonly applied in both these ways. Results on a CSMS question quoted by Hart (1980, 1981) suggest the limitations of a pure 'area' concept. The question was:

> Mary and John both have pocket money. Mary spends ¼ of hers and John spends ½ of his. It is possible for Mary to have spent more than John?............... Why do you think this?..............
>
> (Hart, 1981)

41 per cent of 12 year olds, falling to 19 per cent of 15 year olds, replied that it was impossible for Mary to spend more, since ½ is greater than ¼. One possible explanation for this is that pupils may be too ready to fall back on an 'area' (or 'number line') representation in which ½ and ¼ are seen as disembodied numbers, whereas the 'set' representation may be more powerful in suggesting that the size of the set, quantity or number, to which the fraction relates is of critical importance in many real-life situations.

3.8.1.4 *Fractions as Points on a Number Line*

One might expect that there would be a strong link between the representation of a fraction as a sub-area of a unit area, and that of the same fraction as a sub-length of a unit length, since the latter involves a 1-dimensional analogy of the former i.e.

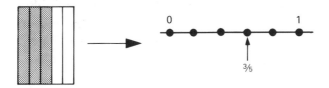

Fig. 3.52

However Payne (1976) reports that the number line model consistently caused difficulty in the various teaching experiments with 8 to 12 year olds in Michigan conducted by Muangnapoe (1975), Williams, H.B. (1975) and Galloway (1975). Novillis (1976), in her investigation of the hierarchical development of the concept of a fraction among 10 to 12 year olds confirmed that the number line model was significantly more difficult than either the 'area part-whole' model or the 'sub-set of a discrete set' model.

One reason for this may be that in Fig. 3.52, which is that used by Novillis, the representation of the fraction is a dot, emphasising the point that ⅗ can be thought of as a *number,* essentially similar to 0 and 1 since they too are represented by dots, but between them in magnitude. Thus unlike the two other representations, the number line does not incorporate the notion that a fraction can be thought of as a part of a concrete object, or as part of a set of objects, but reduces it to an abstract number.

While this clearly causes problems for 8 to 12 year olds, and possibly older children as well, it also possesses several advantages: for example it makes improper fractions appear much more natural, and, probably more important, emphasises that the set of fractions form an extension of the set of natural numbers, helping to 'fill in the holes' in between them. The number line model thus leads naturally to the use of fractions (and hence decimal fractions) in measurement of all types.

Novillis (1976) also found that the difficulty of operating with a number line was aggravated if the number line was extended beyond one:

Fig. 3.53

In this case when 12 year olds were asked to mark the point representing ⅗, many chose to mark a point ⅗ of the way along the whole line segment (e.g. in Fig. 3.53 they might choose to mark the point 3 as being ⅗ of 5). However if the line segment given was only that between 0 and 1, most 12 year olds were able to answer correctly.

The additional problem seemed to arise from the fact that children are unable to decide what constitutes an appropriate 'unit', in this case taking the whole line segment shown as representing a unit, by analogy with the 'area' model, rather than the line segment between 0 and 1 only, which represents a numerical unit. This problem of confusion over the nature of a 'unit' would seem to be less likely to occur if children had had more experience in reading numerical scales involving fractions for measuring, for example, rulers, barometers, thermometers, etc., before moving on to abstract scales. With metrication such experience is less likely to arise naturally but there may be a good case for creating such scales for classroom use.

It thus seems that many children are likely to be uncertain about the aspect of a fraction as a number represented by a point on the number line well into the secondary age range.

3.8.1.5 *Fractions as the Result of a Division Operation*

This aspect of the meaning of fractions associates a fraction with the operation of dividing one whole number by another: thus, for example, ⅗ is identified with the result of 3 ÷ 5, or 'sharing 3 units between 5 people' (see Figures 3.54, 3.55 for illustrations of both the discrete and continuous cases of this).

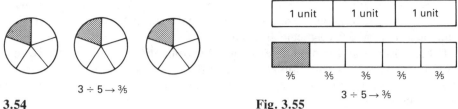

$$3 \div 5 \rightarrow \tfrac{3}{5}$$

Fig. 3.54

$$3 \div 5 \rightarrow \tfrac{3}{5}$$

Fig. 3.55

Payne (1976) notes that it is an aspect of fractions into which research is needed; it was not included with the other aspects in the studies reported either by Payne or by Novillis (1976). However Streefland (1982) uses it as the core of his teaching programme.

The only research which does appear to consider this aspect is the CSMS study on fractions reported by Hart (1980, 1981). As part of this study, the following item was given to representative samples of 12 and 13 year old English children:

> Three bars of chocolate are to be shared equally between five children. How much should each child get?............. (Hart, 1980)

On a separate computation paper, the parallel question '3 ÷ 5' was also included.

In each case, 33 per cent of 12 to 13 year olds gave the correct answer of $\tfrac{3}{5}$ (or 0.6). This suggests that only a third of children in the first two years of secondary school have appreciated that any whole number can be divided by any other to give an exact result expressible as a fraction.

This has some serious implications. For example in order to change a fraction to a decimal or percentage, whether by using a calculator or by some other means, it is generally necessary to convert from, say, $\tfrac{2}{7}$ to 2 ÷ 7. This application is likely to arise in problems which involve scales or ratios, as discussed in the part which follows.

3.8.1.6 Fractions as a Method of Comparing the Sizes of Two Sets, or Two Measurements

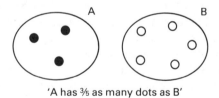

Fig. 3.56

'A has $\tfrac{3}{5}$ as many dots as B'

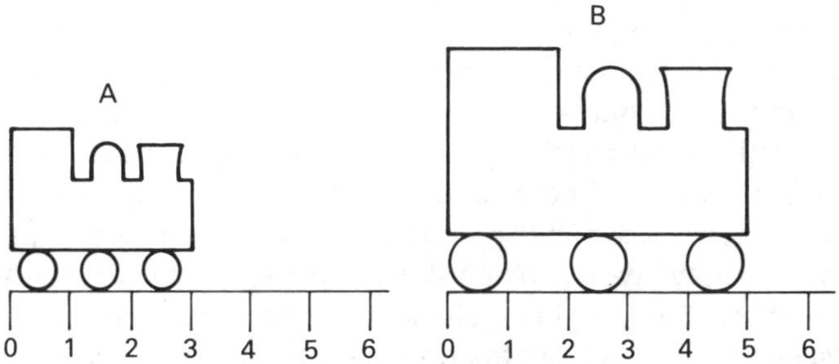

Fig. 3.57

'Train A is $\tfrac{3}{5}$ as long as train B'

This aspect of fractions, summarised in Figures 3.56 and 3.57, is the basis of many of the applications of fractions which occur in real-life, particularly those involving the ideas of ratio or scale (e.g. drawing scale diagrams).

In both discrete and continuous examples, the comparison could easily have been reversed e.g. 'B has ⅔ as many dots as A'; 'Train B is ⅔ as long as train A'. This means that there is no natural 'unit' or 'whole' as in the other aspects of the meaning of fractions already discussed.

Novillis (1976) found in her study of 10 to 12 year olds that this aspect of fractions developed significantly later than the 'area' and 'set' aspects.

This is supported by a number of studies on the understanding of ratio which suggest that the majority of adolescents find difficulty with the notion of expressing one number, or quantity, as a fraction of another. In particular studies by Piaget et al. (1968), Karplus et al. (1977) and Hart (1980) have revealed that children tend to revert to using additive comparisons (e.g. '3 more than') even in inappropriate circumstances.

Unfortunately most of the problems used in the ratio studies involve several steps of which the expression of one number as a fraction of another is only one. It is therefore difficult to obtain an estimate of the difficulty of this step in isolation. Nevertheless it is clear that the ability, and willingness, to compare two numbers in this way is an essential component in solving many such problems.

3.8.1.7   Understanding the Meaning of Decimals

The relationship between fractions and decimal fractions is not unlike that between the system of Roman numerals and the Arabic system which we normally use; in both cases the underlying concepts, whether of whole numbers or rational numbers, are the same. The essential difference between the systems lies in the particular conventions according to which the numbers are recorded. The system of Roman numerals is probably easier to grasp initially than our Arabic 'place value' system, just as the system of recording a fraction as, say, ⅗, is probably easier to cope with in the initial stages than, say, 0.6. But just as the Arabic system of recording whole numbers has considerable pay-offs in easing more sophisticated applications e.g. ordering numbers, recording large numbers, and, especially, complex computations, so does the decimal system of recording fractions have advantages of a similar kind.

For this reason, and, more particularly, since it is the recording system which is adopted on calculators, computers, and so on, it seems likely that the decimal system will be used increasingly in applications and the use of fractions at other than an informal level will gradually die out.

To operate effectively with decimal fractions thus requires both the ability to understand and use the notational conventions of the place value system of recording decimals, and the more basic ability to grasp the underlying concept, of say 'three tenths'. This latter ability involves being able to operate with the various meanings of a fraction, such as those discussed in detail in the previous parts 3.8.1.2 to 3.8.1.6.

It has been generally the case that fractions have been introduced into the curriculum prior to decimals, and that this has been done in a concrete way, using 'area part-whole' models and possibly also some of the models of fractions discussed above.

When decimals have been introduced at a later stage, the emphasis has often been mainly on relating the notational system to that of fractions in a fairly abstract way, on the assumption that the concept of a fraction is firmly grasped. For example the places after the decimal point are likely to be defined as tenths ($\frac{1}{10}$), hundredths ($\frac{1}{100}$) and so on, so that, for example, the number 34.275 is said to have 2 in the 'tenths column' ($\frac{2}{10}$), 7 in the 'hundredths column' ($\frac{7}{100}$) and 5 in the 'thousandths column' and thus be regarded as

'3 tens, 4 units, 2 tenths, 7 hundredths and 5 thousandths' or '3 × 10 + 4 × 1 + 2 × $\frac{1}{10}$ + 7 × $\frac{1}{100}$ + 5 × $\frac{1}{1000}$'.

However there is now evidence which indicates that a majority of children find it diffi-culty to translate decimal notation into fraction notation in this way. For example the APU (1980b) report that for the item:

Write a fraction in the box to complete the statement 6.28 = 6 × 1 + 2 × □ + 8 × $\frac{1}{100}$

the percentage of 15 year olds answering correctly (i.e. $\frac{1}{10}$) was only 10 per cent. Simi-larly, not more than 40 per cent of 11 year olds could rewrite simple decimals like 0.5 as fractions (APU, 1980a, p.40).

Brown (1981b, 1981c) found similar confusions. She quotes the following item statis-tics, obtained from samples of English 12 to 15 year olds which were each representative of the full ability range.

The following results may be on the optimistic side since the first three items follow-ing all appeared at the beginning of the paper directly underneath the solution to a 'trial' question in which 'tenths' and 'hundredths' columns were clearly labelled as the first two after the decimal point.

Brown quotes a number of responses to demonstrate the misunderstandings which arose. For the first question in the table (the value of the 2 in 0.260)

Julie: 'Hundreds, no, hundredths'
MB: 'Are you sure? What is the one after the point *always*? (long pause). Or does it change?'
J: 'It changes.'
MB: 'So what is it there?'
J: 'Hundredths'

For the third question in the table (adding one tenth to 4.254),

Kevin: 'Four point two hundred and sixty four.'

(Brown, 1981c)

Table showing results of selected items related to understanding of place value notation in decimals (data from Brown, 1981c)

Item	Answer	Percentage			
		12 yrs	13 yrs	14 yrs	15 yrs
0.126 The 2 stands for 2 ↑ ↑ *HUNDREDTHS*					
0.260 The 2 stands for 2 ↑ ↑	tenths	53	65	64	72
	(hundredths				
	or hundreds)	(23)	(14)	(14)	(9)
0.412 The 2 stands for 2 ↑ ↑	thousandths	48	58	53	63
	(units/ones)	(21)	(15)	(17)	(11)
Add one tenth	4.354	42	49	48	54
4.254 →	(4.264)	(17)	(12)	(17)	(9)
Six tenths as a decimal is *0.6* How would you write as a decimal three hundredths......	.03	50	60	59	67

For the fourth question in the table (writing three hundredths as a decimal), wrong answers given by children aged 12 to 15 in the interviews included:

.300	(4 children)
3.00	(3 children)
3.0	(1 child)
3.100	(1 child)
00.3	(1 child)
0.3	(1 child)

Brown concluded that:

> The major difficulty that weaker children seemed to have was in understanding that the figures after the point indicated that part of the number which was less than one unit, even though the names of the decimal places in relation to diagrams showing units, tenths and hundredths had been explained to them at the beginning of the interview or test.

> Instead, children seemed to think that the figures after the point represented a 'different' number which also had tens, units etc. (Brown, 1981b)

It seems possible that this confusion may arise from the use of decimal money, where, for example, it is legitimate to read 3.54 as 'three pounds, fifty four', the figures after the point being more readily interpreted as a whole number of pence than a fraction of a pound (see part 2.5.2.4, page 159).

The results given provide evidence that about half of 12 year old children and a third of 15 year olds do not understand decimal notation.

One way of trying to prevent this difficulty may be by treating decimals in a more concrete way, relating the notation to concrete meanings similar to those suggested for fractions rather than merely defining the meaning of decimal notation in rather abstract terms. Bell, Swan and Taylor (1981) also give details of a calculator based teaching programme which produced considerable gains in understanding decimal place-value for less able 14 year olds (see page 319). In fact there is no reason why the teaching of decimals should not precede that of fractions (although it is unlikely that a child will not have some informal knowledge of fractions like ½ and ¼ when decimals are taught in school). Willson (1972) showed that decimals could be taught before fractions with no loss of effectiveness, and Faires (1963) demonstrated that the teaching of decimals first might actually have some advantages, although in the latter case most of the teaching emphasised computation.

Since decimals provide an alternative notation for rational numbers which are also expressible as fractions, it follows that each of the concrete 'meanings' of fractions discussed earlier in parts 3.8.1.1 to 3.8.1.6 can and should be introduced in any programme to teach the meaning of decimals.

However, in everyday life, decimals are probably encountered most frequently in the field of *measurement* as referring to fractions of a standard unit of measurement. (When comparing one number with another, i.e. using fractions in the *operator* sense discussed on page 281, it is more usual in everyday life to use the terminology of percentages rather than decimals, although this is not necessarily the case in more advanced work. For example, '1.2 metres' is used instead of 'one metre and 20 per cent of a metre', but it is usual to say that a wage rise of £16 is 20 per cent of a basic wage of £80, in preference to 0.2 of £80.)

Thus in discussing the 'meaning' of decimals in a concrete way the aspects likely to be emphasised are:

(a) decimals as sub-areas of a unit region
   e.g.

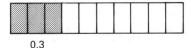

Fig. 3.58   0.3

(b) decimals as points on a number line
   e.g.

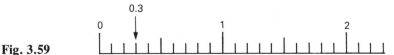

Fig. 3.59

(c) decimals as the result of a division operation
   e.g. $3 \div 10$

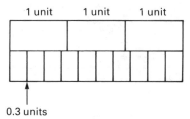

Fig. 3.60   0.3 units

All these three aspects are considered by Brown (1981b,c) but there appears to be very little other research which relates to the basic meaning of decimals.

The 'area' and 'number line' aspects were found to be similar in difficulty; although in each case the use of tenths alone was found to be very significantly easier than that of tenths and hundredths combined. For example the first item in the following table gives results in the 'area' aspect; the second in the 'number line' aspect.

Selected results of items testing the meaning of decimals in the sense of (i) sub-area of a unit region (ii) points on a number line (reproduced from Brown 1981c)

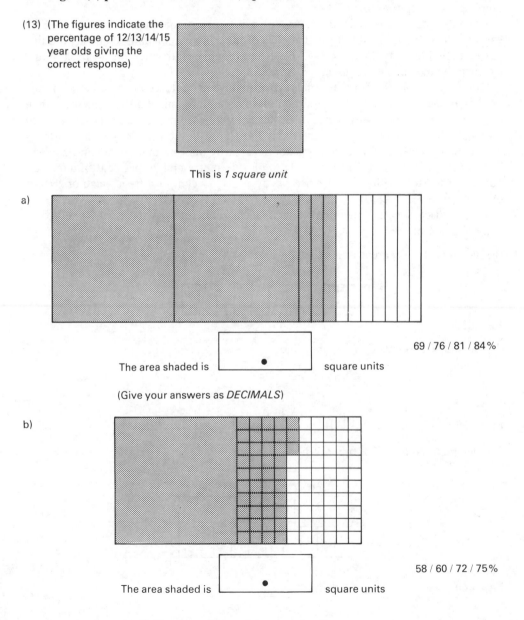

(13) (The figures indicate the percentage of 12/13/14/15 year olds giving the correct response)

This is *1 square unit*

a)

The area shaded is [•   ] square units   69 / 76 / 81 / 84%

(Give your answers as *DECIMALS*)

b)

The area shaded is [•   ] square units   58 / 60 / 72 / 75%

c)

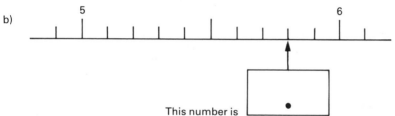

The area shaded is ☐ • ☐ square units

32 / 36 / 48 / 55 %

---

(6)   GIVE THE REST OF YOUR ANSWERS AS *DECIMALS*

b)

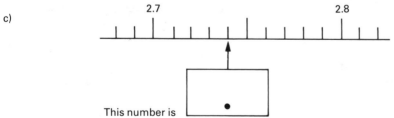

This number is ☐ • ☐

62 / 74 / 83 / 85 %

c)

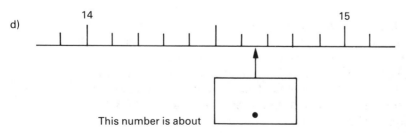

This number is ☐ • ☐

31 / 48 / 66 / 71 %

d)

This number is about ☐ • ☐

24 / 36 / 53 / 61 %

(Parts of these items are discussed in other contexts in sections 2.1.8 page 92 and 4.3.5 on page 367.)

It can be seen that around 65 per cent of 12 year olds, rising to 85 per cent of 15 year olds, can interpret decimals involving tenths alone in each of these concrete ways. However in the case of hundredths, the percentages fall drastically to 25–30 per cent of 12 year olds and 55–70 per cent of 15 year olds. (Item 13b appears to be much easier than the others involving hundredths but in many of the interviews it was found that children obtained the correct answer by merely counting 43 small squares and recording it correctly with very little understanding. For example Brown quotes children who expressed such answers as 'one and 43 tenths' or '4 tens and 3 little ones' but still obtained the correct written answer of 1.43 (Brown, 1981c.)

These results are supported by those of Galloway (1975), as quoted by Payne (1976), which suggest that 9 to 10 year olds can successfully be taught both area and number line meanings for decimals provided that only tenths are involved.

As in the case of fractions, the use of decimals to represent the result of the division of one whole number by another was however found to be extremely difficult:

Results on items relating to decimals as the result of a division of one whole number by another (data from Brown 1981c)

		Percentage			
Item	*Answer*	12 yrs	13 yrs	14 yrs	15 yrs
Divide by 20					
16→..............	0.8	8*	12*	26*	37*
(Write 'NO' if you think					
there is no answer.)	NO	(51)	(58)	(43)	(23)
		(*includes about 1% of sample who gave correct fractional answer)			
Ring the number you think is NEAREST IN SIZE to the answer (do *not* work out the sum): 59 ÷ 190 → .003 / .03 / .3 / 3 / 30 / 300 / 3000	0.3	15	10	13	22

Although neither of these is completely straightforward (e.g. 16 ÷ 20 requires a simplification e.g. from $^{16}/_{20}$ to $^8/_{10}$, 59 ÷ 190 requires an appoximation to be made), nevertheless the results seem to indicate that very few children appreciate that decimals can be used to give the answer to a division of two whole numbers; indeed 60 per cent of 13 year olds consider that 16 cannot be divided by 20 (even on a paper headed 'decimals'). These results are considerably lower than those quoted in 3.8.1.5 which relate to fractions as the result of a division operation.

A further question included in the investigation by Brown was designed to find out whether children appreciated what decimals were in fact used for in everyday life (see part 2.1.8 on page 92). 35 per cent of 12 year olds, rising to 48 per cent of 15 year olds, were able to supply an appropriate context for using decimals (e.g. using lengths, weights, times etc.). Others gave totally inappropriate answers (e.g. involving 6.4 books) or, in some cases, indicated that they did not consider that decimals had any use in everyday life (see interview with Raymond reported in part 2.1.8, page 92).

3.8.1.8   *Understanding the Meaning of Percentages*

A percentage represents a third method of representing a rational number, although it is clearly closely related to the decimal system (e.g. 0.35 is equivalent to 35%).

As noted previously, the use of percentages is normally confined to the *operator* aspect of fractions i.e. 35 per cent is rarely encountered as a point on a number line in the *measurement* sense, but more frequently as expressing the relationship of a sub-area to a unit area, or subset to the size of the whole set, or of one number to another (e.g. 35 per cent of £80, or 35 per cent of a sample of 80 people).

The only reference found on the meaning of percentage notation was a conclusion of the APU that conversion of decimals to percentages and vice versa, in the case of familiar numbers only, was carried out correctly by about 50 per cent of 11 year olds (APU, 1980a).

Similarly there appears to be very little information relating to the concrete meanings of a percentage at an elementary level, except an indication that 70 per cent of 15 year olds, asked what percentage of the square shown in Fig. 3.61 was shaded were able to give the correct answer of 50 per cent.

Fig. 3.61

Other results on percentage incorporate more complex ideas like equivalence, which are now discussed.

3.8.2   **Understanding the Structure of Fractions, Decimals and Percentages**

3.8.2.1   *Equivalence of Fractions*

Many applications of fractions require the knowledge that each rational number can be represented by any member of a family of *equivalent* fractions: thus, for example, two-thirds can be represented by any of the set (⅔, 4/6, 6/9, 8/12,...).

This idea of equivalent fractions is needed in order to compare the sizes of different fractions and decimals (e.g. to distinguish which is the larger of 3/16 or ¼, or of 0.75 or

0.8), to convert between fractions, decimals and percentages, and in performing operations on fractions. (Ordering and conversion are discussed specifically in 3.8.2.2, equivalence of decimals and of percentages in 3.8.2.3, and operations in 3.8.3.) The use of a ratio (e.g. 6 to 8), a fraction or a decimal to compare two numbers also depends on the notion of equivalence.

As with many ideas in mathematics, equivalence can be appreciated in a concrete form by 5 year olds (e.g. half an apple is the same amount as two quarters). Nevertheless in a more abstract form (e.g. $4/14 = 10/?$), it is fully grasped by only a minority of 15 year olds.

The notion of equivalence is often introduced using one or both of the two concrete aspects of fractions which are generally accepted to be the easiest to grasp: the 'area' aspect and the 'set' aspect (see 3.8.1.2 and 3.8.1.3). For example to demonstrate that $2/3 = 4/6$, illustrations of the type shown here can be used:

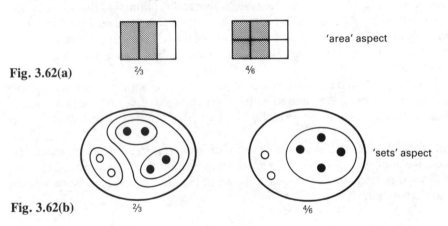

Fig. 3.62(a) $2/3$ $4/6$ 'area' aspect

Fig. 3.62(b) $2/3$ $4/6$ 'sets' aspect

These two approaches appear in questions in the CSMS study (Hart, 1980). The following table shows selected results relating to the equivalence of fractions (data from Hart, 1980).

	Item	Percentage successful	
		12 yrs	13 yrs
Shade in two-thirds:			
Fig. 3.63 (a)		61	66
Fig. 3.64 (b)		58	64

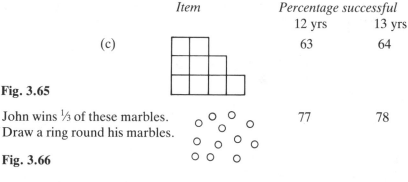

	Item	Percentage successful	
		12 yrs	13 yrs
(c)		63	64

Fig. 3.65

John wins ⅓ of these marbles.   77   78
Draw a ring round his marbles.

Fig. 3.66

Jane wins ⅔ of the marbles.
How many marbles does she win?...........   71   73

The above results may suggest that the 'sets' aspect may be an easier one than the 'area' aspect in which to introduce equivalence. However it is not clear whether the items are strictly comparable; the strategies used are probably different in the two cases.

A number of studies conducted in the United States and reviewed by Payne (1976) have included the aspect of equivalence. Many of these found that equivalence was a source of particular difficulty, and especially reduction to 'lowest terms' e.g. changing $\frac{8}{12}$ to $\frac{2}{3}$, even more than moving from $\frac{2}{3}$ to $\frac{8}{12}$ (e.g. Hinkleman, 1956; Bidwell, 1968; Anderson, 1966; Green, 1970).

Bohan (1971) compared three teaching sequences for their effectiveness in teaching the idea of equivalence to 11 year old children. Although much material was common, the sequences approached the idea from different points of view:

a.   the use of area ('region') diagrams, as discussed above
b.   the use of paperfolding related to area

    e.g.

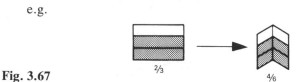

Fig. 3.67

c.   the use of multiplication by 1
     e.g. $\frac{2}{3} = \frac{2}{3} \times 1 = \frac{2}{3} \times \frac{2}{2} = \frac{4}{6}$

The 'paper-folding' sequence was found to be the most effective, both as regards learning and in producing favourable attitudes, and the more abstract numerical approach was least effective. However even after the 'paper-folding' sequence, less than half the children were successful in simplifying fractions to their lowest terms (e.g. going from $\frac{8}{12}$ to $\frac{2}{3}$), whereas 75 per cent were able to generate equivalent fractions using large numbers (e.g. going from $\frac{2}{3}$ to $\frac{8}{12}$).

A study by Coburn (1974) on 10 year olds compared the introduction of equivalent fractions through an 'area' model with their introduction through a 'ratio' model, (see 3.8.1.6), as shown in this figure:

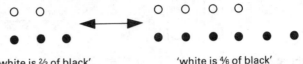

Fig. 3.68   'white is ⅔ of black'   'white is ⁴⁄₆ of black'

In both teaching sequences however the other aspect was covered later in the sequence. Coburn found the 'area' model to be generally more effective, especially in leading to addition and subtraction, but strangely enough both groups found questions on the ratio aspect easier to cope with than those on the regions aspect, which perhaps supports the CSMS results given above.

Steffe and Parr (1968) had previously found the opposite result to Coburn i.e. that introducing the ratio aspect first was most effective, but Coburn attributed this to the problems children had in actually drawing the regions.

Thus research evidence is by no means conclusive as to the best way of introducing the idea of equivalence. It may even be that, in a calculator age in which decimals are emphasised, a more useful approach in the long term might be to link the 'division' and 'number line' aspects (see 3.8.1.4 and 3.8.1.5) in using calculators to demonstrate that, for example,

$$⅔ = 2 ÷ 3 = 0.6666....$$

and

$$⁴⁄₆ = 4 ÷ 6 = 0.6666....$$

Introductory experience of equivalence with concrete aspects of fractions later leads to the ability to manipulate fractions in a symbolic form.

There is evidence that in reasonably straightforward cases most children can do this successfully. For example Hart reports the results shown in the following table as part of the CSMS study:

Selected results on equivalence of fractions (data from Hart, 1980)

Item	*Percentage successful*			
	12 yrs	13 yrs	14 yrs	15 yrs
⅓ = ²⁄?	72	77	78	79
⁴⁄₁₂ = ¹⁄?	56	52	61	63
²⁄? = ⁵⁄₁₄	57	55	63	74

These percentages are very similar to those obtained in comparable surveys in America (NAEP, 1980) and Britain (APU, 1980a,b).

It is however difficult to tell whether children obtaining correct answers in these numerical items are demonstrating any real understanding of equivalence as opposed to 'pattern-spotting'. For example, Hart quotes responses like:

'²⁄? = ¹⁰⁄₁₅ because 2 is 5 less than 7 and we want a number 5 less than 15' (Hart, 1981).

Also the results of an item in a national survey in America (NAEP, 1980) suggests that there may be considerable underlying confusion about the notion of equivalence. See the next table:

Responses to item on equivalence (NAEP, 1980)

	Percentage responding		
Item	9 yrs	11 yrs	17 yrs
Suppose x/y represents a number. If the values of x and y are each doubled, the new number is			
(a) ½ as large as x/y	17	10	8
(b) equal to x/y	15	18	41
(c) double x/y	47	65	46
(d) I don't know	21	7	5

Also, in a more demanding situation in which the spotting of simple number problems is insufficient, the percentage of children able to cope appears to fall steeply. For example in the last part of the CSMS item

$$\tfrac{2}{7} = \tfrac{\square}{14} = \tfrac{10}{\triangle}$$

the success rate varied from only 24 per cent of 12 year olds to 31 per cent of 15 year olds (Hart, 1980).

Verbal examples may also provide a better test of equivalence ideas. Some examples which relate to various different aspects in which fractions are applied are given in the next table:

Verbal items testing the concept of equivalence (data from Hart, 1980, pp. 217–240; Ward, 1979, p.96)

Item	*Percentage successful*
'Part-whole' aspect (¼ = 5⁄20)	
Two boys have the same amount of pocket money. One decides to save ¼ of his pocket money; the other decides to save 5⁄20 of his pocket money. Tick the answer you think correct:	
(a) 5⁄20 is more than ¼	76 per cent of 12 year olds
(b) ¼ is more than 5⁄20	rising to 84 per cent of 15
(c) 5⁄20 and ¼ are the same	year olds.

(Hart, 1980)

'Set' aspect (⅗ = 24⁄40)

There are 40 children in a class. ⅗ of them are girls. How many girls are there in the class?	22 per cent of 11 year olds.

(Ward, 1979)

| *Item* | *Percentage successful* |

'Number-line' aspect ($\frac{2}{8} = \frac{1}{4}$ or $\frac{6}{8} = \frac{3}{4}$)

A relay race is run in stages of $\frac{1}{8}$ km each.
Each runner runs one stage. How many runners   48 per cent of 12 year olds
would be required to run a total distance of   rising to 57 per cent of 15
$\frac{3}{4}$ km?..............   year olds.

(Hart, 1980)

It is again difficult to draw detailed conclusions from the above results; in particular the strategies used by children may be very different in each case.

The following text concerns applications of the concept of equivalence to ordering fractions, and conversions between fractions, decimals and percentages.

3.8.2.2   *Applying Equivalence Ideas to Ordering Fractions and Converting from Fractions to Decimals and Percentages*

The idea of equivalence is also important in the ability to order fractions (and ratios). It has already been seen (Section 3.8.1.4) that children do not always readily appreciate that fractions are numbers, and as such fill in some of the gaps in the number-line between whole numbers. It may thus not be clear to them that, given two fractions, either they are equivalent and hence represent the same number, or that one of them represents a larger number than the other. Except in simple cases (e.g. comparing $\frac{3}{8}$ with $\frac{5}{8}$, or $\frac{1}{3}$ with $\frac{1}{5}$) the mechanism for this comparison is normally to find appropriate equivalent forms for one or both of the fractions. For example, $\frac{5}{8}$ can be seen to be smaller than $\frac{3}{4}$ if $\frac{3}{4}$ is converted to $\frac{6}{8}$. However to compare $\frac{5}{8}$ with $\frac{2}{3}$ it is necessary to either convert into equivalents to arrive at $\frac{15}{24}$ and $\frac{16}{24}$ respectively, or to use some other equivalence-based informal reasoning e.g. '$\frac{5}{8}$ is $\frac{3}{8}$ less than 1, whereas $\frac{2}{3}$ is $\frac{1}{3}$ less; $\frac{1}{3}$ is $\frac{3}{9}$ which must be smaller than $\frac{3}{8}$, hence $\frac{5}{8}$ is smaller than $\frac{2}{3}$'.

It is thus clear that the difficulty of comparing two fractions can vary greatly depending on the numbers involved in the numerators and denominators.

For example, Hart (1980) noted that 66 per cent of 15 year olds could recognise that $\frac{3}{10}$ was larger than $\frac{1}{5}$, whereas in an American Survey (NAEP) only 3 per cent of 13 year olds were able to decide which, of $\frac{1}{4}$, $\frac{5}{32}$, $\frac{5}{16}$, $\frac{3}{8}$, was nearest to $\frac{3}{16}$.

Noelting (1978, 1980), working in Quebec designed an extremely thorough experiment to examine the relative difficulty of comparing different ratios. He used the context of mixing orange juice, for example asking children to choose which of the combinations in Fig. 3.69 will produce the strongest mixture:

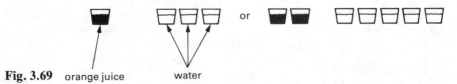

Fig. 3.69   orange juice   water

He found that a number of informal methods were used. Questions like that illustrated in Fig. 3.69 in which the number of orange juice (or water) units on one side is a simple multiple of the number of orange juice (or water) units on the other were first answered by the average child at age $12\frac{1}{2}$. However some children in the sample were able to do this at age 8, while others had not reached this level at 16 years. (This task is equivalent to comparing the fractions $\frac{1}{3}$ and $\frac{2}{5}$, or $\frac{1}{4}$ and $\frac{2}{7}$, depending on whether the quantity of orange juice is seen in relation to that of water, or to the whole volume). More difficult were situations in which there was no simple relationship, e.g. that illustrated here:

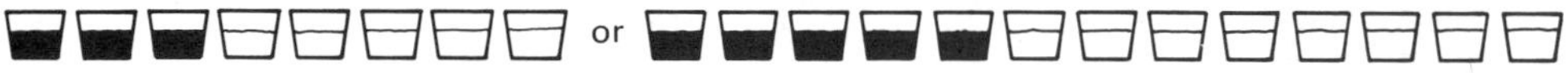

Fig. 3.70

which is equivalent to a comparison between $\frac{3}{5}$ and $\frac{5}{8}$ (or $\frac{3}{8}$ and $\frac{5}{13}$ taking all the units in each side). In these cases the 'average' child could only solve the problem at 17 to 18 years of age, although one of the thirty-two 10 year olds managed to cope with tasks of this type.

Noelting quotes examples of children successfully using both fractions and percentages to solve these problems e.g:

Fig. 3.71

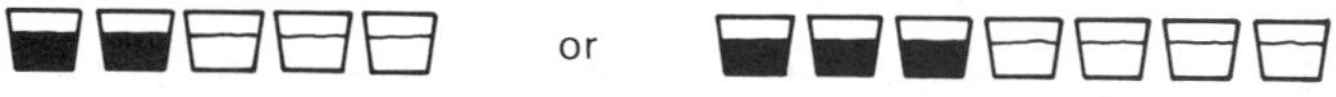

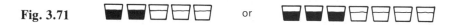

Sylvie (14 years): 'At the right, there is $\frac{3}{7}$ of juice for $\frac{4}{7}$ of water, that is $\frac{15}{35}$ of juice; at the left there is only $\frac{14}{35}$ juice.'

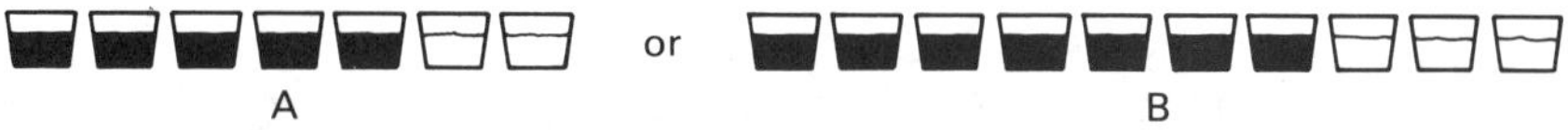

Fig. 3.72

Réjean (13 years): 'A = $71\frac{3}{7}$% because $\frac{5}{7}$ orange juice;
B = 70% because $\frac{7}{10}$ orange juice'

Children who failed the tasks were more likely to compare the total number of glasses of water (or orange juice) e.g.
Diane (8 years): 'It is that there are less glasses of water' or to look at the differences between the number of glasses of water and orange juice instead of at their ratio e.g.
Louise (11 years): 'Because the left side has one glass of water more, while the right side has two of them more' (Noelting, 1978).

Thus Noelting confirmed that the comparison of ratios, or fractions, can range enormously in difficulty depending on the relationship of the numbers.

A related use of equivalence is in finding a fraction between two other fractions e.g. between $\frac{1}{2}$ and $\frac{2}{3}$. In this particular case, which requires conversion of each of these fractions to appropriate equivalent forms (e.g. $\frac{6}{12}$ and $\frac{8}{12}$, leading to the choice of $\frac{7}{12}$ as an intermediate fraction), Hart (1980) found that only 21 per cent of 15 year olds were successful (although only 12 per cent gave $\frac{7}{12}$). Perhaps a more serious point is that it was not generally recognised that it *is* always possible to find more fractions on the number line between any two fractions.

'How many fractions lie between ¼ and ½?'

Hart found that the replies in the case of 15 year olds were as in the following table:

Results for item 'How many fractions lie between ¼ and ½?' (Data from Hart, 1980)

Response	Percentage (15 yrs)
Infinitely many, lots, etc.	16
One	30
A number between 1 and 20	22
Other answers	15
Omit	17

These percentages are remarkably similar to those obtained by Brown (1981b) when children were asked:

'How many different numbers could you write down which lie between 0.41 and 0.42?'

Thus it appears that very few children even at 15 years have a picture of the number line being packed with an infinite number of rational numbers, whether expressed as fractions or decimals, between any two whole numbers.

A final application of equivalent fractions is in converting from fractions to decimals or percentages, particularly in simple cases like ¼, ⅗ which can be readily converted into ²⁵⁄₁₀₀ (0.25 or 25%), or ⁶⁄₁₀ (0.6 or 60%) respectively. There is very little evidence on children's ability to do this, except that the APU found in their first survey of 11 year olds that:

50% of pupils correctly wrote the fraction ¼ as a percentage while about 25% knew its decimal equivalent ...Facilities *(i.e. percentages successful)* of just over 40% were obtained for the conversion of tenths to decimals and less for converting hundredths to decimals (30%). (APU, 1980a)

The notion of equivalence as it applies to decimals and percentages is now considered.

3.8.2.3   Equivalence of Decimals and Percentages

Decimals   Since the decimal representation of number is based on the concept of fractions (tenths, hundredths, and so on), the understanding of the idea of equivalence is just as important in using decimals as using fractions although the place value notation makes the notion of equivalence less explicit.

For example, 0.21 may be thought of as 'two tenths and one hundredth' according to a strict place value definition. However in comparing 0.21 with, say, 0.07, it is more helpful to regard 0.21 as also equivalent to 21 hundredths, which is more readily seen as either 3 times as much as 7 hundredths, or 14 hundredths (0.14) more than 7 hundredths. The ability to see a number like 0.21 as *either* 2 tenths and 1 hundredth *or* 21 hundredths depends on an intuitive acceptance that $²⁄₁₀ = ²⁰⁄₁₀₀$ (although this formal symbolism may not be used by the child), which again rests on the notion of equivalence of fractions.

Brown quotes an example of an 11 year old child who is on the brink of learning to use the idea of equivalence:

> Frances had given as the answer to 'Is there a difference between 4.90 and 4.9?' 'Yes, 4.90 is more', but in the next question decided that 0.8 was bigger than 0.75 because: 'Oh, it is eight tenths which equals 80 hundredths.' This was said with great satisfaction, as if somehow it had all come together. (Brown, 1981b)

She gives a further example of a 14 year old who is able to call on this idea to reason out the answer to the following simple multiplication which was set in a non -standard form:

$$\boxed{\text{Multiply by 10}}$$
$$5.13 \; \rightarrow \; \dots..$$

Billy: ...(pause)...'You can't put a nought on the end of there as it's a decimal'...(long pause)...
Interviewer: How big, roughly...?'
B: '50 ... 51.3'
I: 'How...?'
B: 'I multiplied that (5) by ten first, and then that one (1)... ten tenths are one... and then multiplied that (3) by ten and put it in three tenths.'

...(Billy and Frances) seemed to be at a similar transitional stage of development which seemed to be common among secondary school children. (Brown, 1981b)

Results of several survey items are quoted in the next table to back up this conclusion:

Results and items which involve the notion of decimal equivalence (data from Brown, 1981c)

		Percentage responding			
Item	*Answer*	12 yrs	13 yrs	14 yrs	15 yrs
Ring the bigger of the numbers 0.75 or 0.8 Why is it bigger?..............	0.8	51	65	69	75
Write down any number between 0.4 and 0.5..............	(correct)	42	54	71	75
0.41 and 0.42..............	(correct)	37	49	66	71
Ring the number nearest in size to: $\boxed{0.18}$ → 0.1/10/0.2/20/0/1/2	0.2	44	48	61	59
$\boxed{\text{Multiply by 10}}$ 5.13→......	51.3	37	42	58	65
$\boxed{\text{Add one tenth}}$ 2.9→......	3.0 or 3	38	44	51	59

Item	Answer	Percentage responding			
		12 yrs	13 yrs	14 yrs	15 yrs

Item	Answer	12 yrs	13 yrs	14 yrs	15 yrs
This number is	3.2 (3.1)	23 (48)	37 (42)	50 (33)	58 (21)
Four tenths is the same as ..... hundredths	40	28	31	42	40
Six tenths as a decimal is *0.6*. How would you write a a decimal: eleven thousandths.....	0.011 (0.0011)	30 (17)	33 (24)	33 (28)	31 (36)
eleven tenths............	1.1 (.11)	28 (49)	35 (47)	36 (48)	34 (45)

These results are supported by data from the APU (1980a). For example 23 per cent of 11 year olds were able to correctly place the following decimals in order of size, with the smallest first:

$$0.07 \qquad 0.23 \qquad 0.1$$

Similarly only about 35 per cent of 15 year olds could select the smallest number from:

$$0.625 \qquad 0.25 \qquad 0.375 \qquad 0.125 \qquad 0.5$$

    The above items include many aspects which are extremely important in applying decimals; for example approximation, reading of scales, comparing the sizes of different numbers expressed as decimals, and so on. All these aspects require an understanding of the idea of equivalence of decimals, which is in turn based on the notion of equivalence of fractions. However it is clear from the above results that the ability to view a given decimal flexibly, using ideas of equivalence, is not something which springs naturally from the way in which decimals are defined, but which is learned over a long period, and which is harder to develop in some contexts than in others.

Percentages Equivalence ideas are also implicit in almost all applications of percentages. In fact most applications relate to the 'subset' or 'ratio' aspects of fractions e.g.

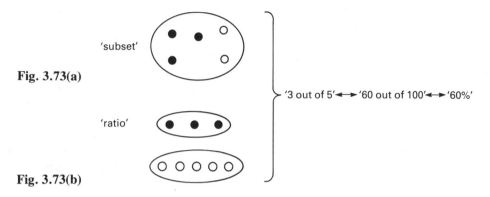

'subset'

Fig. 3.73(a)

'3 out of 5' ←→ '60 out of 100' ←→ '60%'

'ratio'

Fig. 3.73(b)

In view of the obvious importance that the understanding of percentages has in everyday life and commercial activity (e.g. wage rises, unemployment rates, mortgage rates, etc.) there has been relatively little research relating to the ability of children to use them. A survey of adults commissioned by the Cockcroft Committee of Enquiry (Sewell, 1981; ACACE, 1982), revealed a widespread inability to understand percentages. Almost a third of a representative sample of about 100 adults

indicated, sometimes indirectly, that *all* percentages were meaningless: 'They're beyond me', 'You've got me there' and 'I'm completely lost with percentages' were typical replies (Sewell, 1981).

When a smaller sample of 50 adults were asked to calculate 15 per cent of £60, 32 obtained correct answers but there was a great variety of methods used. These included:

	Number of People	Method
Mental	15	10% + 5%
	2	scaling down 15 per 100
	1	6×15
Paper	7	60×15
	4	$^{15}/_{100} \times 60$
	2	10% + 5%
Calculator	1	
Wrong	1	90p (mentally)
	1	15p in the pound, giving 4.50
	1	$\begin{array}{r} \times\ 15 \\ \underline{60} \\ 65 \\ \underline{350} \\ 415 \end{array}$
Unsuccessful	1	$\frac{\cancel{60}^{3}}{\cancel{100}_{\,5}} \times \frac{1}{\cancel{15}} = 3 \times 11 = 33$ (faulty cancelling)

(Sewell, 1981)

This prevalence of informal methods suggests a considerable degree of understanding. It was not clear to what extent the results would be similar if less straightforward figures had been chosen, although comments included:

'That's easy but I couldn't do 12%.'
'I can do 10% or 15% but when it comes to 8½% I'm stuck.'

The APU (1980b) report that about half of the 15 year old sample were successful with each of the two straightforward percentage items. The first item required the new cost of a suit to be calculated given the original price and a 30 per cent reduction; the second gave a number and its percentage of a total number, and asked for the total to be calculated. (Neither item was released). At 11 years old, only about 15 per cent could work out what percentage 50 is of 250, and on no percentage question were more than a third of the sample successful (APU, 1980a). This latter item requires the equivalence $^{50}/_{250} = ^{?}/_{100}$, but seems to be rather harder put in the context of a percentage than many of the items reported on equivalence of fractions, thus suggesting that perhaps it is the idea of a percentage itself which is not understood.

Hart reports results on three items dealing with straightforward percentage usages, each of which depend on the idea of equivalence:

Results on percentage items (data from Hart, 1981)

Item	Percentage successful		
	13 yrs	14 yrs	15 yrs
% means *per cent* or *per 100* so 3% is 3 out of every 100.			
6% of children in a school have free dinners. There are 250 children in the school. How many children have free dinners?...............	37	46	58
The newspaper says that 24 out of 800 Avenger cars have a faulty engine. What percentage is this?..............	32	40	48
The price of a coat is £20. In the sale it is reduced by 5%. How much does it now cost?...............	20	27	35

The extra difficulty of the final item may be due to the fact that it involves an extra step of subtraction after the 5 per cent of £20 has been calculated.

Hart reports than many children attempted to manipulate the numbers in any way which would produce a small decrease; thus in the final example 41 per cent of children either subtracted 5 from 20 to obtain £15, or subtracted one fifth of 20 to obtain £16. Even when children did attempt to involve the number 100, this seemed sometimes to be done without any understanding e.g. in the second part, 192 was obtained by around 10 per cent of children by dividing 800 by 100 and multiplying by 24.

The above results show once again that a minority of children in secondary schools really understand the principle of equivalence of fractions, including equivalence arising in the context of ratios, decimals and percentages. The problem appears to be one of bridging the gap between concrete examples involving, for example, paper-folding and shading, which appear to be reasonably successful, and knowing how to proceed in a problem which just involves numbers, especially when these numbers are not easily related to each other.

3.8.3 Operations on Fractions and Decimals

3.8.3.1 Operations on Fractions – 'Meaning' and 'Computation'

In part 3.8.1.3, page 281, it was noted that when all the various aspects of the meaning of fractions are considered, they fall into two major categories:

  a. fractions in *measurement*;
  b. fractions as *operators*.

The *measurement* use (e.g. ¾ hour, 1⅔ bucketfuls) involves a fraction as expressing part of a physical unit. This relates closely to the 'area' and 'number line' aspects detailed earlier in parts 3.8.1.2 and 3.8.1.4.

The *operator* use (e.g. ¾ of 12) involves a comparison of two numbers, one being expressed as a fraction of another. This relates closely to the 'subset' and 'ratio' aspects described in parts 3.8.1.3 and 3.8.1.6.

One of the problems with operations on fractions, as is pointed out in parts 3.8.3.2 and 3.8.3.3, is that *addition* and *subtraction* are most easily related to the *measurement* aspect (e.g. 1½ hours + ¾ hour), while *multiplication* and *division* are most meaningfully understood and applied in terms of the *operator* aspect (e.g. ¼ of ⅔ of 24). This means that if only one concrete model of fractions is emphasised, then this is likely to prove unsatisfactory for serving to illustrate either addition/subtraction or multiplication/division, depending on which one is chosen.

In both types of operation, the distinction needs to be clearly made between understanding the *meaning* of, say, ¾ + ½ or ¾ × ½, which relates to an ability to apply such operations to solve realistic problems, and the ability to actually carry out the *computations* successfully. That these two abilities are to some extent separate is illustrated by Hart, who set parallel 'problems' and 'computations' as part of the CSMS testing.

e.g. *Problem*   *Computation*
                                          ⅙ of ¾ = ...

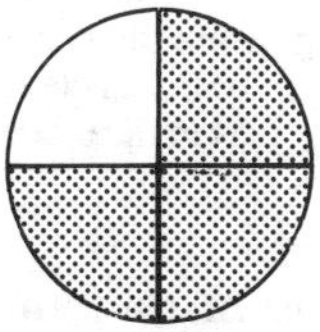

Fig. 3.74

Shade in ⅙ of the dotted section of the disc.
What fraction of the whole disc have you shaded?

It was noted that:

> The ability to solve addition and subtraction computations *declines* as the child gets older. The ability to solve the problems *does not decrease* with age and one is left with the hypothesis that the problems are solved without recourse to the computational algorithms. Many children do not in fact seem to connect the algorithms with the problem solving and use their own methods... (Hart, 1981)

Hasemann (1980, 1981) combined some of the CSMS questions with further items and tested them with 13 year old German students in the lowest half of the ability range. He observed that:

> ...we learned from informal interviews with the students that many of them are not unduly concerned at getting different aswers from using different methods of solving the problem (as here by calculation and by a diagram)... (Hasemann, 1980)

He cites several examples, including one of a 13 year old who gave, as the answer to ¼ + ⅙, the following:

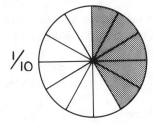

Fig. 3.75

Thus the student had used a wrong method of calculation, reasoning that ¼ + ⅙ = ¹⁄₁₀, but had correctly shaded in ¼ + ⅙ of the circle by shading ¼ (³⁄₁₂) and then ⅙ (²⁄₁₂) without apparently realising the discrepancy between the two answers.

Similar results were found on practical tests with 15 year olds by the APU. Here also:

> Testers commented that very few pupils seemed familiar with pictorial representation of operations with fractions (APU, 1981b).

It seems likely that computational procedures for manipulating fractions are thus introduced rather prematurely and before a sufficient concrete foundation is laid.

Hasemann also noted of his lower ability 13 year olds:

> ...our students use, if it is at all possible, algorithms and rules which they have learnt; and they are certain that they will get the right solution for their problem (even when the algorithm they have used is quite unsuitable for that problem). Most of them are also not in a position to explain why a rule is so and not otherwise .... In other words, the rules and algorithms seem to play a too large part in the mathematical thinking of the poorer students... (Hasemann, 1980)

Even with this concentration on computational rules, Hasemann found that only 19 per cent of his sample could correctly add ⅙ + ¼, which suggests once again that an over-emphasis on 'computation' at the expense of 'meaning' may be counter-productive.

Suydam (1978), in a comprehensive review of previous research on fractions and ratio, quotes the similar conclusions of Carpenter et al. (1978) on the poor performance on operations with fractions reported in the American National Survey (NAEP):

> Results indicate that many students have little computational skill with fractions and probably little conceptual understanding. An increase in the amount of time spent on operations with fractions is not necessarily an appropriate remedy. The development of algorithms should be placed so as to connect firmly with the main ideas in the initial development. (Carpenter et al., 1978, as quoted in Suydam, 1978)

Bright (1978), also quoted in Suydam (1978), suggests that the relationship between 'meaning' and 'computation' may be rather different in the cases of addition and multiplication:

> It should be noted in passing that the multiplication algorithm, because it can be performed without writing the fractions with a common denominator, is mechanically an easy one. Conceptually, however, multiplying fractions is more complex than adding them... this hierarchy of difficulty is supported empirically. Apparently the mechanical difficulties of adding fractions without a common denominator outweigh the conceptual difficulties of multiplication. (Bright, 1978, as quoted in Suydam, 1978)

The following text deals with points specific to addition/subtraction (part 3.8.3.2) and multiplication/division (part 3.8.3.3) of fractions.

3.8.3.2   Addition and Subtraction of Fractions

The applications of addition and subtraction of fractions are mainly in the field of *measurement* e.g. working out how long it takes to prepare and cook a dish if the preparation is estimated to take ½ hour and the cooking time is 1¾ hours. However to an increasing extent measurement is on a metric scale in which it is conventional to use decimal fractions for recording e.g. 1.25 metres rather than 1¼ metres.

It is therefore important to question the relevance for practical purposes of teaching anything other than informal methods of adding common related fractions (e.g. sixteenths, eighths, quarters and halves; maybe also thirds and sixths; or fifths and tenths.) Indeed Freudenthal (1973) has argued that the process of adding fractions should not be introduced until children are able to view the rational numbers abstractly as an example of a formal number system with specific algebraic properties; it seems likely that this would involve a delay until late in the secondary school for most children, and many might not reach it at all.

The 'meaning' of addition and subtraction is traditionally developed using an 'area' representation e.g.

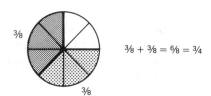

$$ \tfrac{3}{8} + \tfrac{3}{8} = \tfrac{6}{8} = \tfrac{3}{4} $$

Fig. 3.76

However if the child decides to represent the two fractions to be added on separate diagrams, this may cause problems, e.g.

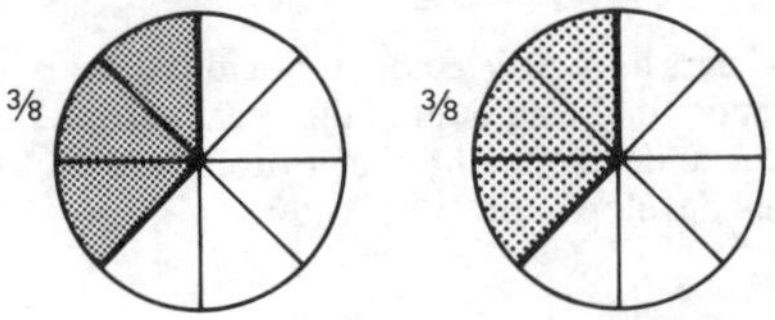

Fig. 3.77

the child may interpret the answer in relation to the whole area of both circles and therefore arrive at $\frac{6}{16}$, rather than $\frac{6}{8}$. (This could even be used in this case to 'justify' the common mistake of adding numerators and denominators separately.)

Similar problems occur when the total exceeds a whole unit e.g. $\frac{5}{8} + \frac{7}{8}$ can be represented as in Fig. 3.78.

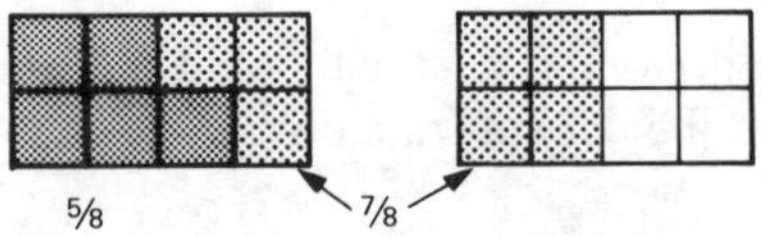

Fig. 3.78

but may easily be interpreted as $\frac{12}{16}$ rather than $1\frac{4}{8}$ or $1\frac{1}{2}$.

The number line model, although initially more difficult to grasp, does avoid these problems; for example here $\frac{5}{8} + \frac{7}{8}$ can be illustrated:

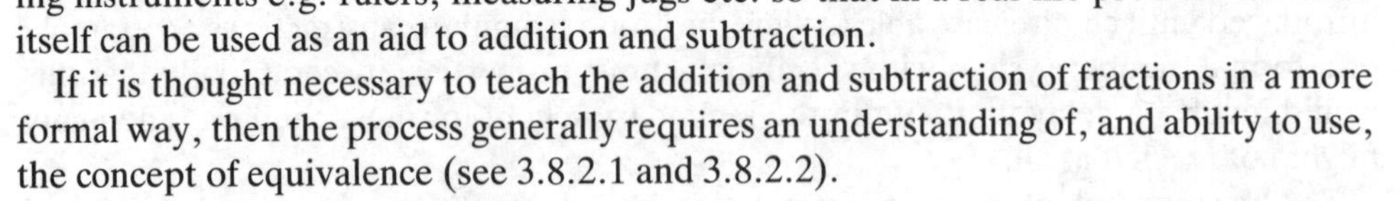

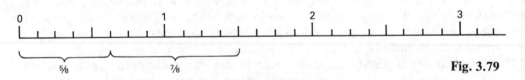

Fig. 3.79

This also has the advantage of relating immediately to the scales shown on most measuring instruments e.g. rulers, measuring jugs etc. so that in a real-life problem the scale itself can be used as an aid to addition and subtraction.

If it is thought necessary to teach the addition and subtraction of fractions in a more formal way, then the process generally requires an understanding of, and ability to use, the concept of equivalence (see 3.8.2.1 and 3.8.2.2).

e.g. $\frac{2}{3} + \frac{1}{6} = \frac{4}{6} + \frac{1}{6} = \frac{5}{6}$
$\frac{2}{3} + \frac{1}{4} = \frac{8}{12} + \frac{3}{12} = \frac{11}{12}$

The results reported in sections 3.8.2.1 and 3.8.2.2 suggest that many children find the idea of equivalence difficult unless a concrete context (e.g. a diagram) is provided. Many children may therefore be taught the procedures for adding fractions by rote, with little basis of understanding. However survey results for performance in adding fractions are generally poor, suggesting that this is not a reliable strategy.

For example Lankford (1972) in a comprehensive study on computation with fractions with 176 13 year olds in the United States, found that of the possible answers, 35 per cent were right, 33 per cent were wrong, and 32 per cent were omitted. He gives details of the most common errors, including in particular the 'adding numerators and adding denominators' strategy for addition.

e.g. $\frac{1}{3} + \frac{1}{4} = \frac{2}{7}$

This strategy has also been well-documented elsewhere; for example Hart (1981) found 30 per cent of English 13 year olds making this error in the above example. APU note that in the case of 11 year olds,

> more complex questions had facilities around 35%, with about 20% adding numerators and denominators... although success rates at the secondary level were around 10–15% higher than among primary pupils, the strategy of adding numerators and denominators was almost as widely used by 15 year olds as with 11 year olds. (APU, 1980b)

It is clear that the more complex the example becomes, the higher the frequency becomes of this type of error. An example of this is given by the APU results for 11 year olds, as shown in the following table.

Selected results in addition and subtraction of fractions for 11 year old sample (data from APU, 1981a)

Item	Correct	Omitting item	Adding numerators and denominators
	%	%	%
$\frac{6}{10} + \frac{3}{10}$	72	2	8
$\frac{2}{5} + \frac{3}{5}$	60	7	13
$\frac{1}{4} + \frac{1}{2}$	58	5	13
$\frac{3}{4} - \frac{1}{2}$	48	18	12
$\frac{8}{9} - \frac{1}{3}$	31	19	17
$\frac{1}{2} + \frac{1}{3}$	33	9	24

The APU also report other wrong strategies; for example

> a few pupils cancelled before attempting to add: $\frac{2}{4} + \frac{2}{7}$ or $\frac{3}{4} + \frac{4}{3}$ or both.

(APU, 1980b)

These results suggest a high degree of confusion. As Brown, J.S. and Vanlehn (1982) propose (see p.261), it seems likely that a majority of pupils forget part of the taught procedures, and attempt to 'repair' them. Since there is in many cases, and especially the more abstract examples, little conceptual understanding to provide guidance, the 'repairs' often contain steps which are inconsistent with the mathematical structure and hence lead to errors.

The likelihood of this happening is illustrated by work by Brueckner and Grosssnickle (1947) who identified 160 different types of question on addition of fractions, depending on factors like the presence of mixed numbers, the relationships of the denominators, and so on. It is perhaps not surprising that children taught by rote find this variety confusing.

Erlwanger gives insight into what happens with individual children by reporting on the case of 'Benny'. Benny has evolved a number of incorrect personal 'procedures' for coping with different types of questions involving fractions, which he confidently applies and is convinced that he has been taught. His view of mathematics becomes clear in the following excerpt:

> B: In fractions we have 100 different kinds of rule…
> E: Would you be able to say the 100 rules?
> B: Yes…maybe, but not all of them.
>
> (Erlwanger, 1973)

This illustrates the point made by Ginsburg that

> Errors are seldom capricious or random. Often children think of mathematics as an isolated game with peculiar sets of rules and no evident relation to reality. (Ginsburg, 1977).

3.8.3.3   Multiplication and Division of Fractions

Evidence was given earlier in the context of whole numbers (see page 236) that multiplication is the most difficult of the four operations to which to attach a concrete meaning. In particular, the majority of 12 year olds were found to be somewhat limited in their grasp of the meaning of multiplication even with whole numbers. However in the case of fractions, the problem is likely to be that much greater. For, as Hart notes:

> The meaning of multiplication is firmly rooted in the child's experience of whole numbers where the operation can always be replaced by repeated addition. If the child sees 4×3 as four groups of three objects (which can be spaced out and counted), the meaning he attaches to $\frac{1}{3} \times \frac{6}{7}$ is unclear… (Hart, 1981)

One concrete picture of multiplication of fractions which is sometimes used in teaching is that of the product of two fractions as represented by the area of a rectangle:

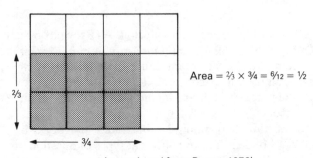

Area $= \frac{2}{3} \times \frac{3}{4} = \frac{6}{12} = \frac{1}{2}$

Fig. 3.80

(reproduced from Payne, 1976)

Green (1969), as reported in Payne (1976) compared the introduction of multiplication using this 'product as area' approach with an approach combining the operator idea (e.g. '⅔ of 5') with the 'sub-area as part of a whole area' model as illustrated in Fig. 3.81.

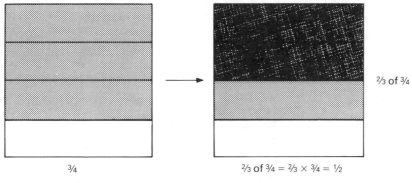

⅔ of ¾

¾

⅔ of ¾ = ⅔ × ¾ = ½

Fig. 3.81 (reproduced from Payne, 1976)

It is clear that although this 'operator'/'area' approach works reasonably well in the example given in Fig. 3.81, it is less easy to carry out in the example ⅔ × ⁴/₅, in which case presumably the resulting diagram would resemble Fig. 3.82 (although this is not made clear in the reference).

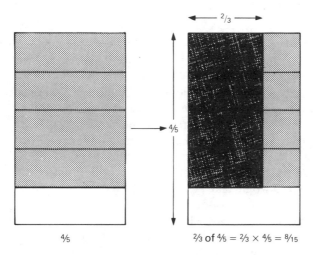

Fig. 3.82 ⁴/₅ ⅔ of ⁴/₅ = ⅔ × ⁴/₅ = ⁸/₁₅

Green (1970) found, with a sample of 481 11 year olds in Michigan, that both aspects were equally successful with the high achievement group, but with middle and lower achievement groups the 'product as area' model as shown in Fig. 3.80 was more effective than the 'operator/area' model of Figs. 3.81 and 3.82. However both groups experienced difficulty with:

a.   finding a fractional part of a set,

b. multiplying a fractional number by a whole number,
c. multiplying fractions given as mixed numbers e.g. 1¾ × 2⅓.

Although Green reports some success in teaching the 'product as area' approach with 11 year olds, there must be some doubt whether this is effective in the long term. For example Hart used the following item with a representative sample of English 14 and 15 year olds:

What fraction of the square
metre is shaded?........

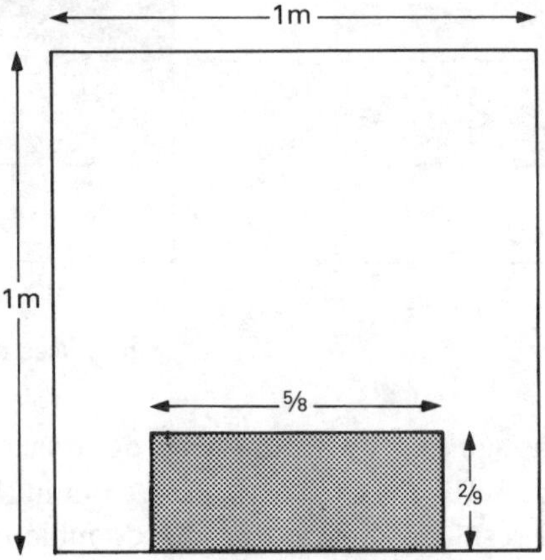

Fig. 3.83   (reproduced from Hart, 1980)

Only 13 per cent of 14 year olds and 19 per cent of 15 year olds obtained a correct answer (involving ⁵⁄₃₆ or ¹⁰⁄₇₂), although just under 50 per cent in each case were able to obtain the correct answer to a comparable computation (⅔ × ⅝).

Even though the 'product as area' model was found to be more successful than an 'operator' model by Green, perhaps partly because it avoids the difficult transition from 'of' to '×', it does not necessarily mean that it is the best model to teach. This is because very few direct applications of multiplication of fractions are easily translated into the idea of finding areas by multiplying fractional lengths; as noted in 3.8.3.1, applications of multiplication of fractions are more likely to occur in terms of ratio. For example, working out how long it would take to walk 7 miles if it takes 3½ hours to walk 10 miles requires the use of ⁷⁄₁₀ as an 'operator' i.e. ⁷⁄₁₀ of 3½.

Another everyday example, used by Sewell with a sample of adults, relates to a genuine jam recipe using 3½lb of sugar for 4lb of strawberries. The problem was to determine the amount of sugar if only 3lb of strawberries had been bought. Here again the 'operator' approach is the most natural one; as one person put it:

'I need ¾ of 3½. I'd get my son to do it on his calculator.'

In fact, out of a sample of 50 adults, Sewell found that:

only 3 solved the problem by multiplying fractions
    (1 mentally, 2 on paper);

6 used a method involving finding a rate
    (e.g. dividing by 4 to obtain the sugar per pound and then multiplying by 3);
1 used a calculator;
12 gave a reasonable estimate (between 2¼ and 3lb);
1 gave a wrong answer;
27 could make no attempt.

(Sewell, 1981)

This supports a finding by Hart (1980) using a similar 'ratio-type' problem with more complex fractions (⅔ × ⅝), that under 10 per cent of 14 to 15 year olds were able to obtain a correct answer, although as reported earlier, about 50 per cent were able to do the same operation when it was presented symbolically out of context.

This suggests that the concept of multiplication of fractions is not widely used or understood, and indeed may not be necessary in many real-life problems when the process can be avoided by finding a 'rate' by successively dividing and multiplying by whole numbers. This is in any case the easiest method to employ with a standard calculator.

A possible way of making the operation of multiplication of fractions more concrete using the 'operator' idea which leads naturally into ratio is that due to Dienes (1971). He presents fractions from the beginning as 'operators', so that ¾ is, for example, seen as a '3 for 4' operator e.g. '3 boxes for every 4 children', or '3 green counters for every 4 blue counters', or, more abstractly, '3cm for every 4cm'.

This basically introduces fractions using an approach similar to that of 'subsets', since ¾ of 12 is solved by dividing 12 into groups of 4, and giving '3 for every 4' as in Fig. 3.84.

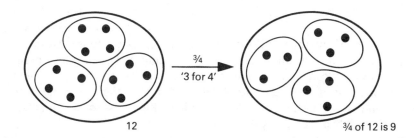

If the effect of two operators, one after the other, is explored then we have, say:

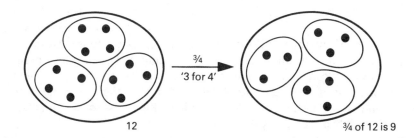

this can be expressed as ⅗ of 2/7 of 35, or ⅗ × 2/7 × 35.
However the two operators can be replaced by the single operator '6 for 35'

e.g.

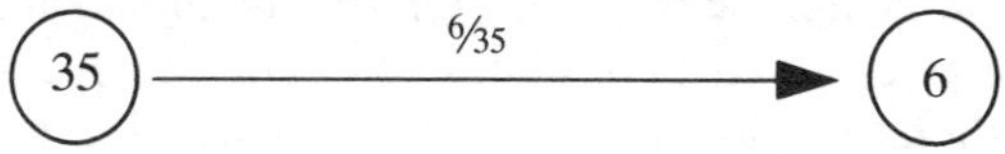

Hence '⅗ of ²⁄₇' is equivalent to '⁶⁄₃₅'
     i.e.   ⅗ × ²⁄₇ = ⁶⁄₃₅

While this appears to be a rather complex approach it can be made into a concrete activity, as suggested by Dienes (1971), and does, unlike the 'area' approach, readily translate into the use of multiplication of fractions for enlargement, and rate and ratio, at a more abstract level.

Floyd (1980) also proposes an operator approach going via

    '¼ of ⅓ of 24 counters is 2 counters'
    to '¼ of ⅓ = ¹⁄₁₂'.

APU (1980a) results suggest that children find the operator idea more straightforward than multiplication when a fraction is operating on a whole number. For example items of the form '¹⁄ᵦ of c', with b and c particular whole numbers, obtained success rates of 60 per cent of 11 year olds, compared with around 40 per cent for comparable items of the form '¹⁄ᵦ × c'. Similarly if the numerator was not one, items of the type 'ᵃ⁄ᵦ' were answered correctly by about 40 per cent of 11 year olds in comparison with 20 to 35 per cent for 'ᵃ⁄ᵦ × c'.

However again there is evidence that these ideas are less easy to apply in problem situations, especially involving two fractions. For example although the APU (1981b) found that 53 per cent of 15 year olds were able to correctly work out ¾ × ⅓ = ¼, only 24 per cent were able to demonstrate this concretely using a circle divided into twelfths.

Thus all the research on multiplication of fractions suggest that straightforward computation 'sums' (i.e. those not involving mixed numbers), set out of context, are not necessarily more difficult than addition and subtraction calculations. (Although even so only 50 per cent of 15 year olds and about 30 per cent of 11 year olds obtain correct answers.) However the meaning of multiplying fractions is clearly very difficult to either understand or apply in everyday problems; and indeed there is some doubt as to whether a competence in the operation itself is really necessary, since most realistic problems can readily be solved by other means.

The same is true, but to an even greater extent, for the division of fractions. Here possible concrete meanings of, for instance, ⅔ ÷ ⅝ become even more difficult to provide since the only meanings are in terms of the inverse of multiplication, either in the 'product as area' aspect or in the 'ratio aspect'. This is far removed from the intuitive way the operation of division is introduced as 'sharing' or 'grouping' when whole numbers are concerned (see page 236). Even the calculation itself becomes correspondingly more complex than in the case of multiplication.

For example, Hart gave a problem situation involving the inverse of the 'product as area', together with the comparable computation to her sample of 14 and 15 year olds. The results are reported in the table on page 313.

With a more complex calculation 3⅓ ÷ 2⅕, the success rates were reduced to around 14 per cent. There were no clear patterns of errors in the case of the division computations, and in each case large numbers of children were unable to attempt the questions.

Thus once again these results cast considerable doubt on the usefulness of teaching the meaning of, or methods for, dividing fractions to most children, especially when it is

Results on selected items involving division of fractions (data from Hart, 1980)

Item	*Success rates*	
	14 yrs	15 yrs

Problem

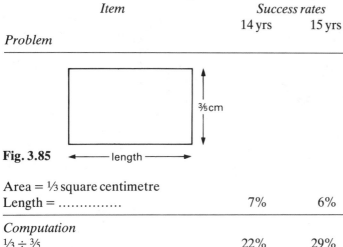

Fig. 3.85 ◄──── length ────►

Area = ⅓ square centimetre
Length = ...

	7%	6%

Computation

⅓ ÷ ⅗	22%	29%

so difficult to find everyday examples which cannot better be solved by some other method.

3.8.3.4   *Operations on Decimals*

Addition and subtraction   The meaning of a rational number expressed as a decimal (e.g. 0.75) is more difficult to grasp than that of a rational number expressed as a fraction (e.g. ¾). However the base-ten place value representation of decimals has considerable advantages as compared with fractional notation when the meaning of number operations is considered, since the analogy with whole numbers is emphasised by the similarity of the notation.

Number operations are always defined for children initially with respect to whole numbers and thus to sets of discrete objects. An extension of this meaning must be made in order to define operations on either fractions or decimals, since, for example, addition can no longer be thought of as simply 'combining two sets' or division as 'sharing out objects into piles'.

Nevertheless, particularly in the case of addition and subtraction, the analogy between decimals and large numbers is close. For example addition of large numbers can be made more concrete by the representation of numbers in terms of volumes, areas or lengths, and each transfer across into decimals as demonstrated in the tables on pages 314 and 315.

Similar representations can be made for subtraction, although the process is slightly more complex.

Thus, provided children understand the meaning of decimal notation, and of addition/subtraction with large numbers, the step to transferring these to the addition/subtraction of decimals would seem to be relatively small.

Addition: analogies between whole numbers and decimals

(a) *Volume representation*

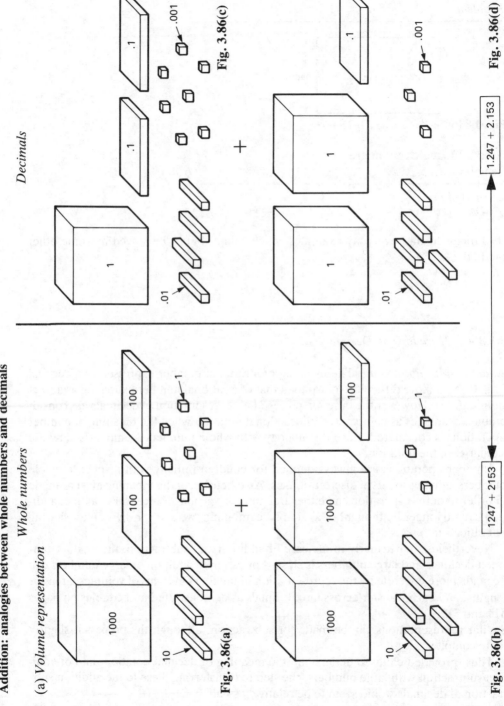

Whole numbers

Decimals

Fig. 3.86(a)

Fig. 3.86(b)

Fig. 3.86(c)

Fig. 3.86(d)

1247 + 2153

1.247 + 2.153

(b) *Area representation*

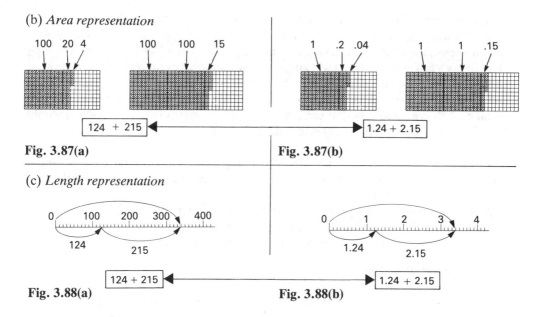

Fig. 3.87(a)   Fig. 3.87(b)

(c) *Length representation*

Fig. 3.88(a)   Fig. 3.88(b)

Also the computational methods for addition and subtraction are essentially similar for whole numbers and decimals. This is demonstrated by the APU finding that:

> Adding was hardly less well done with decimals than it was with whole numbers – as long as the format was vertical and the decimal points and places were aligned for each number involved (APU, 1980a).

The actual success rates quoted for such addition questions are around 90 per cent for 15 year olds and 80 per cent for 11 year olds. In the case of subtraction items set in standard vertical form the success rates for both whole numbers and decimals were in the 80–90 per cent range for 15 year olds, and 50–65 per cent for 11 year olds (APU, 1981b; 1980a). These results suggest that many children may treat decimal 'sums' just like whole numbers 'sums', and simply ignore the presence of the decimal point.

The major difference between success rates for decimal and whole number addition and subtraction computations came when the items were set in horizontal format and especially in which the decimals had varying numbers of digits after the point, as is shown in the following table.

Success rates for selected APU items involving whole numbers and decimals (data from APU, 1981b)

Item	*Success rate (percentages)*	
	11 yrs	15 yrs
9417 – 283	67	83
5.07 – 1.3	37	64

However the problem in the decimal item quoted above is probably more due to a lack of understanding of decimal notation (see page 00) than to any problem concerned specifically with subtraction.

In fact there must be some question over the legitimacy of setting decimal computations with varying numbers of decimal places. The major application of decimals is to *measurement*, for example, working out the weight of Coca-Cola in a can by weighing the can both full and empty, or working out how long it takes (in seconds and tenths of seconds) to perform an operation by subtracting the times shown on a digital clock beforehand and afterwards. In all such cases it would be expected that the number of significant figures would be the same in both measurements (i.e. the problem would arise in the context of 5.07 – 1.30 rather than 5.07 – 1.3).

Multiplication and division   The conceptual step from whole numbers to decimals would seem to be much greater in the case of multiplication and division than in the case of addition and subtraction.

Multiplication can no longer be thought of in terms of repeated addition ('4 lots of 3' cannot be extended to '4.7 lots of 3.2', and even less to '0.47 lots of 0.32'). Thus, as in the case of fractions, the major 'meanings' of multiplication must be seen as

a.   'product as area' e.g.

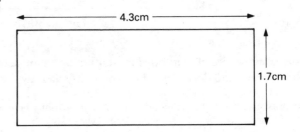

Fig. 3.89

b.   'rate/ratio' e.g. finding the cost of 26.7 litres of petrol at 48.2 pence per litre.

As in the case of fractions, the 'area' model may be easier to understand but is less used in practice.

Similarly 'division' can no longer be thought of as simple 'sharing' or 'grouping', and can only be understood as the inverse of multiplication in the two aspects given above. (However in the case of rate/ratio the division operation is used to find the rate or ratio itself and thus directly compare two quantities e.g. finding a share or unit cost. Thus it may be easier to understand than multiplication in this aspect.)

Brown examined children's understanding of the number operations on decimals by asking them to select which 'expression' matched a problem in the various cases shown in Fig. 3.90, for which the results for a representative sample of English children are given in the table following the problem.

Fig. 3.90 Selected items on the recognition of number operations involving decimals (reproduced from Brown, 1981d)

(19) Ring the (CALCULATION) you *would* need to do to find the answer:

A.	A table is 92.3 centimetres long. About how many inches is this? (1 inch is about 2.54 centimetres.)	2.54 + 92.3 92.3 ÷ 2.54 2.54 ÷ 92.3 92.3 – 2.54 2.54 – 92.3 92.3 × 2.54

B. The car. My car tank was full after I put in 6.44 gallons of petrol. The tank holds 8.37 gallons. How much petrol was in it to start with?

8.37 + 6.44	6.44 ÷ 8.37
8.37 ÷ 6.44	6.44 − 8.37
8.37 − 6.44	6.44 × 8.37

C. The price of minced beef is shown as 88.2 pence for each kilogram. What is the cost of a packet containing 0.58 kg of minced beef?

88.2 + 0.58	0.58 ÷ 88.2
88.2 ÷ 0.58	0.58 − 88.2
88.2 − 0.58	0.58 × 88.2

D. The cost of the 6.44 gallons of petrol was £4.86. What would the price of one gallon be?

6.44 + 4.86	4.86 ÷ 6.44
6.44 ÷ 4.86	4.86 − 6.44
6.44 − 4.86	4.86 × 6.44

E. My car can go 41.8 miles on each gallon of petrol on a motor-way. How many miles can I expect to travel on 8.37 gallons?

41.8 + 8.37	8.37 ÷ 41.8
41.8 ÷ 8.37	8.37 − 41.8
41.8 − 8.37	8.37 × 41.8

(NB   It should be noted that the prices given were realistic at the time of the testing)

Results for items shown in Figure 3.90 (reproduced from Brown, 1981d)

Question	Response	Percentage of children giving response			
		1st year	2nd year	3rd year	4th year
A	92.3 ÷ 2.54 (2.54 ÷ 92.3)	27% (17%)	35% (15%)	44% (19%)	45% (12%)
B	8.37 − 6.44 (6.44 − 8.37)	52% (11%)	52% (15%)	68% (8%)	63% (6%)
C	0.58 × 88.2	18%	17%	21%	29%
D	4.86 ÷ 6.44 (6.44 ÷ 4.86)	19% (34%)	23% (39%)	30% (43%)	28% (40%)
E	8.37 × 41.8	32%	42%	54%	53%

It will be seen that with the exception of the subtraction problem in part B, the remaining four parts are all concerned with the use of multiplication or division to either calculate a rate, or to use a given rate to work out an unknown quantity.

The following conclusions can be drawn:

i.   The total numbers recognising the correct operation to use, regardless of the order of the numbers, were highest for subtraction, with division next and multiplication being the hardest. (This was similar to the order found with whole numbers, but in the case of subtraction and division the percentages above were about 25 per cent lower than for the comparable 'whole number' items, and the gap was even greater for the multiplication items.)

ii.   The items involving only numbers greater than one in both the question and the answer were easier than those which involved at least one number less than one (i.e. parts A and E were easier than the corresponding pairs D and C, respectively). The reasons for this difficulty with numbers less than one seemed to be different in the cases of multiplication and division.

In part C, the children interviewed tended to recognise that the answer should be less than 88.2, and considered that the way to obtain a smaller answer was to use the operation of division, on the basis that 'multiplication makes it bigger, division makes it smaller'. Hence the percentage who selected a division answer for part C ranged from 37 per cent of 1st year (12 year old) children to 42 per cent of 4th year (15 year olds), which was substantially greater than the percentage choosing a multiplication answer (and was also much larger than in part E).

Support for this reasoning came from a further item used by Brown:

> Ring the one in each pair which gives the (BIGGER) answer:
> (a) 8×4 or $8 \div 4$
> (b) 8×0.4 or $8 \div 0.4$
> (c) 0.8×0.4 or $0.8 \div 0.4$

In this case 13 per cent of 12 year olds, rising to only 18 per cent of 15 year olds, correctly recognised that multiplication by a number less than one produced a smaller number than division by a number less than one; about half the sample ringed the multiplication expression for each part (Brown, 1981c).

In the case of division, the problem was not so much of selecting the wrong operation, but of selecting the numbers in the wrong order i.e. $6.44 \div 4.86$ rather than $4.86 \div 6.44$. The problem was generally that children did not consider it legitimate to divide by a larger number, e.g.

Julie: ' 'Cos you can get that into that (i.e. 4.86 into 6.44) but you can't get that into there.'
(Brown, 1981d)

However some children chose at random between the two expressions $4.86 \div 6.44$ and $6.44 \div 4.86$, believing them to be equivalent. Thus the percentage of success for this question is almost certainly an over-estimate of those children who fully understood what they were doing.

Brown notes that if the answers to all the five parts of the question are taken together, probably less than 10 per cent of 15 year old children were consistently successful, and thus could be said to have a sound understanding of the meaning of number operations in real-life applications involving decimals. She notes that none of the 39 children interviewed obtained a correct answer to all five parts, and in a top stream class of 15 year olds doing 0-level examinations that year, only two out of 22 achieved this.

This suggests that the meaning of the multiplication and division operations in the case of decimals is very difficult for children, particularly since false generalisations like those referred to are made from the 'whole number' situations (e.g. 'multiplication makes it bigger', 'you can't divide by a bigger number'). The strategies used by successful children seemed to fall into two categories:

a.   approximating the decimal numbers by whole numbers, and working out what operation would be appropriate in this case e.g., for part A

   Peter: 'See how many 2's go into 92'.

b.   identifying the rate structure e.g.

   Paul: 'Cos price is going to be times whatever you buy of it'. (Brown, 1981c)

Bell et al (1981) in a study referred to on page 296, used the former strategy, together with other activities, as part of a teaching programme which achieved some success in assisting less able 14 year olds to select the correct operation. Many of their observations confirm those of Brown.

It is clear from the above results that children find it difficult to recognise what operation to apply in real-life problems involving decimals. This suggests that teaching the algorithms for multiplication and division of decimals out of context may often be a waste of time. However this does not mean that the computation involved is itself difficult, for once again the analogy with whole numbers is of considerable assistance here and the extra difficulty comes only in the positioning of the decimal point. However this in itself is not necessarily so trivial as it might seem.

For example, although the APU (1981b) report that 64 per cent of 15 year olds could work out the correct answer to

$$2.54 \times 12$$

only 35 per cent could do so for

$$40 \div 0.8$$

The problem here would again seem to be that of using the false generalisation that 'division makes it smaller', since 29 per cent gave either 5 or 0.5 as an answer.

The methods taught for determining the position of the decimal point tend to focus either on the use of rules (e.g. counting from the right the total number of decimal places in both numbers to be multiplied) or on the use of estimation.

Brown (1981c) found that only one child out of 35 interviewed was able to use a rule correctly for determining the answer to 0.2×0.4. In fact only 15 per cent of 12 year olds rising to 31 per cent of 15 year olds obtained a correct answer to this, the majority giving 0.8. And in the parallel division problem $60 \div 0.3$, all those interviewed who attempted to use a taught rule to determine the position of the point in the answer made a mistake; in fact all four children interviewed who obtained correct answers did so by reasoning without depending on a rule. (Altogether 13 per cent of 12 year olds and 44 per cent of 15 year olds obtained correct answers in a large representative sample.)

This suggests that although it may be possible to teach children a rule, in the long term many of them either do not retain it, or apply it incorrectly.

However there are also problems with the estimation method. For example Brown asked children to select the number nearest in size to the answers from the selection of numbers shown on the right:

| 2.9×7 | $\rightarrow$ | .002 / .02 / .2 / 2 / 20 / 200 / 2000 |
| 0.29×7.1 | $\rightarrow$ | .002 / .02 / .2 / 2 / 20 / 200 / 2000 |

Although 44 per cent of 12 year olds, rising to 62 per cent of 15 year olds, were successful with 2.9×7, only 15 per cent of 12 year olds, rising to 31 per cent of 15 year olds, chose correctly for 0.29×7.1 (Brown, 1981c). Hence neither the 'rule' nor the 'estimation' method of placing the decimal point is very successful when the numbers involved are less than one.

Nor are the results much more promising in the case of adults. Sewell (1981) asked 50 adults to work out the cost of 3.4 metres of cloth at £2.45 a metre, using any method they wished (including calculators). Only four of the fifty attempted it, all of whom had passed GCE O-level mathematics at school, and the three who obtained a correct answer used informal methods rather than school-taught rules or algorithms. The researcher reported,

This sum was widely viewed with horror and alarm (Sewell, 1981).

The results quoted above suggest that for many children the only realistic method of multiplying or dividing decimals is to use a calculator. Even this may not be effective unless their understanding of the nature of decimals, and in particular of the nature and effect of multiplication and division by decimals, is sufficiently developed to enable the child to recognise which operation should be applied, and, in the case of division, in which order the numbers should be entered. This implies a shift in teaching towards expertise in using calculators in problems, and away from pencil-and-paper techniques used out of any real-life context.

References for Section 3

Advisory Council for Adult and Continuing Education – ACACE (1982) *Adults' Mathematical Ability and Performance.*

Anderson, R.C. (1966) A Comparison of two Procedures for Finding the Least Common Denominator in the Addition of Unlike, Related Fractions. (Doctoral dissertation, 1975). *Dissertation Abstracts,* **26,** 5901. University of Iowa.

Assessment of Performance Unit (APU) – See references under Department of Education and Science.

Ashlock, R.B. (1976) *Error Patterns in Computation : A Semi-programmed Approach.* Columbus, Ohio: Charles E. Merrill.

Barclay, T. (1980) Buggy. *Mathematics Teaching,* **92,** 10–12.

Beckwith, M. and Restle, F. (1966) Processes of Enumeration. *Psychological Review,* **73,** 437–444.

Bednarz, N. and Janvier, B. (1979) The Understanding of Place-Value (Numeration); Unpublished paper presented at the Third International Conference for the Psychology of Mathematics Education, Warwick University (1979).

Bednarz, N. and Janvier, B. (1982) The Understanding of Numeration. *Educational Studies in Mathematics,* **13(1),** 33–57.

Bell, A; Swan, M. and Taylor, G. (1981) Choice of Operation in Verbal Problems with Decimal Numbers, *Educational Studies in Mathematics,* **12(4),** 399–420.

Bever, T.G; Mehler, J. and Epstein, J. (1968) What Children Do in Spite of What They Know. *Science,* **162,** 921–924.

Bidwell, J.K. (1968) A Comparative Study of the Learning Structures of Three Algorithms for Division of Fractional Numbers. (Doctoral dissertation, 1968). *Dissertation Abstracts,* **29,** 830A. University of Michigan.

Biggs, E. (1970) *Mathematics for Younger Children.* London: Macmillan.

Biggs, J.B. (1967) *Mathematics and the Conditions of Learning.* Windsor: N.F.E.R.

Bohan, H.J. (1971) A Study of the Effectiveness of Three Learning Sequences for Equivalent Fractions. (Doctoral dissertation, University of Michigan 1970). *Dissertation Abstracts International,* **31,** 6270A.

Botsmanova, M. (1972a) The Forms of Pictorial Visual Aids in Instruction in Arithmetic Problem-Solving. In *Soviet Studies in the Psychology of Learning and Teaching Mathematics,* 6 (Ed) Kilpatrick, J. & Wirszup, I. Stanford, California: School Mathematics Study Group.

Botsmanova, M (1972b) The Role of Graphic Analysis in Solving Arithmetic Problems. In *Soviet Studies in the Psychology of Learning and Teaching Mathematics,* 6 (Ed) Kilpatrick, J. and Wirszup, I. Stanford, California: School Mathematics Study Group.

Brainerd, C.J. (1979) *The Origins of the Number Concept.* New York: Praeger. (Available from Holt, Rinehart and Winston, Eastbourne.)

Bright, G. (1978) Assessing the Development of Computation Skills. In *Developing Computational Skills: 1978 Yearbook* (Ed) Suydam, M.N. Reston, Virginia : National Council of Teachers of Mathematics.

Brown, J.S. and Burton, R.B. (1978) Diagnostic Models for Procedural 'Bugs' in Basic Mathematical Skills. *Cognitive Science,* **2,** 155–92.

Brown, J.S. & Van Lehn, K. (1982) Towards a Generative Theory of 'Bugs'. In *Addition and Subtraction : A Cognitive Perspective* (Ed) Carpenter, T.P. et al. New Jersey: Laurence Erlbaum.

Brown, M. (1978) *CSMS Number Operations Test.* Windsor: N.F.E.R.

Brown, M. (1981a) Number Operations. In *Children's Understanding of Mathematics : 11–16* (Ed) Hart, K.M. John Murray.

Brown, M. (1981b) Place-Value and Decimals. In *Children's Understanding of Mathematics : 11–16* (Ed) Hart, K.M. John Murray.

Brown, M. (1981c) *Levels of Understanding of Number Operations, Place-Value and Decimals in Secondary School Children* – Ph. D. Thesis. University of London, Chelsea College.

Brown, M. (1981d) Is it an Add, Miss? Part 3. *Mathematics in School,* **10,** (1), 26–28.

Brown, M. and Kuchemann, D. (1977) Is it an Add, Miss? Part 2. *Mathematics in School,* **6,** (1), 9–10.

Brown, P.G. (1969) *Tests of Development in Children's Understanding of the Laws of Natural Numbers* – M. Ed. Thesis. Manchester University.

Brown, S.I. (1974) Musing on Multiplication. *Mathematics Teaching,* **69,** 26–30.

Brown, S.I. (1981) Sharon's 'Kye'. *Mathematics Teaching,* **94,** 11–17.

Brownell, W.A. (1941) *Arithmetic in Grades I and II: a Critical Summary of New and Previously Reported Research.* Durham, U.S.A: Duke University Press.

Brownell, W.A. (1967) *Arithmetical Abstractions: The Movement Towards Conceptual Maturity Under Differing Systems of Instruction (Co-operative Research Project No. 1676).* Office of Education, U.S. Dept. of Health, Education and Welfare. Berkeley: University of California Press.

Brownell, W.A. and Moser, H.E. (1949) Meaningful vs. Mechanical Learning: a Study in Grade III Subtraction – Duke University Studies in Education, no. 8. Durham, N.C; Duke University Press.

Brueckner, L. and Grossnickle, F. (1947) *How to Make Arithmetic Meaningful.* New York: Holt, Rinehart and Winston.

Bryant, P.E. (1974) *Perception and Understanding in Young Children.* Methuen.

Burton, R.B. (1981) DEBUGGY: Diagnosis of Errors in Basic Mathematical Skills. In *Intelligent Tutoring Systems* (Ed) Sleeman, D.H. and Brown, J.S. New York: Academic Press.

Callahan, L.G. and Glennon, V.J. (1975) *Elementary School Mathematics: A Guide to Recent Research.* Washington: Association for Supervision and Curriculum Development.

Carpenter, T.P. (1980a) Research in Cognitive Development. In *Research in Mathematics Education* (Ed) Shumway, R.J. New York: National Council of Teachers of Mathematics.

Carpenter, T.P. (1980b) Heuristic Strategies used to Solve Addition and Subtraction Problems. In *Proceedings of the Fourth International Conference for the Psychology of Mathematics Education* (Ed) Karplus, R. California: University of Berkeley.

Carpenter, T; Coburn, T.G; Rays, R.E. and Wilson, J.W. (1978) *Results from the First Mathematics Assessment of the National Assessment of Educational Progress.* Reston, Virginia: National Council of Teachers of Mathematics.

Carpenter, T.P. and Moser, J.M. (1979) The Development of Addition and Subtraction Concepts in Young Children. In *Proceedings of the Third International Conference for the Psychology of Mathematics Education.* (University of Warwick, Mathematics Education Research Centre.)

Carpenter, T.P. and Moser, J.M. (1982) The Development of Addition and Subtraction Problem-Solving Skills. In *Addition and Subtraction: A Cognitive Perspective* (Ed) Carpenter et al. New Jersey: Laurence Erlbaum.

Coburn, T.G. (1974) The Effect of a Ratio Approach and a Region Approach on Equivalent Fractions and

Addition/Subtraction for Pupils in Grade Four. (Doctoral dissertation, University of Michigan 1973). *Dissertation Abstracts International*, **34**, 4688A–4689A.

Collis, K. (1975) *A Study of Concrete and Formal Operations in School Mathematics: A Piagetian Viewpoint.* Victoria, Australia: Australian Council for Educational Research.

Copeland, R.W. (1974) *How Children Learn Mathematics: Teaching Implications of Piaget's Research.* New York: Macmillan.

Cox, L.S. (1975) Systematic Errors in the Four Vertical Algorithms in Normal and Handicapped Populations. *Journal for Research in Mathematics Education*, **6**, (4), 202–220.

CSMS (Concepts in Secondary Maths and Science), See Hart, K.M. (Ed, 1981).

Davis, R.B. (1973) The Misuse of Educational Objectives. *Educational Technology,* November 1973.

Davydov, V.V. and Andronov, V.P. (1981) *Psychological Conditions of the Origination of Ideal Actions (Project Paper 81–82).* Wisconsin: Research and Development Center for Individualized Schooling, University of Wisconsin.

Department of Education and Science, APU – Assessment of Performance Unit (1980a) *Mathematical Development, Primary Survey Report No. 1.* HMSO.

Department of Education and Science, APU – Assessment of Performance Unit (1980b) *Mathematical Development, Secondary Survey Report No. 1.* HMSO.

Department of Education and Science, APU – Assessment of Performance Unit (1981a) *Mathematical Development, Primary Survey Report No. 2.* HMSO.

Department of Education and Science, APU – Assessment of Performance Unit (1981b) *Mathematical Development, Secondary Survey Report No. 2.* HMSO.

Denvir, B; Stolz, C; Brown, M. (1982) *Low Attainers in Mathematics 5–16: Policies and Practices in Schools* (Schools Council Working Paper 72) London: Methuen Educational Ltd. for the Schools Council.

Descoeudres, A. (1921) *Le Development de l'Enfant de Deux a Sept Ans.* Paris: Delachaux et Niestle C.A.

Dickson, L. (in preparation) Secondary School Pupil's Understanding of the Symbolic Representation of Natural Number and Computation with Particular Reference to Place-Value. Unpublished Ph. D. Thesis, University of London.

Dienes, Z.P. (1959) The Growth of Mathematical Concepts in Children through Experience. *Educational Research*, **2**, 9–28.

Dienes, Z.P. (1960) *Building Up Mathematics.* London: Hutchinson.

Dienes, Z.P. (1971) *The Elements of Mathematics.* New York: Herder and Herder.

Dodwell, P.C. (1962) Relation Between the Understanding of the Logic of Classes and of Cardinal Number in Children. *Canadian Journal of Psychology*, **16**, 152–160.

Donaldson, M. (1978) *Children's Minds.* Fontana.

Donaldson, M. and Balfour, G. (1968) Less is More: A Study of Language Comprehension in Children. *British Journal of Psychology*, **59**, 461–471.

Donaldson, M. and Wales, R.J. (1970) On the Acquisition of Some Relational Terms. In *Cognition and the Development of Language* (Ed) Hayes, J.R. New York: Wiley.

Duffin, J. (1976) Left-Right-Left: A Source of Confusion. *Mathematical Education for Teaching, **2**, 14–18.

Easterday, K. (1964) An Experiment with Low Achievers in Arithmetic. *Mathematics Teacher, **57**, 462–468.

Elkind, D. (1964) Discrimination, Seriation and Numeration of Size. *Journal of Genetic Psychology,* **104,** 275–296.

Engelhardt, J.M. (1977) Analysis of Children's Computational Errors: A Qualitative Approach. *British Journal of Educational Psychology,* **47,** 149–154.

Erlwanger, S.H. (1973) Benny's conception of Rules and Answers in IPI Mathematics. *Journal of Children's Mathematical Behaviour,* **1,** (2), 7–26.

Faires, D.M. (1963) Computation with Decimal Fractions in the Sequence of Number Development. *Dissertation Abstracts,* **23,** 4183. Wayne State University.

Fennema, E. (1972) Models and Mathematics. *The Arithmetic Teacher,* **19,** 635–640.

Fitzgerald, A. (1976) An Excursion into Subtraction and Long Multiplication. *Mathematics Teaching,* **74,** 30–34.

Flournoy, F; Brandt, D., & McGregor, J. (1963) Pupil Understanding of the Numeration System. *Arithmetic Teacher,* **10,** 88–92.

Floyd, A. (1980) Towards Thinking Mathematically, Part II: Combining Fractions. In *Proceedings of the Fourth International Conference for the Psychology of Mathematics Education* (Ed) Karplus, R. California, U.S.A: University of Berkeley.

Freudenthal, H. (1973) *Mathematics as an Educational Task.* Dordrecht: D. Reidel.

Fulkerson, E. (1963) Adding in Tens. *The Arithmetic Teacher,* **10,** 139–140.

Fuson, K.C. (1980) The Counting Word Sequence as a Representational Tool. In *Proceedings of the Fourth International Conference for the Psychology of Mathematics Education* (Ed) Karplus, R. California, U.S.A: University of Berkeley.

Fuson, K. (1982) An analysis of the Counting-On Solution Procedure in Addition. In *Addition and Subtraction: A Cognitive Perspective* (Ed) Carpenter, T.P. et al. New Jersey: Laurence Erlbaum.

Galloway, P.J. (1975) *Achievement and Attitude of Pupils Toward Initial Fractional Number Concepts at Various Ages from Six through Ten Years and of Decimals at Ages Eight, Nine and Ten,* (Doctoral dissertation). University of Michigan.

Gelman, R. (1972) Logical Capacity of Very Young Children: Number Invariance Rules. *Child Development,* **43,** 75–90.

Gelman, R. and Gallistel, C.R. (1978) *The Child's Understanding of Number.* Harvard University Press.

Gibb, E.G. (1954) Take-Away is Not Enough. *Arithmetic Teacher,* **1,** (2), 7–10.

Gibb, E.G. (1956) Children's thinking in the process of subtraction. *Journal of Experimental Education,* **25,** 71–80.

Ginsburg, H.P. (1977) *Children's Arithmetic: how they learn it and how you teach it.* Austin, TX: PRO-ED, 1982.

Girling, M. (1977) Towards a Definition of Basic Numeracy. *Mathematics Teaching,* **81,** 4.

Glennerster, A. (1980) Trubloons. *Mathematics Teaching,* **93,** 42–43.

Gonchar, A.J. (1975) *A Study in the Nature and Development of the Natural Number Concept : Initial and Supplementary Analyses* (Wisconsin Research and Development Center Technical Report no. 340.) Madison: The University of Wisconsin.

Gréco, P. & Morf, A. (1962) *Structures Numeriques Elementaires: Etudes d' Epistemologie Genetique, Vol. 13.* Paris: Presses Universitaires de France.

Green, G.A. (1970) A Comparison of Two Approaches, Area and Finding a Part of, and Two Instructional Materials, Diagrams and Manipulative Aids, on Multiplication of Fractional Numbers in Grade Five (Doctoral dissertation, University of Michigan, 1969). *Dissertation Abstracts International*, **31,** 676A–677A.

Grenier, M – A.C. (1975) An Investigation of Summer Mathematics Achievement Loss and the Related Fall Recovery Time. Unpublished doctoral dissertation, University of Georgia, 1975.

Groen, G.J. and Resnick, L.B. (1977) Can Pre-School Children Invent Addition Algorithms? *Journal of Educational Psychology,* **69,** 645–652.

Grossnickle, F.E. (1959) Discovering the Multiplication Facts. *Arithmetic Teacher,* **6, (4)**, 195–198, 208.

Gunderson, A.G. (1955) Thought-Patterns of Young Children in Learning Multiplication and Division. *Elementary School Journal,* **55,** 453–461.

Hasemann, K. (1980a) On the Understanding of Concepts and Rules in Secondary Mathematics: Some Examples Illustrating the Difficulties. In *Proceedings of the Fourth International Conference for the Psychology of Mathematics Education* (Ed) Karplus, R. California, U.S.A: University of Berkeley.

Hasemann, K. (1980b) On difficulties with fractions. *Educational Studies in Mathematics,* **12, (1)**, 71–88.

Hart, K.M. (1980) *Secondary School Children's Understanding of Mathematics – Research Monograph* (A report of the Mathematics Component of the Concepts in Secondary Mathematics and Science Programme). Chelsea College, University of London.

Hart, K.M. (Ed, 1981) *Children's Understanding of Mathematics 11–16*. John Murray.

Hartung, M.L. (1958) Fractions and Related Symbolism in Elementary School Instruction. *Elementary School Journal,* **58,** 377–384.

Hatano, G. (1982) Learning to Add and Subtract : a Japanese Perspective. In *Addition and Subtraction : A Cognitive Perspective* (Ed) Carpenter, T.P. et al. New Jersey : Laurence Erlbaum.

Hendrickson, A.D. (1979) Why do Children Experience Difficulty in Learning Mathematics? *FOCUS On Learning Problems in Mathematics,* **1, (2)**, 31–37.

Hiebert, J. and Carpenter, T.P. (1982) Piagetian Tasks as Readiness Measures in Mathematics Instruction : a Critical Review. *Educational Studies in Mathematics,* **13,** 329–345.

Higginson, W. (1980) 'Berry Undecided' : a Digital Dialogue. *Mathematics Teaching,* **91,** 8–13.

Hill, E.H. (1952) Study of Third, Fourth, Fifth and Sixth Grade Children's Preferences and Performances on Partition and Measurement Division Problems – State University of Iowa. *Dissertation Abstracts,* **12,** 703.

Hinkleman, E.A. (1956) A study of the Principles Governing Fractions Known by Fifth- and Sixth-Grade Children. *Educational Administration and Supervision,* **42,** 153–161.

Howell, A; Walker, R. and Fletcher, H. (1979) *Mathematics for Schools, Teacher's Resource Book, Level 1.* London : Addison Wesley.

Hughes, M. (1981) Can Pre-School Children Add and Subtract? *Educational Psychology,* **1, (3)**, 207–219.

Hutchings, L.B. (1975) Low-Stress Subtraction. *The Arithmetic Teacher,* **10,** 139–140.

Inner London Education Authority – Abbey Wood Mathematics Centre (1975) *The Development of Place-Value.* Inner London Education Authority.

Karplus R; Karplus, E; Formisano, M. and Paulsen, A. (1977) A Survey of Proportional Reasoning and Control of Variables in Seven Countries. *Journal of Research in Science Teaching,* **13,** 411–417.

Kieren, T.E. (1976) On the Mathematical, Cognitive and Instructional Foundation of Rational Numbers. In *Number and Measurement* (Ed) Lesh, R.A. and Bradberd, D.A. Columbus, Ohio: ERIC/SMEAC.

Kurtz, R. (1973) Fourth Grade Division : How Much Is Retained in Grade Five. *Arithmetic Teacher,* **20,** 65–71.

Lankford, F.G. (1972) *Some Computational Strategies of Seventh Grade Pupils.* Charlottesville, Va: University of Virginia (ERIC Document Reproduction Service no. ED069469).

Leeb-Lundberg, K. (1979) Zero. *Mathematics Teaching,* **78,** 24–25.

Lesh, R; Landau, M. and Hamilton, E. (1980) Rational Number Ideas and the Role of Representational Systems. In *Proceedings of the Fourth International Conference for the Psychology of Mathematics Education* (Ed) Karplus, R. Berkeley : University of California.

Lowry, W.C. (1965) Structure and the Algorithms of Arithmetic. *Arithmetic Teacher,* **12,** 146–150.

Lunzer, E.A; Bell, A.W. and Shiu, C.M. (1976) *Number and the World of Things.* University of Nottingham, Shell Centre for Mathematics Education.

Luriya, A.R. (1969) On the Pathology of Computational Operations. In *Soviet Studies in the Psychology of Learning and Teaching Mathematics Vol. I.* (Ed) Kilpatrick, J. & Wirszup, I. Stanford California, School Mathematics Study Group.

McIntosh, A. (1978) Some Subtractions: What do you think you are doing? *Mathematics Teaching,* **87,** 14–18.

McIntosh, A. (1979) Some Children and Some Multiplications. *Mathematics Teaching,* **87,** 14–15.

Magne, O. (1978) The Psychology of Remedial Mathematics, *Didakometry,* **10,** 59: (Oslo).

Matthews, J. (1980) 'Five More Pages Fewer …'. *Mathematics in School,* **9,** 24–25.

Menchinskaya, N.A. (1969) Intellectual Activity in Solving Model Arithmetic Problems. In *Soviet Studies in the Psychology of Learning and Teaching Mathematics Vol. VI.* (Ed) Kilpatrick, J. and Wirszup, I. Stanford, California, School Mathematics Study Group.

Muangnapoe, C. (1975) *An Investigation of the Learning of the Initial Concept and Oral/Written Symbols for Fractional Numbers in Grades Three and Four.* (Doctoral dissertation). University of Michigan.

National Assessment of Educational Performance – NAEP (1980) *Mathematics Technical Report : Summary Volume.* Washington : NAEP.

Nesher, P. (1982) Levels of Description in the Analysis of Addition and Subtraction Word Problems. In *Addition and Subtraction : A Cognitive Perspective* (Ed) Carpenter, T.P. et al. New Jersey : Laurence Erlbaum.

Nesher, P. and Katriel, T. (1977) A Semantic Analysis of Addition and Subtraction Word Problems in Arithmetic. *Educational Studies in Mathematics,* **8,** 251–269.

Noelting, G. (1978) The Development of Proportional Reasoning in the Child and Adolescent through Combination of Logic and Arithmetic. In *Proceedings of the Second International Conference for the Psychology of Mathematics Education* (Ed) Cohors-Fresenborg, E. and Wachsmuth, I. West Germany: University of Osnabruck.

Noelting, G. (1980) The Development of Proportional Reasoning and the Ratio Concept, Part I : Determination of stages. *Educational Studies in Mathematics,* **11,** 217–251.

Novillis, C.F. (1976) An Analysis of the Fraction Concept into a Hierarchy of Selected Subconcepts and the Testing of the Hierarchical Dependencies. *Journal of Research in Mathematics Education,* **7,** 131–144.

Oesterle, R.A. (1959) What about the 'Zero Facts'? *The Arithmetic Teacher,* **6,** (2), 109–111.

Osburn, W.J. and Foltz, P.J. (1931) Permanence of Improvement in the Fundamentals of Arithmetic. *Educational Research Bulletin,* **10,** 227–234.

Payne, J.N. (1976) Review of Research on Fractions. In *Number and Measurement: Papers from a Research Workshop* (Ed) Lesh, R.A. Ohio, U.S.A: ERIC/SMEAC.

Piaget, J. (1952) *The Child's Conception of Number.* Humanities Press Inc. and Routledge and Kegan Paul.

Piaget, J. (1968) Quantification, Conservation and Nativism. *Science, 162,* 976–979.

Piaget, J; Inhelder, B. and Szeminska, A. (1960) *The Child's Conception of Geometry.* London : Routledge and Kegan Paul.

Piaget, J; Grize, J.B; Szeminska, A. and Bang, V. (1968) *Epistemologie et Psychologie de la Fonction.* Paris: Presses Universitaires de France.

Plunkett, S. (1979) Decomposition and All That Rot. *Mathematics in School,* **8,** (**3**), 2–5.

Potter, M.C. and Levy, E.I. (1968) Spatial Enumeration Without Counting. *Child Development,* **39,** 265–273.

Resnick, L.B. (1982) Syntax and Semantics in Learning to Subtract. In *Addition and Subtraction: A Cognitive Perspective* (Ed) Carpenter, T.P. et al. New Jersey : Laurence Erlbaum.

Richards, J. (1980) Counting and the Constructivist Program. In *Proceedings of the Fourth International Conference for the Psychology of Mathematics Education* (Ed) Karplus, R. Berkeley : University of California.

Roberts, G.H. (1968) The Failure Strategies of Third Grade Arithmetic Pupils. *The Arithmetic Teacher,* **15,** 442–446.

Romberg, T.A. & Collis, K.F. (1980) Cognitive Level and Performance on Addition and Subtraction Problems. In *Proceedings of the Fourth International Conference for the Psychology of Mathematics Education* (Ed) Karplus, R. Berkeley : University of California.

Ronshausen, N. (1978) Introducing Place-Value. *Arithmetic Teacher,* **25,** (**4**).

Schaeffer, B; Eggleston, V.H; Scott, J.L. (1974) Number development in young children. *Cognitive Psychology,* **6,** 357–379.

Schell, L.M. and Burns, P.C. (1962) Pupils' Performance with Three Types of Subtraction Situations. *School Science and Mathematics,* **62,** 208–269.

Sewell, B. (1978) *Use of Mathematics by Adults in Daily Life.* Advisory Council for Adult and Continuing Education.

Siegel, L.S. (1978) The Relationship of Language and Thought in the Pre-operational Child – a Reconsideration of Non-Verbal Alternatives to Piagetian Tasks. In *Alternatives to Piaget : Critical Essays on the Theory* (Ed) Siegel, L.S. & Brainerd, C.J. New York : Academic Press.

Skemp, R.R. (1976) Relational Understanding and Instrumental Understanding. *Mathematics Teaching,* **77,** 20–26.

Starkey, P. & Gelman, R. (1982) The Development of Addition and Subtraction Abilities Prior to Formal Schooling in Arithmetic. In *Addition and Subtraction : A Cognitive Perspective* (Ed) Carpenter, T.P. et al. New Jersey : Laurence Erlbaum.

Steffe, L.P. and Parr, R.B. (1968) *The Development of the Concepts of Ratio and Fraction in the Fourth, Fifth and Sixth Year of Elementary School.* Washington, D.C: Department of Health, Education and Welfare.

Steffe, L.P; Thompson, P.W; Richards, J. (1982) Children Counting in Arithmetical Problem Solving. In *Addition and Subtraction : A Cognitive Perspective.* New Jersey : Laurence Erlbaum.

Streefland, L. (1982) Subtracting Fractions with Different Denominators. *Educational Studies in Mathematics* **13** (**3**), 233–255.

Suydam, M.N. (1978) Review of Recent Research Related to the Concepts of Fractions and Ratio. In *Proceedings of the Second International Conference for the Psychology of Mathematics Education* (Ed) Cohors-Fresenborg, E. and Wachsmuth, I. West Germany : University of Osnabrück.

Suydam, M.N. and Dessart, D.J. (1976) *Classroom Ideas from Research on Computational Skills*. Reston, Virginia: National Council of Teachers of Mathematics.

Turnbull, W.P. (1903) *The Teaching of Arithmetic*. Newmann.

Vergnaud, G. (1982) A Classification of Cognitive Tasks and Operations of Thought Involved in Addition and Subtraction Problems. In *Addition and Subtraction : A Cognitive Perspective*. New Jersey : Laurence Erlbaum.

Vergnaud, G. and Durand, C. (1976) Structures additives et complexite psychogenetique. *La Revue Francaise de Pedagogie*, **36**, 28–43.

Vergnaud, G. et al. (1979) *Acquisition des 'structures multiplicatives' dans le premier cycle du second degre* (Report RO No. 2) Institut de Recherche sur l'Enseignement de Mathematiques, Orleans/Centre d'Etudes des Processes Cognitifs et du Langage, Paris.

Wang, M.C; Resnick, L.B. and Boozer, R.F. (1971) The Sequence of Development of Some Early Mathematical Behaviours. *Child Development*, **42**, 1767–1778.

Ward, M. (1979) *Mathematics and the 10-year old, Working Paper 61*, Schools Council. Evans/Methuen.

Weaver, F. (1955) Big Dividends from Little Interviews. *Arithmetic Teacher*, **11**, (2), 40–47.

Williams, H.B. (1975) *A Sequential Introduction of Initial Fraction Concepts in Grades Two and Four and Remediation in Grade Six* (Ed. S. Research Report). University of Michigan.

Williams, J.D. (1963) The Teaching of Mathematics (VI): Arithmetic and the Difficulties of Calculative Thinking. *Educational Research*, **5**, 216–218.

Williams, J.D. (1964) Some Remarks on the Nature of Understanding. *Educational Research*, **7**, 15–36.

Williams, M. and Moore, W. (1979) *Nuffield Maths 2 Teacher's Handbook*. London : Longman.

Willington, G.A. (1967) *The Development of the Mathematical Understanding of Primary School Children –* M.Ed. Thesis. Manchester University.

Willson, G.H. (1972) Decimal-Common Fraction Sequence Versus Conventional Sequence. *School Science and Mathematics*, **72**, 589–592.

Wilson, G.M. (1951) *Teaching the New Arithmetic*. New York : McGraw-Hill.

Zweng, M.J. (1964) Division Problems and the Concept of Rate. *Arithmetic Teacher*, **11**, 547–556.

SECTION 4: *Language – Words and Symbols*

SECTION 4: *Language – Words and Symbols*

4.1 WHAT IS THE ROLE OF LANGUAGE IN THE DEVELOPMENT OF MATHEMATICAL CONCEPTS?

4.1.1 Language and Thought

Can a child acquire the concept of seven before he learns the names of the counting numbers?

Should a child be familiar with the experience of combining two collections and working out the number of objects in the combined set before the specialised vocabuary of 'two add three' is introduced?

Or on the other hand should a child be positively encouraged to use words like 'a million' before he has any real notion of their exact meaning?

In teaching any mathematical topic, decisions have to be made about the optimum time for introducing relevant vocabulary and symbolism. Yet this happens against a background of psychological uncertainty about the role of language in concept acquisition.

> The relationship of language and thought has provoked a long and extensive debate in linguistics, philosophy and psychology (Moore and Harris, 1978).

At one extreme, the American behaviourist tradition treated thought simply as 'unvocalized speech', and hence thought without language was inconceivable. At the other extreme Piaget considered thought to comprise 'internalized actions' and thus, at least in his early writings, stated that language could only reflect, and not determine, cognitive development:

> Linguistic progress is not responsible for logical or operational progress. It is rather the other way round. The logical or operational level is likely to be responsible for a more sophisticated language level.
>
> (Piaget, 1972)

These two extremes, one holding thought to be verbal in nature and the other holding thought to be essentially spatial, are reminiscent of the 'two sides of the brain' hypothesis discussed in Section 1.1 (p.6). There the point was made that spatial and verbal aspects of thought are essentially complementary, although it should be recognised that some individuals may find one type of activity much more accessible than the other.

Choat stresses the close interdependence of language and conceptual development:

> Even if the learner interacts with the physical aspect of the learning situation, i.e. objects, the verbal element is necessary both as a means of communication and as an instrument of individual representation … in the acquisition of mathematical knowledge, a new concept brings a new word. Devoid of the conception, a child will not understand the word; without the word he cannot as easily assimilate and accommodate the concept. (Choat, 1974. Mathematics Teaching, **69**, 11)

This reflects the views of some psychologists, e.g. Vygotsky (1962), that thought and language are interdependent. Even Piaget in his later work, e.g. his preface to Ferreiro (1971), accepted that there might be parallel development in the linguistic and cognitive areas, both perhaps related to the child's development of more general underlying strategies for making sense of the world

The experimental evidence is equivocal. A particular problem in testing whether language precedes or succeeds related concept development is that many of the recognised tests of concept development, in particular those of Piaget, themselves depend heavily on the child's understanding and use of language.

Siegel points out the inconsistency of Piaget's position in claiming that thought precedes language, yet choosing to rely on verbal interviews to judge concept attainment. She then goes on to describe a number of attempts to evolve non-verbal tasks of concept development, some but not all related to Piagetian experiments. For instance she tried to train 3 and 4 year olds to consistently select the larger of two collections of dots simultaneously presented to them, (see Fig. 4.1 for a typical configuration), by rewarding them with a sweet each time they chose the larger set.

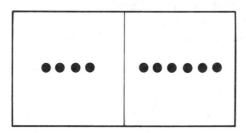

Fig. 4.1   (reproduced from Siegel, 1978)

A second group of children were trained to choose the smaller collection. The children were also asked to identify which of two sets was 'the big one' (or, for the second group, 'the little one').

The results were as shown in the following table:

Concept and Language Performance on Comparison Tasks

Group	Pass language, pass concept	Fail language, pass concept	Pass language, fail concept	Fail language, fail concept	Total
3 year olds	15	20	2	8	45
4 year olds	37	17	1	2	57

(Siegel, 1978)

This would seem to suggest that the ability to learn the concept of comparison of numerosity ('larger'/'smaller') in the non-verbal task precedes the acquisition of the related language ('big'/'little').

The experiment was repeated with the concept of equality, and the children were trained in the same way to select whichever lower collection was equal to the top one in configurations such as Fig. 4.2

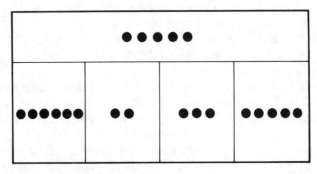

Fig. 4.2   (reproduced from Siegel, 1978)

They were also asked verbally to select the set which was 'the same' as the top set, in the verbal equivalent. This time the results were as shown in the following table.

Concept and Language Performance on Equality Tasks

Group	Pass language, pass concept	Fail language, pass concept	Pass language, fail concept	Fail language, fail concept	Total
3 year olds	7	4	0	34	45
4 year olds	40	8	0	9	57

(Siegel, 1978)

Here the correspondence between language and concept success is much closer, perhaps because both tasks rely on a 'matching' or 'counting' technique.

Siegel also looked at the abilities of the children to justify their choices verbally when asked to do so. (Statements counted as correct included e.g. 'the big one', 'the one that was more big' etc. for the comparison experiment and e.g. 'it's the same', etc. for the equality experiment. The proportion of those able to make the *correct* choice who were *unable* to give an appropriate verbal explanation were as shown in the next table for the two tasks of comparison and equality respectively.

Proportion of those passing concept attainment task who failed to give any appropriate verbal justification:

Group	Comparison	Equality
3 year olds	69%	100%
4 year olds	32%	79%

(Siegel, 1978)

This seems to demonstrate quite clearly that at least in this case the children's ability to *produce* the relevant language lags well behind their ability to attain the concept. The question of whether or not the ability to *understand* the language develops after or in parallel with attainment of the concept is less clear.

It is not certain whether this finding will generalise to other concepts or to older children, but a similar point relating to the greater difficulty of production of the appropriate language is certainly noted in the first APU secondary report in relation to 15 year olds:

> ...nearly all the testers commented that the ability of many pupils to express themselves clearly was the main stumbling block (APU, 1980b).

Wheatley stresses the weaknesses of reliance on verbal explanations as an indication of understanding:

> Words are not the only medium for knowing, although a study of our educational practices would belie this statement. I was raised on the adage, 'You don't know it if you can't explain it'. I now know this to be false.... Children can 'know' without being able to state their thoughts in words. (Wheatley, 1977)

Nevertheless, in spite of Wheatley's warning against over-dependence on verbal facility, it is clear that a central aim of mathematical education must be to enable children to express their mathematical ideas verbally. This includes the ability to listen and to talk about mathematics, as well as to read and to write about it.

4.1.2   Learning the Language of Mathematics

The acquisition of language and concepts is a dynamic process. It is not an all-or-nothing passive type of learning. The child's understanding and use of language varies with the involvement of the child in the situation in which it is used, and with the relevance it holds for him. Since language development is dynamic in nature it is essential that the child and teacher should discuss various meanings and interpretations of words and phrases so each becomes aware of what the other means and understands by particular linguistic forms. The teacher is then in a position to help the child express himself more coherently.

Nicholson wondered to what extent this negotiation of meaning can be accomplished by using less specialised terminology than is typical of the language of mathematics. He gave the following example:

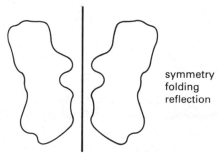

symmetry
folding
reflection

Fig. 4.3   (Nicholson, 1980)

The relationship which is perceived between the two parts of this diagram is one which can be described by the term 'symmetry'. However the child will probably first encounter it within a practical context involving perhaps the painting of a shape and folding it over to make a print. It may then be called a 'folding'. Later on when he comes to learn about transformational geometry, he will hear the term 'reflection'. Perhaps the child needs to employ the less specialised term of 'folding' because it has more meaning for him. However if and when he does come to acquire the other linguistic forms the negotiation of meanings and interpretations between teacher and pupil should certainly include reference to the everyday language being used by the child. According to Nicholson,

> This will lead to the consolidation of the concept in relation to the word, and to the confirmation of the word as the name of the concept. (Nicholson, 1980).

The first APU Primary Report found that,

> In the practical interview words could be used to probe and draw attention to particular features of the situation under discussion and the pupils' understanding of the terms and expressions to describe it. In cases where the meaning of a term (e.g. 'estimate') was not understood, or needed to be negotiated, (e.g. 'different' in 'different number pattern') then prompts improved performance.

> For example, one pupil's response to the first question (Estimate the length of line *l*) was, 'Does that mean when you say how long it is?' He was assured that it did and went on to complete the topic with ease.
> (APU, 1980a)

In a booklet entitled *Children Reading Maths* edited by Rothery (1980) it is considered that some technical terms in mathematics can be avoided and replaced by phrases such as 'top number' for 'numerator' etc. However many specialised terms have an essential and rightful place in mathematics and it is necessary that they are incorporated into the learning and teaching of the subject.

Austin and Howson suggest that in the last twenty years curriculum development in mathematics has made it increasingly necessary for pupils to be able to read with understanding. Although there is now an awareness of the need to introduce new vocabulary with greater care, they feel that not sufficient attention is directed toward ensuring that it is in fact understood. Furthermore, they point out that there are very real dangers in simplifying verbal forms in order to facilitate readability. They mention studies which suggest, for example, that individualised learning schemes involving workcards may use a basic vocabulary in a very limited fashion thus creating barriers to both language and concept development. Also a specific and restricted vocabulary, often used in books and examinations, tends to encourage pupils

> to learn connotations derived from the problems and not through experience with natural language. Such artificiality is unlikely to contribute either to the student's motivation or his ability to apply mathematics. (Austin and Howson 1979)

Austin and Howson advocate that rather than just simplifying written material in mathematics children should be specifically taught to read it. They argue that, particularly with less able children, it is best for them to read and discuss a mathematics question as an exercise in English in the first instance. Sweet also maintains that:

> If we were to teach mathematics as we teach English we would perhaps begin to have a few less negative attitudes to cope with. No child is taught to read before he can speak, nor is he expected to write before he can read…. (Sweet, 1972)

In a similar vein Trivett asserts that the language of mathematics is really never taught as a language and often, particularly where the teaching of number is concerned, the emphasis is on:

> the written codings only, with perhaps a hastily passed over superficial and temporary attachment to meaning. (Trivett, 1978)

Later on in this section there is a more detailed look at specific difficulties children experience with the words and symbols involved in the language of mathematics. Before this there is a section on analysis of the different aspects of the use of language which may give rise to low attainment in mathematics and is particularly related to the sequence of steps involved in solving problems.

Overall the research literature is rather limited for, as Austin and Howson (1979) point out, interest in the role of language in mathematics learning is a relatively recent development.

4.2 HOW DOES LANGUAGE AFFECT PROBLEM-SOLVING ABILITY?

Recently research from Australia has looked into the errors children make during the problem solving process. Much of this analysis has been carried out by using a particular model for classifying errors put forward by Newman. Although its use is limited to problems involving only a single step to solution it is nonetheless very useful as a guide for outlining the role of language in the problem solving process. The Newman classification of errors is described by Watson (1980) as follows:

1. *Reading Ability*
   Can the pupil read the question?
   (i) Word recognition: (ii) Symbol recognition.

2. *Comprehension*
   Can the pupil understand the question?
   (i) General comprehension: (ii) Understanding of specific terms and symbols.

3. *Transformation*
   Can the pupil select the mathematical processes which are required to obtain a solution?

4. *Process Skills*
   Can the pupil perform the mathematical operations necessary for the task?

5. *Encoding*
   Can the pupil write the answer in an acceptable form?

Two other types of error can occur at any stage throughout the sequence of these five error categories.

1. *Motivation*
    The pupil could have correctly solved the problem had he tried.

2. *Carelessness*
    The pupil could do all the steps but made a careless error which is unlikely to be repeated.

Another category of error is:

Question form
This is where the pupil makes an error because of the way the problem has been presented – it may be ambiguous for example.

Clements (1980) gives examples for each type of error:

Reading Error
i. Jane, aged 12 years, was given the following question:

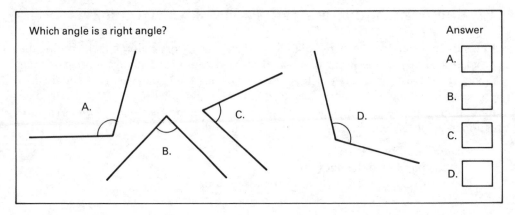

Fig. 4.4

She asked,

> When it says here, 'Which angel is a right angel, does it mean that the wings should go this way, or that way?'

ii. Billy, aged 12 years, gave the answer, 96 hours, to the following problem:

> What does fifty-six minus forty equal?

He had read it as:

> What does fifty-six minutes forty equal?

and then reasoned:

> It didn't tell me what I had to do, so I added and got ninety-six. Now ninety-six is more than sixty, so the answer must be in hours.

Comprehension Error

Charles, aged 12 years, could read the words but had difficulty with understanding the meaning of the following question:

> Sam goes to bed at 10 minutes to 9. John goes to bed 15 minutes later than Sam. What time does John go to bed?

He gave 15 as his answer because:

> It says John goes to bed fifteen minutes later, so the answer must be fifteen.

Transformation Error

Take the problem:

Here are some children.

I have 24 lollies and I want each child to have the same number of lollies. How many lollies will I give each child?

Fig. 4.5

John was able to read and understand the problem but could not transform it into an appropriate ordering of mathematical concepts and procedures. He said:

> There are twelve children, and twenty-four lollies; 12 into 24 goes 2, so we have two twelves; you multiply these two twelves; 12 times 12 is 144.

For this question Percy, aged 12 years, gave the answer 'one' because:

> I would give each child one lolly and keep 12 for myself.

From the wording of the question this should be regarded as a correct reply. His error, if it can be regarded as such, is due to the *Question Form*.

Process Skill Error
For the question:

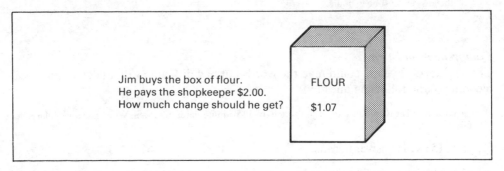

Jim buys the box of flour.
He pays the shopkeeper $2.00.
How much change should he get?

FLOUR

$1.07

Fig. 4.6

Elaine used a faulty algorithm and gave as her answer $1.93.

Encoding Error
On the same question Kelvin was marked wrong as he just wrote down '93' which was judged to be an unacceptable written form within the context of this question.

   All these errors came to light through oral discussion between teacher and pupil. However on a purely written test it becomes impossible to pinpoint the exact source of difficulty. One of the items on Ward's (1979) survey of 10 year olds here in Britain high-lights this. It was a written test so it is not possible to give an accurate analysis of the errors committed, but from the replies it seems that children were responding incorrectly to the item for reasons associated with at least one, if not all the categories of error outlined above. They were required to:

Write <, >, or = in the circles to make these correct.

$$3 \times 2 \quad \bigcirc \quad 3 + 4$$

$$90p \quad \bigcirc \quad £1$$

$$2 \times 6 \quad \bigcirc \quad 4 \times 3$$

Some children seemed to interpret these as a 'write the answer in the box' type question and would just work out the left hand side as in: $3 \times 2 \textcircled{6} 3 + 4$.
Some wrote 'or' in the circle, and others, two of the symbols such as: '<>'.
   Clements (1980) cites a study by Newman where she investigated the errors made by twelve year olds on a 40-item test. In her investigation Newman (1977) gave a written test to 917 children who were drawn from 31 different classes. She interviewed the four lowest performers from each class, i.e. 124 twelve year olds in all. Each interview was structured according to her error classification list. In the interview the pupil would be asked to attempt for a second time each item in which he was unsuccessful in the original

test. Once the pupil had attempted it, whether successfully or not, Newman then proceeded according to the following guideline:

1. Please read the question to me. If you don't know a word leave it out.
2. Tell me what the question is asking you to do.
3. Tell me how you are going to find the answer.
4. Show me what to do to get the answer. Tell me what you are doing as you work.
5. Write down the answer to the question.

The 124 low attainers had made a total of 3002 errors on the original test. It appeared that over 70 per cent of these errors were repeated during the interviews. (Special care had been taken in compiling the questions and it appeared that no errors could be attributed to Question Form.)

From Newman's results it could be seen that almost half the errors occurred before reaching the stage where children had to implement process skills e.g. performing a calculation. Also more than half the items did not require any transformation process since they were already in a mathematical form such as $\begin{array}{r} 554 \\ -108 \\ \hline \end{array}$ and many of the errors involved

the reading and comprehension of these symbols.

Clements has also collected interview data from 92 low attainers and 92 average attainers (all 13 years old) and has classified their errors according to the Newman classification. (See the following table – each entry relates to the stage at which the first error occurred.)

A classification of 1981 errors made by 92 low achievers and 92 average achievers (grade 7), 1977–1979 (Clements 1980)

Error category	Low achievers (n = 92)		Average achievers (n = 92)	
	No. of errors	% of errors	No. of errors	% of errors
Reading	117	8	18	3
Comprehension	225	16	32	6
Transformation	401	28	150	28
Process skills	351	24	126	23
Encoding	37	3	12	2
Carelessness or motivation	306	21	206	38
Total	1437	100	544	100

It is interesting that although the average achievers are committing 9 per cent of their errors at the reading and comprehension levels, the *low attainers are making a much higher proportion of their errors at these two initial stages* in the problem solving process; about a quarter of their total errors occurring here.

The Newman and Clements data seem to indicate that reading and comprehension difficulties with words and symbols play a particularly crucial role in childrens' low attainment in mathematics.

Clements comments that the Newman classification has been found to be a useful diagnostic strategy to follow by a teacher when helping individual pupils in the classroom. *Rather than showing a child how to do a problem, it is of far more use to both pupil and teacher to ask the child to read the problem and for him to say what the question is asking him to do, and so on.* However, as Clements himself points out, much deeper probing is also necessary in order to clarify more precisely the nature of their difficulties within each error category.

He cites the case of John, a 14 year old, who was given the task:

Here are three fractions: $\frac{1}{3}$ $\frac{1}{4}$ $\frac{2}{5}$
Write these fractions in order of size, from smallest to largest.

John:	(attempts question – mumbles to himself, and after about a minute writes down: $\frac{1}{3}$ $\frac{2}{5}$)
	'I'm no good at fractions. Is that right?'
Interviewer:	'Read the question to me.'
John:	'Here are three fractions... Ah... one-third... Ah...one-fourth... Ah... two-five. Write down the fractions in order of size from smallest to largest.'
Interviewer:	'Good. How are you going to do that John?'
John:	'I haven't a clue. Can't do fractions.'
Interviewer:	'How did you get that answer?'
John:	'Dunno. Guessed. Was I right?'

(Clements 1980)

John's error was classified as a comprehension error. Although he had not read $\frac{2}{5}$ correctly, Clements felt that this had not prevented him from proceeding. John's main problem seemed to be that he did not know the meaning of the symbol $\frac{2}{5}$. Clements continued interviewing him. He recounts:

After preliminary discussion on the meaning of 'one-third' and 'one-fifth', I wrote the numeral $\frac{2}{7}$ and asked him to read it. He read: 'two-seven'. I told him that it should be read 'two-sevenths', and that $\frac{1}{4}$ was usually read 'a quarter'. After a while he was able to read numerals such as $\frac{2}{5}$, $\frac{4}{7}$, $\frac{3}{8}$ correctly. I then asked him 'to find the number in the box'

$$\frac{3}{8} \text{ of } 16 = \square$$

He said he could not do it because he did not know what $\frac{3}{8}$ meant. I drew a rectangle and asked him to shade three-eighths of it. He had no idea what to do. I explained $\frac{3}{8}$ is 'three lots of $\frac{1}{8}$', and showed him how to shade three-eighths of the rectangle. He appeared to be happy with the explanation, but when asked to shade five-sixths of another rectangle he could not do so. He could not shade one-eighth of a circle. When showed sixteen marbles he could separate out one-eighth of them from the others, but did not know how many marbles he would have if had three-eighths of the 16 marbles. My strong impression was that any numeral of the form $\frac{m}{n}$ where $m \geq 2$ evoked no visual imagery in John's mind. (Clements, 1980)

The remainder of this section is concerned with literature and research which gives a similar deeper insight into the nature of the difficulties experienced with the words and symbols of the language of mathematics. Parts of the section on 'Number' have already given many examples of errors in the category of 'Process Skills' so the emphasis that follows will be on the other kinds of errors in the Newman classification.

4.3   WHAT ARE SOME OF THE SPECIFIC DIFFICULTIES CHILDREN EXPERIENCE WITH THE LANGUAGE OF MATHEMATICS?

When looking at the literature concerning a more detailed account of the difficulties children experience with the language of mathematics it becomes apparent that although the Newman classification is a useful way to organise the data, its simplicity is not totally appropriate to the reality of the situation. For instance, since language and thought are so closely intertwined, the way a child reads a mathematical symbol may depend upon his understanding of it and vice versa. For example if he reads '3 + 2 = 5' as 'three add two makes five', he may not fully understand '=' as representing equality or equivalence. More is said of this later. Apart from an interdependence between the reading and comprehension of symbols, a child's understanding of what symbols represent may influence the way in which he transforms a verbal problem into a symbolic expression. Again further discussion of this occurs later. Because of this inter-relatedness between categories of errors the headings provided by the Newman classification have been adapted in this review to incorporate these aspects. Therefore what follows is concerned with reflecting the difficulties children experience with:

(a)   the reading and comprehension of words;
(b)   the reading and comprehension of symbols;
(c)   transforming problems into mathematical operations;
(d)   attaching meaning to symbols;
(e)   difficulties with encoding.

4.3.1   The Reading and Comprehension of words

4.3.1.1   *Words which are Similar in Appearance or Sound*

The previous quotations from Clements (1980) include examples of children who made errors in reading words such as 'angel' for 'angle' and 'minutes' for 'minus'. Another source of confusion apart from the visual similarity between written words is the similarity of certain spoken words in terms of their pronunciation. For example 'ten' and 'tenth'; 'hundred' and 'hundredth' etc. Brown in her study of secondary school children found that for the item:

5214   The 2 stands for 2 HUNDREDS
521   The 2 stands for 2 _____

> The major error was to write 'tenths' instead of 'tens', which indicates a confusion over words but suggests that children basically understood the idea of 'tens'. (Brown, 1981c)

Also the 'teens' may often be confused with the '...ty's' – for instance 'fourteen' and 'forty'. Haigh gives an example of this kind of error which was made by a 7 year old who was counting in *ones* from 'ten' saying:

> ten, twenty, thirty, forty, fifty, sixty, seventy, eighty, ninety, twenty, twenty-one. (Haigh, 1979)

Other words are the same but depending upon the context or the use of definite and indefinite articles mean very different things, e.g. 'It is tenth', 'It is the tenth', and 'It is a tenth'. The first two phrases imply order whereas the third is to do with a fractional quantity, but all involve the word 'tenth'.

In Rothery (1980) three broad categories of words are distinguished. These are:
(i) Words which are specific to mathematics and not usually encountered in everyday language;
(ii) Words which occur in mathematics and ordinary English but involve different meanings within these two contexts, and
(iii) Words which have the same or roughly the same meaning in both contexts.

4.3.1.2   Words Specific to Mathematics

Such words include for example: hypotenuse, parallelogram, coefficient, multiply etc. It is pointed out in Rothery (1980) that many of the difficulties caused by such words occur because children only come across them in their mathematics lessons where they may be defined once and never again. Children usually do not have easy access to looking them up, as few school textbooks have an index. It is suggested that the onus should be on the teacher to repeat definitions, discuss them and index them in some way. Krulik (1980) suggests children should compile their own dictionaries using illustrative sentences and examples. Ginsburg (1977) comments that:

> The artificial (albeit carefully defined) language of mathematics can present the child with considerable difficulties. 'Plus', 'congruent', 'minus' and the like are unfamiliar words that children frequently misunderstand. Defining the artificial words does not guarantee comprehension.

Ginsburg cites the case of Patty, aged 9, who only associated the word 'plus' with the standard algorithm for addition, which she always did incorrectly in any case. For example, for '10 plus 1' she wrote:

$$\begin{array}{r} 10 \\ +1 \\ \hline 20 \end{array}$$

When asked to draw marks on her paper to find what 10 and 1 made she then said:

Patty:	'Altogether it would be 11.'
Interviewer:	'OK, what about 10 plus 1, not altogether, but plus?'
Patty:	'Then you'd have to put 20.'
Interviewer:	'What if we write down on paper, here's 20, now I write down another 1, and you want to find out how much the 20 and 1 are altogether.'
Patty:	'It's 21.'
Interviewer:	'OK, now what would 20 plus 1 be?'
Patty:	'20 plus 1?' She wrote

$$\begin{array}{r} 20 \\ +1 \\ \hline 30 \end{array}$$

Following a lengthy interview with Patty, Ginsburg concluded:

> For Patty, language was crucial. Given the word 'plus' she applied an incorrect addition method to both objects and written numbers. Given 'altogether', she used a sensible counting procedure, again for both objects and written numbers. 'Altogether' is a natural word for addition: Patty probably used it in everyday life to talk about adding things. 'Plus' is a school word that Patty seems to have associated with a wrong algorithm that she did not understand. (Ginsburg, 1977)

Another clear example of the difficulty an unknown word can cause occurs in the national monitoring programme in the United States (National Assessment of Educational Progress). Carpenter et al. (1981) report the results in the following table concerning the perimeter of a rectangle:

		Per cent responding	
	Exercise	*Age 9*	*Age 13*
1.	What is the DISTANCE ALL THE WAY AROUND this rectangle?		

10 ft.
6 ft.

		Age 9	*Age 13*
●	32 feet	40	69
○	16 feet	39	12
○	60 feet	4	13

2. What is the PERIMETER of this rectangle?

10 ft.
6 ft.

		Age 9	*Age 13*
●	32 feet	8	49
○	16 feet	66	25
○	60 feet	3	17

3. Mr. Jones put a rectangular fence all the way round his rectangular garden. The garden is ten feet long and six feet wide. How many feet of fencing did he use?

		Age 9	*Age 13*
●	32 feet	9	31
○	16 feet	59	38
○	60 feet	14	21

These results demonstrate not only the extra difficulty caused by the use of the word 'perimeter' but also the effect of setting a long verbal problem rather than a short instruction with an accompanying diagram.

4.3.1.3   *Words which have Different Meanings in Mathematics and Ordinary English*

Words which are used in everyday language and have different meanings when used in mathematics can often be a particular source of difficulty to children.

For example, Matthews (1980) in investigating the problems of comprehension of 'subtraction words', considers the word 'difference'. She gave the following question, taken from the ILEA Checkpoints Assesment Cards (ILEA, 1979) to 81 children aged six or seven years, drawn from three different London schools:

> (I gave the child 9 fir-cones and gave myself 7.) I've got 7 fir-cones and you've got 9. What is the difference between the number you've got and the number I've got? (Matthews, 1980).

30 out of all the 81 children gave a correct reply, and 19 were judged to make an error due to misunderstanding of the meaning of the word 'difference'.

> To explain 'the difference between the numbers' gave some of the children much food for thought and pro-
> duced some thoughtful replies such as: 'well, if I were 9 and you were 7 I would be older than you'. And
> 'well if I had 9 sweets and you had 7 it wouldn't be fair'. Or... the child who, although she could not give the
> difference as 2, nevertheless offered 8. I feel sure you can work out her reasons! (Matthews, 1980)

In Rothery (1980) it is pointed out that there are varying degrees of relatedness between words in their mathematical and everyday contexts. For example, 'gradient' has
more similarity of meaning in the two contexts than does 'product'. Krulik (1980) gives
other examples of words with dual meanings, e.g. volume, count, odd, prime, power,
mean.

Sometimes words are met in their everyday usage before their mathematical usage
e.g. 'difference'. Other words are often encountered within a mathematical context before they are used in ordinary language e.g. 'parallel'. Some words which have a specific
mathematical meaning and an everyday meaning occur in mathematics in both senses
e.g 'similar', 'shape'.

There can also be different mathematical meanings for the same words e.g. 'base',
'figure'. Even the word 'into' can be used to indicate either multiplication or division
e.g. 5 (x+3) is sometimes read as '5 into x plus 3' and $5\overline{)12}$ can be read '5 into 12'. In
Rothery it is also noted that some words carry derogatory overtones:

> Fractions may be *vulgar* or *improper,* a *mean* is a rather underhand average, a *negative* number feels less
> good than a *positive* one. (Rothery, 1980).

Again meanings should be discussed and children encouraged to catalogue and illustrate meanings in sentences. Krulik (1980) gives the following examples for the word
'prime'.

(Mathematical)   A prime number is a number that is divisible only by itself and
                        one.

(Non-mathematical)   The most often viewed T.V. programme is always placed in
                        prime time to attract a large audience.

(Non-mathematical)   The water pump will not begin to work unless you prime it.

There are some words used in certain contexts which may cause conceptual confusions
and should probably be avoided altogether. This is particularly the case where everyday
connotations are used to draw attention to superficial symbol manipulation thus obscur-
ing rather than emphasising the all important underlying concepts e.g. 'borrow' in subtraction: 'add a nought' to multiply by ten – this is particularly misleading because of
the different sense in which 'add' is used but also because of the difficulties this creates
when multiplying decimal quantities by ten. Speer points to the matter of 'reducing' a
fraction, saying:

> If 'reduce' connotates making something smaller, then doesn't the process of reducing ⁶⁄₈ to ¾ yield a
> 'smaller fraction?' (Speer, 1979).

Examples of children's usage of these sorts of terms were given in section 3.

4.3.1.4   *Words Meaning the Same for Mathematics and Ordinary English*

In Rothery (1980) it is mentioned that because many everyday words have a different meaning when used in mathematics, the main problem with words which have the same meaning in both contexts is knowing that they do in fact mean the same. Sometimes children may think that an ordinary word takes on some mystical meaning when used in a mathematical setting or maybe they do not fully understand its everyday meaning in any case. Brown found instances where secondary children did not understand 'between' in the following item:

Write down any number between 4000 and 5000.

Kim said: 'What do you mean "between?"'

Other children, during interview on this item, also had to have 'between' explained to them. Later on Kim was asked:

'Is anything in between? … Suppose it was between forty and fifty.'

She immediately replied,

'Forty-five.'   (Brown, 1981c)

It may be that the phrase 'in between' is more easily understood than just 'between' because it has overtones of some physical action which children can more readily grasp. This is particularly the case with their interpretation of symbols which is discussed later.

A well-documented group of words are those used for comparison, probably since they occur in the Piagetian conservation tasks. For example Lawson, Baron and Siegel (1974) found that 3 to 5 year old children confuse terminology related to number ('more') with that related to length ('longer'), and that this was one of the reasons for lack of conservation. Donaldson and Balfour (1968) also found in the same age group that children were interpreting 'less' as meaning 'more'; when asked to indicate which now contained less sweets they pointed to the same row as they did when asked to indicate which now contained more. That this confusion is verbal rather than conceptual was demonstrated by Bryant (1974), who showed that 4 year old children could be trained by a non-verbal method to choose either the larger or smaller of two collections with equal ease.

Matthews, in the article referred to on page 343, also found problems with the word 'fewer'. Eighty-one 6 and 7 year olds were given the following task, again taken from the ILEA Checkpoint Assessment Cards (ILEA, 1979).

(I gave the child a box of about 25 crayons.) Take 9 crayons out of the box for yourself. Now give me 3 fewer than you've got. How many do I have? (Matthews, 1980)

She found that only 13 out of 81 children obtained the correct answer, 61 appearing to misunderstand the word 'fewer'. However when she translated it to 'less' or 'not so many', many children still gave her 3, rather than 3 fewer.

It can be seen then, that even with ordinary English vocabulary, there is still a need to discuss meanings with children.

4.3.1.5   Which Words are not Readily Understood?

Otterburn and Nicholson (1976) report a study which investigated 300 pupils' under-standing of certain words commonly used in CSE Mathematics courses. The children were drawn from approximately the middle 50 per cent of the ability range. They were presented with a list of 36 such words and were required to fill in four columns of infor-mation for each term as in this example.

Word	1 *Understand it:* *Yes/No*	2 *Symbol:*	3 *Draw* *Diagram*	4 *Describe in* *words*
e.g. Plus	Yes	+	∴ + ∴ = ∷	Add, eg. four plus five are nine.

Responses were classified into three categories:

Correct: If any replies in the last three columns indicated a correct interpretation of the word.

Blank: If nothing was given in any of the last three columns to indicate understanding, even if 'yes' had been written in column 1.

Confused: If any of the responses to the last three columns were in disagreement or not at all clear, they were classified as 'confused'.

The results are given in the following table:

Reproduced from Otterburn and Nicholson (1976)

		Percentages	
Word	Correct	Blank	Confused
Minus	99.7	0.3	0
Multiply	99.7	0.3	0
Square	94	2	3
Remainder	92	8	1
Fraction	91	8	0.3
Rectangle	88	4	8
Parallel	77	19	3
Area	72	11	17
Decimal Fraction	68	29	2
Square Number	65	24	10
Radius	64	25	11
Perimeter	64	29	8
Volume	58	35	7
Prime Number	52	34	13
Kite	49	20	32
Reflection	45	51	4

	Percentages		
Word	Correct	Blank	Confused
Average	43	37	19
Intersection	41	50	9
Square Root	40	44	16
Rotation	37	60	3
Parallelogram	37	41	22
Perpendicular	35	61	4
Factor	32	62	6
Rhombus	31	47	22
Union	26	65	9
Ratio	25	71	4
Gradient	23	73	4
Symmetry	22	75	3
Product	21	59	20
Multiple	20	45	34
Similar	19	67	15
Congruent	18	77	5
Index	16	78	5
Mapping	16	81	3
Integer	15	76	9
Trapezium	11	79	10

(Note: the figures above are generally given to the nearest whole number.)

The authors point out that it should be borne in mind that responses classified as Blank do not necessarily mean the words are not understood – they may instead reflect the inability of pupils to express their meaning.

They give examples of some responses to several of the words. For instance:

Product
Common errors were to confuse it with 'sum' or 'difference'. A number of pupils described it in terms of its everyday usage as in 'something produced'. Other confused comments included:

The product of 6 could be 3 x 2.
To bring or take away.

Multiple:
Common causes of confusion for this word seemed to derive from pupils thinking it said or meant 'multiply', e.g:

It is really long multiplying but with two numbers.

Of the 103 confused responses for this word, 85 muddled it with factor e.g:

Multiple = factors of eight = two fours or eight ones.
2 is a multiple of 6 and 10.
6 and 8 are multiples of 48.

Many descriptions gave rise to other language difficulties particularly in the misuse of words such as 'even, evenly', 'equal and equally' for 'exact and exactly'. e.g.:

6 = 3, 2, 1, e.g. numbers that will divide into a number evenly.
4 |20   A number that can be divided equally into another number.
 $\underline{\quad 5}$

Other confused replies for *multiple* included:

7 + 8 = 15   When numbers are increasing.
A number which is and divided by itself.
Multiple choice question – more than one choice.

Parallelogram:
There were many confusions with this term. Children were sometimes muddled with ideas of 'parallel lines', 'square' and 'rhombus'. Also diagrams were often very poorly drawn and classed as unacceptable. e.g.

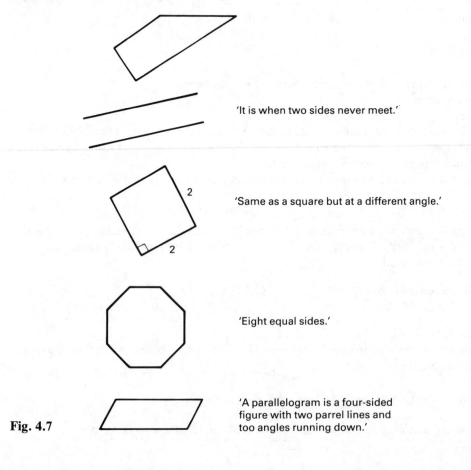

'It is when two sides never meet.'

'Same as a square but at a different angle.'

'Eight equal sides.'

'A parallelogram is a four-sided figure with two parrel lines and too angles running down.'

Fig. 4.7

Nicholson (1977) reports further investigations to study children's understanding of mathematics vocabulary. Two other approaches were employed. The first involved setting the word within a specific context which was kept as simple as possible, e.g.

Multiply 12 by 6.
What is true about the sides of a rhombus?

A child was judged to understand the word if he produced a correct answer; 'Blank' indicates an omission, and a wrong answer was classified as 'confused'. The results were as given in the following table:

(Reproduced from Nicholson (1977))

		Percentages	
Word	Correct	Blank	Confused
Multiply	99.5	0	0.5
Remainder	98	0	2
Factor	91	5	4
Reflection	86.5	0.5	13
Parallel	86	5	9
Square Root	81.5	3	15
Intersection	75.5	3	21
Square Number	75	4	21
Prime Number	68.5	9.5	21.5
Parallelogram	65.5	5.5	29
Rectangle	64.5	4.5	31.5
Union	55	5	40
Mapping	41	32.5	26.5
Rotation	34.5	6	59.5
Rhombus	32.5	24	44
Product	20.5	4.5	75
Multiple	11.5	2	86.5
Integer	9.5	18.5	72

Again there was evidence of confusion with each of the terms *Product* and *Multiply*, e.g.

Product – 'What is the product of 15 and 3?'

38 of 185 gave an acceptable answer. 139 gave confused responses. Of these 66 replied '5' i.e. '15 ÷ 3' and 54 gave the answer '18' i.e. '15 + 3'.

Multiple – 'Give one example of a multiple of 30.'

21 of 185 gave an acceptable answer. 160 gave confused replies of which 92 gave factors in the form 5, 10, 6, 3, 15 or 2 and another 62 answers were of the form 5 x 6, 10 x 3, 2 x 15.

Nicholson regards such a test as being very helpful for teachers in assessing their pupils difficulties with mathematical vocabulary.

The other approach to studying children's comprehension of mathematical words involved an exercise where pupils were required to fill in a missing word which named the concept being described, e.g: 'When 15 is multiplied by 3 the answer is 45. We call 45 the .... of 15 and 3.'

The responses were classified as before but with an additional category of 'Neutral' which comprised instances where the choice of word was not totally correct but may have indicated a step in the right direction e.g. for the item:

'if two straight lines are at right angles one of them is said to be .... to the other.'

Responses classified as 'Neutral' were for example 'horizontal', 'vertical.'

Nicholson (1977) found the results given in the following table:

SCHOOL A — Classes 5CA and 5CB
Responses (46 pupils)

Word	Acceptable	Neutral + Blank	Confused
Multiply	45	0 + 0	1
Rectangle	43	1 + 0	2
Parallel	42	3 + 0	1
Average	39	1 + 5	1
Radius	36	0 + 3	7
Remainder	34	1 + 0	11
Square root	25	5 + 2	14
Factors	23	2 + 1	20
Parallelogram	22	0 + 8	16
Prime (numbers)	18	5 + 7	16
Rhombus/Diamond	17	7 + 6	16
Symmetrical	16	19 + 10	1
Square (numbers)	14	11 + 12	9
Multiples	13	8 + 9	16
Product	8	10 + 10	18
Congruent/Identical	7	18 + 13	8
Perimeter	5	21 + 2	18
Perpendicular	1	12 + 8	25

The following are examples of some of the confused responses.

For: '3 and 7 are ...... of 21'

a very common answer was 'multiples' instead of 'factors/divisors'.

'The numbers 2, 3, 5, 7, 11, 13, 17, 19 are all examples of .... numbers.'

Instead of 'prime' a common mistake was to write 'odd'.

Nicholson concludes:

> These pupils who enter for CSE mathematics (broadly the middle 50% of the whole ability range) have significant difficulties in understanding some of the mathematical terms in common use. It is important that teachers should recognise the extent of their difficulties and work continuously for better understanding. To that end it is relatively simple to devise and administer short diagnostic tests, and ... possible alternative forms are suggested above. Once difficulties have been accurately diagnosed the teacher can then attempt to remedy them – perhaps on an individual or small group basis. (Nicholson, 1977).

4.3.1.6 Some Suggestions for Helping Children Understand Mathematical Vocabulary

In Rothery there is a selection of ideas to help children read and understand mathematical vocabulary. Many of the suggestions emanate from the U.S.A. where there seems at present to be a more active concern over the importance of the role of language and the readability of mathematical material in mathematics education. Specific activities such as the following are quoted from an article by Cribb.

Selecting best answer:
Bisect means   (a)   cut into pieces
                (b)   cut into two pieces
                (c)   cut into two equal pieces
                (d)   cut twice

Matching words with phrases:
Evaluate   Put in a table
Expand ⎯⎯⟶ Find the value of
Tabulate ⎯⎯⟶ Remove the brackets

Words, symbols and clues:
The idea here is for the pupil to match the scrambled words with the correct symbolic form and also the correct description.

Word/letters	*Symbols*	*Clue Description*
U E B C	□	Has three sides
U A R Q E S	△	Has 6 faces
V D E I I D	 ⬡ (cube)	Opposite to multiply
T A E I N R G L	÷	Has 4 vertices

Fig. 4.8   (Examples from Cribb as quoted by Rothery, Ed. 1980).

They also suggest such activities as simply structured crosswords where the clues give a description of the meaning of the word to be filled in. Further ideas are presented in the next section concerned with the reading and comprehension of symbols.

4.3.2 The Reading and Comprehension of Symbols

4.3.2.1 Introduction

In Rothery it is stated that the most striking feature of the language of mathematics is the symbolism where much of it is a very short and precise code for words and groups of words. Much of this symbolism has meanings which vary according to the spatial presentation of the symbols e.g.

3^2 means 3 squared
32   means thirty-two or $(3 \times 10) + 2$
(3, 2)   coordinates, or number pairs depending upon the context.

                                                      (Rothery, Ed. 1980)

Children, particularly younger ones, often confuse number symbols which look similar e.g., 5 and 2, 6 and 9, 12 and 21 etc. may often be muddled. In addition to words with dual meanings, there are symbols which are not specific to mathematics, and these may be confused with their role in ordinary language. For example as Trivett (1978) points out, 'X' may mean a kiss or a wrong answer: certain punctuation marks take on very specific meanings in the language of mathematics such as '−' and '.'.

One of the main problems with the spatial structuring of symbols in mathematics is that the rules for it are not consistent, this being particularly the case with algebraic expressions. This is illustrated in Rothery by the example that:

> 3x, when x = ½, is not the same as 3½. (Rothery, Ed. 1980).

Booth cites the case of a 4th year secondary pupil in a top stream who was asked the meaning of 'y' in '5y'.

> WA: 'y could be a number, it could be a 4, making that 54. Or it could be 5 to the power of 4, making it 20.' (Writes 54, 5^4).
>
> LB: 'Do you think it *could* be either, or do you think it's one of those only your're not sure which?'
>
> WA: 'No it could be either, you can't really say.'
>
> LB: 'So y could be any number.' (Pupil nods). 'Suppose I made it 23. What would you write down then?'
>
> WA: 'Oh!' (laughs) 'Well!' (laughs) 'five hundred and twenty-three? But I dunno – it doesn't seem very promising. Wait, it could be 28, 5 plus 23 ... yes ... There again, it could be 5 to the power 23.' (Writes 5 + 23, 5^{23}).
>
> (Booth, 1982)

Booth suggests that it would be advisable not to omit the multiplication sign too soon when working with algebraic products and then to make frequent checks that a pupil is interpreting such a term correctly.

One of the crucial aspects concerned with the reading of mathematical symbolism is that it is quite often not a straightforward left to right matter. Krulik (1980) gives the example:

$$\tfrac{3}{4} \;+\; \tfrac{1}{4} \;=\; 1 \qquad \text{which is read}$$

in the direction shown by the arrows. He suggests children should do such arrow diagrams as an exercise in reading symbolic expressions.

The following example is taken from Rothery:

$$(3 + x)^2 = 49$$
$$3 + x^2 = 49$$

It is pointed out that both of these may be read as 'three plus x squared is 49' but particularly in the first instance

> the reader must scan the phrase repeatedly to build up an overall picture of meaning. (Rothery, Ed. 1980).

Kieran investigated the methods used by six American children in the 12 to 14 age range for evaluating arithmetical expressions. All six evaluated all expressions by working from left to right, regardless of the 'order of operations' conventions that they had all been taught.

e.g. Greg: '5 + 2 x 3 = '
Interviewer: 'What is the value of that left side?'
G: '21'
I: 'So you're adding the 5 and 2 first, are you?'
G: 'Yes.'

(Kieran, 1979)

When asked if they could insert brackets, two children put them only round the first two numbers in the expression e.g. (5 + 2) x 3 which retained the left-to-right order they had followed anyway.

When Patricia was asked if she could put them anywhere else, she put them round the whole expression i.e. (5 + 2 x 3). Her usage was at least consistent with her explanation:

I: 'Why do you put in brackets?'
Pa: 'To show which one you do first.'

(Kieran, 1979).

Austin and Howson refer to literature which points out that apart from the normal level of abstraction inherent in mathematical symbolism, algebra takes this abstraction a stage further to the point where it,

is rich in syntax, but weak in meaning. (Austin and Howson, 1979)

The structuring of brackets, which is usually first encountered in algebra, is a complication which does not generally exist in ordinary language.

Confusion about the use of brackets in algebra is also reported by Booth. She gave the question in Fig. 4.9 to sixteen middle-ability 12 to 14 year olds drawn from five different schools in Greater London as part of the project 'Strategies and Errors in Secondary Mathematics' (SESM).

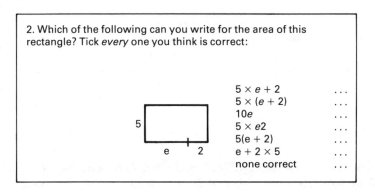

Fig. 4.9 (reproduced from Booth, 1982)

Twelve of the 16 children included 5 × e2 in their answers. A typical illustration of the reasoning leading to this was:

CC: '5 × e2, and 5 (e + 2). That's all.'
LB: 'Why do you say 5 × e2?'
CC: '5 times e2. So that means the plus 2, the e and 2 put together, times 5. That's what the answer should be, 5 times e2.'
LB: 'What does e2 mean?'
CC: 'The answer to e add 2...5 times e2...the e2 means you have to add the e and the 2 before. So it's...I think I'll just stick to that one (5 × e2) actually because you've got to add the e and the 2 first.'
LB: 'And e2 tells you to do that.'
CC: 'Yes, to get the answer.' (Booth, 1982)

This illustrates an understanding of the operations needed, but the use of idiosyncratic symbol conventions to avoid the introduction of brackets. However as in Kieran's example, these untaught rules appear to be common to a number of children, and not confined to a single school or age group.

The first APU Secondary Survey also found a high proportion of 15 year olds ignoring the effect of brackets as shown in the following item:

Which one of the following is not equal to any of the other three?

A.	$a - b + c$	12% chose A
B.	$(a - b) + c$	5% chose B
C.	$a - (b + c)$	30% chose C
D.	$a + (c - b)$	46% chose D

(APU, 1980b)

D was probably chosen the most frequently because children attended only to the different order of the letters.

Two other APU secondary items also illustrate how the increased abstraction inherent in algebra causes confusion when it cannot readily be related to a more meaningful context (see the next table). The APU mathematics team suggest that the discrepancy between the two items presumably occurred because it is easier to expand a^2 and a^3 to a x a and a x a x a respectively. However it is possible that some children may have considered that 'a + b' could be written as 'ab' as in the extract from Booth quoted previously.

Item	Percentage correct (15 year olds)	Other responses
$2^a 2^b = 2^\square$	20%	2^{ab} (45%)
$a^2 a^3 = a^\square$	55%	a^6 (10%)

4.3.2.2 Difficulties Children Experience with Number Symbols

This section is concerned with the difficulties children experience in reading and understanding the symbolisation of number. It does of course involve place value considerations and should therefore be read in conjunction with parts 3.4 and 3.8.1.7 of the 'Number' section.

Brown (1981c) studied secondary school children and found that it was easier for them to read number symbols by translating them into words than it was for them to transform the written word to symbols. Those who did have difficulty in reading the symbols were generally the least able. For 8030 they gave such incorrect responses as:

'Eight hundred and thirty' or
'Eight thousand three hundred'.

Brown points out that difficulties in translating numbers to words and vice versa may occur because the pronunciation of numbers is bounded by different rules than those for their symbolisation: e.g. 23456 : symbolically this is 2 ten thousands, 3 thousands, 4 hundreds, 5 tens and 6, usually worked out from right to left, but it is read as: 23 thousand, 4 hundred, 56, from left to right, thus condensing the columns to thousands, hundreds and units. Often children read the 2 in 23456 as 2 million since this is the next pronounced name to be used for numbers with more digits than thousands.

Brown also mentions that there are alternative forms of reading numbers up to 2000. e.g. 1300 as 'one thousand, three hundred' or 'thirteen hundred'.

She concluded from her study of secondary pupils that:

...children are generally able to recognize on sight numbers under a thousand. With numbers above that size at least 20 per cent do not recognize them and have no systematic method of pronouncing the numbers. However the proportion who cannot apply the symbolic, as opposed to the verbal, system of place value to writing and evaluating numbers seems likely to be considerably greater than this and it probably reaches 50 per cent or more of the secondary population. (Brown, 1981c).

This is well illustrated by Frances who was trying to determine which was bigger 20100 or 20010. Brown found her to be the only one during interview who, having failed in reading the numbers, then turned to attend to the symbolism, albeit rather dubiously.

Frances: 'Twenty thousand and one hundred, and two hundred thousand ...two...'
Writes down: u t h t u   u t h t u
                2 0 1 0 0   2 0 0 1 0
Frances: 'That's units, tens, hundreds, thousands, units (!) Oh that's the bigger.' (rings 20100).
                                                                                  (Brown, 1981c)

One point which recurs throughout Brown's study, which was part of the 'Concepts in Secondary Mathematics and Science (CSMS) Project, is the confusion children show over the symbolic representation of decimal fractions. Often their understanding of decimals in their symbolic form is based on inappropriate analogy to whole numbers. For instance, seventeen of the secondary pupils interviewed were asked to read out '0.29'. Eleven gave the reply 'point twenty-nine', one said, 'twenty-nine', and six replied '(nought) point two nine'. In the written test where pupils were required to write down how they would *say* '0.29' the following results occurred:

Response	*Percentage correct*			
	12 yrs	13 yrs	14 yrs	15 yrs
(Nought) point two nine	26	32	41	41
(Nought) point twenty nine	25	32	30	27
twenty-nine	19	13	8	10

(Brown, 1981b)

Brown also found that about a third of this sample considered 0.75 to be greater than 0.8 because 75 is bigger than 8 e.g.

> Jane's response was: 'This is nothing before and seventy-five; this is nothing before and just eight.' (Brown, 1981b).

A similar, but less frequent, mistake was made in comparing 4.06 and 4.5, 'because 406 is greater than 45'.
When children were required to ring the number nearest in size to:

$$\boxed{0.18} \;\rightarrow\; 0.1 \quad 10 \quad 0.2 \quad 20 \quad 0 \quad 1 \quad 2$$

over a quarter of first and second year secondary pupils chose '20', again reflecting confusion with whole number notation.

From her study Brown (1981c) concluded that the proportion of children who use a whole number analogy as a basis for their reasoning on decimal items, and who apparently do not appreciate the significance of the decimal point and place value, ranges from around 40 per cent at 11 years of age to about 25 per cent of 15 year olds.

4.3.2.3   How do Children Interpret the Operation Signs: $+, -, \times, \div$ *and* $=$?

Much of the research relating to children's reading and understanding of the arithmetic symbols $+, -, \times, \div$ emphasises the finding that they are often interpreted in terms of physical actions.

Brown (1981c) asked secondary pupils to write stories to fit a given symbolic expression, e.g. '84 – 28', '9 ÷ 3', '84 x 28', '9 + 3'. The vast majority of stories involved physical actions such as uniting things or adding-on for '+'; physical removal, or taking away for '–'; sharing something out (often sweets) for '÷'. The multiplication sign caused the most difficulty. Brown found that children often read it as 'times' which she points out is not a verb involving any straightforward physical action. Those children who did manage to compose a story for the 'x' expressions usually interpreted it in terms of repeated addition or in terms of 'rate' (e.g. for 3 x 9, 3 packets with 9 sweets in each packet).

Brown points out that although most children probably first encounter the symbolism for arithmetic operations in a concrete setting involving physical actions on objects, this limited interpretation has serious consequences later on. For example she found that children were unable to interpret an expression such as '16 ÷ 20' since, to many of them, 16 things cannot be shared between 20 people. In each of the age groups the following percentage of children indicated that they considered there was *no answer* to 16 ÷ 20.

12 yr olds	*13 yrs*	*14 yrs*	*15 yrs*
51%	47%	43%	23%

Confusion and difficulty often stem from associating physical action with the '=' sign. Austin and Howson (1979) point to the frequent conflict between natural language and mathematical symbolism with these examples:

$$8 - 2 = 6 \qquad\qquad 6 + 2 = 8$$
$$\downarrow \qquad\qquad\qquad\qquad \downarrow$$
'leaves' \qquad\qquad\qquad 'makes'

Ginsburg (1977) quotes Donna, an 8 year old who, for $3 + 4 = \square$, said,

the = sign mean that it adds up to.

He points out that when children interpret '=' in this way they find it difficult to make sense of perfectly valid forms such as: $\square = 3 + 4$. Tommy was given this expression and changed it to: $4 + 3 = \square$ because,

It's backward ... Do you read backwards?

Similarly, Behr, Erlwanger and Nichols (1980) found 6 and 7 year olds resisting sentences of this form.

Eve, given: $\square = 3 + 5$, counts on her fingers and writes: $8 = 3 + 5$, but reads, '5 plus 3 equals 8'. When given $3 = 2 + 1$ she says, 'You should put a 5 here (i.e. at 1). But when asked to read it she says,

'There...yeah, that's ...I was reading the wrong way again.' Eve explains that $2 + 3 = 5$ is easier than $5 = 2 + 3$ 'because it's (5=) on this side, and I'm used to having it on that side (=5)'.

(*Mathematics Teaching*, **92**, 14).

Children commonly changed an expression of the form $\square = a + b$ to either $a + b = \square$ or $\square + a = b$.

Behr et al. also found that 6 and 7 year olds do not view sentences like '$3 + 2 = 2 + 3$' as being about number relationships. They fail to appreciate such an expression as indicating the 'sameness' of two sets of objects. They are still interpreted as 'do something' sentences. Kay, aged 6 years when given $3 + 2 = 2 + 3$ says,

... you forgot to put the 5.

She writes $3 + 2 = 5$ and $2 + 3 = 5$. However Mel, aged 7 years, accepts $3 + 2 = 2 + 3$

'because they both have the same numbers. Only $2 + 3$ is backwards.' Mel notes a difference between $3 + 2 = 2 + 3$ and $4 + 1 = 2 + 3$ explaining that 'Two four... and one three doesn't rhyme. They are sort of equal because they both equal five. They don't go together; not made the same... they both equal five, but they're not the same.' (*Mathematics Teaching*, **92**, 15).

Another child, Dee, would not accept '$3 + 2 = 2 + 3$' at all but instead changed it to '$3 + 2 + 2 + 3 = 10$'.

Behr et al. conclude:

There is a strong tendency among ... the children to view the = symbol as being acceptable in a sentence only when one (or more) operation signs ($+, -$ etc.) precede it. Some children, in fact, tell us that the answer must come after the equal sign. We observe in children's behaviour an extreme rigidity about written sentences, an insistence that statements be written in a particular form, and a tendency to perform actions (e.g., add) rather than to reflect, make judgements and infer meanings. Moreover, we have some evidence to suggest that children do not change in their thinking about equality as they get older.

(*Mathematics Teaching*, **92**, 15).

This certainly seems to be the case, particularly when older children meet algebraic terms such as '2 + m' or 'x + y', for as we saw earlier Booth (1982) has found evidence that secondary children feel the need to 'do something' with such an expression and may condense it to, say, '2m' or 'xy'. It seems likely that the tendency noted by Behr et al. is due to the fact that children are introduced to written sums at too early an age, when the teacher is forced to translate 2 + 3 = 5 to '2 and 3 make 5' in order to make the symbolism meaningful. The child then persists with this verbalisation, since it is unlikely that at any later point he will be consciously forced to examine and re-interpret the meaning of the symbols.

4.3.2.4 A Suggestion to Help Children Read and Comprehend Symbols

An idea, which occurs in the literature, for assisting children in their reading and comprehension of symbols, is based on a matching activity or domino game. See Rothery (Ed. 1980) and Krulik (1980). Krulik's example is typical of this suggestion. A child is given the following cards and is required to match them.

$=$	equals	eight $?$ 4×2
$<$	is less than	$6 ? 10$
$\equiv$	is not equal to	$7 ? 5 + 6$
$>$	is greater than	$2^3 ? 5$
$\times$	multiplied by	$4 ? 3 = 12$

Krulik (1980)

This kind of activity is easily adapted to incorporate different symbols or different aspects to the task of reading and understanding them.

4.3.3 Transforming Problems into Mathematical Operations

4.3.3.1 Transforming Word Problems

It was seen earlier that a major source of difficulty experienced by children in the problem-solving process is transforming the written word into mathematical operations and the symbolisation of these. Even at the level of translating words into figures, particularly where zeros are required as place holders, Brown has found that secondary pupils, of average, as well as low ability, have problems. One of her items was: 'four hundred thousand and seventy-three'. The percentage correct for each year group was:

1st yr	*2nd yr*	*3rd yr*	*4th yr*
42%	51%	57%	57%

(Brown, 1981b)

Further evidence of such difficulties was outlined in Section 3.3.

At a slightly more complex level of solving a word problem involving simple operations or numbers, the expected procedure might be:

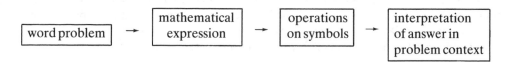

However it is clear from several studies that in many cases children solve the word problem in the context in which it is set, without ever making a translation into symbolism.

In section 3 (pages 191 and 222) the study by Carpenter and Moser (1979) was quoted in which they found that pre-school children were able to solve many word problems involving addition and subtraction before they had been formally taught any symbolism or number operations. The strategies used mainly involved counting, but some were more sophisticated. Matthews (1981), in the investigation quoted in section 4.3.1 (pages 343 and 345) found that all the eighty-one 6 and 7 year olds tested obtained the correct answer to the following problem from the ILEA Checkpoints Assessment Cards (1979),

Give yourself 11 pegs and now if you gave me 4 of yours how many would you have left? (Matthews, 1981).

However when children were asked:

'Now write a number sentence to show what you did', it became apparent that the phrase 'a number sentence' was not understood. Even when the instructions were changed to 'Now write the figures and put in the signs to show what you did to get the answer',

only 20 of the 76 were able to write $11 - 4 = 7$. In the case of a subtraction problem involving the word 'difference', only 10 children were able to write correctly $15 - 6 = 9$. Thus the ability to obtain the correct answer is by no means dependent on the ability to change the problem into a formal 'sum' and then solve it.

Brown (1981a) showed that even among 12 year olds a substantial minority of children were not able to select the correct expression to 'match' a simple word problem. In the case of nine such problems, similar to that given in Fig. 4.10, the average percentage of 12 year old children *unable* to select an expression which even contained the correct operation sign $(+, -, \times, \div)$ was 21 per cent. However the percentage ranged from 7 per cent to 38 per cent depending on the operation and the size of the numbers.

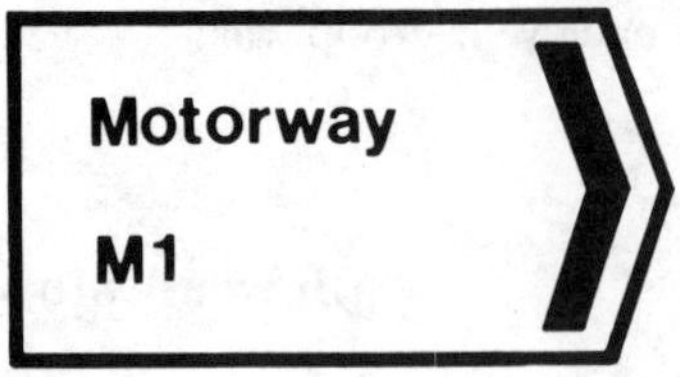

6.		87×3	$261 + 87$
The Green family have to drive 261 miles to get from London to Leeds. After driving 87 miles they stop for lunch.		$87 \div 261$	$261 - 87$
		261×87	$261 \div 87$
How do you work out how far they still have to drive?		$87 - 261$	$87 + 174$

Fig. 4.10 (reproduced from Brown, 1978)

As with the 6 and 7 year olds, some children who could not select the correct expression could nevertheless solve the problem; as is the case with Tony (aged 13) their strategy did not easily translate into a symbolic expression:
(The problem is the one in Fig. 4.10 with the numbers replaced by 228 and 87).

Tony: 'You add it on again... you add...three on to make 80(!) and then another 20 to make 100, then 128 from that is er... one hundred and forty something.'
Interviewer: 'Which of these (expressions) do you think...?'
Tony: 'That one'. (87 + 228).
Interviewer: 'Are you sure it's that one? Did you add 87 onto 228?'
Tony: 'No, I built it up.'
Interviewer: 'You built it up? Do you think it's any of these (expressions)?'
Tony: 'I think its this one (87 ÷ 228).'
Interviewer: '87 divided by 228?'
Tony: 'No... I don't know the sign for adding it on.'

                                                                                                      (Brown, 1981a)

Tony used an 'adding on' strategy for this complementary addition model of subtraction. He could not transform it to the mathematical form $228 - 87 = 141$. However it should be stressed that his approach to the problem allowed him to maintain an overall 'feel' for the situation and approach a reasonable solution... 'one hundred and forty something'. Ideally this beneficial aspect of his strategy should be preserved and built upon in helping him to progress towards a more mathematical structuring of the problem, leading eventually to the symbolic representation. A calculator may provide a useful stimulus for doing this, since only symbols are available on the keyboard.

It was also found in the CSMS study that instead of transforming a problem into its exact symbolism, children may structure it according to an inverse operation e.g. 3×4 instead of $12 \div 3$ for the item shown in Fig. 4.11.

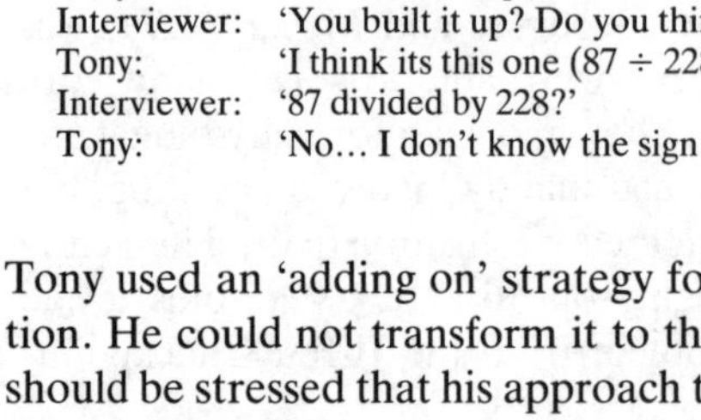

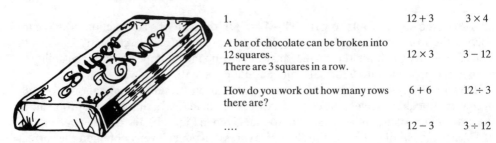

1.		$12 + 3$	3×4
A bar of chocolate can be broken into 12 squares. There are 3 squares in a row.		12×3	$3 - 12$
How do you work out how many rows there are?		$6 + 6$	$12 \div 3$
....		$12 - 3$	$3 \div 12$

Fig. 4.11 (reproduced from Brown, 1978)

Thirty-four per cent of 12 year olds gave 3×4, as opposed to 32 per cent choosing $12 \div 3$. For example Michael (aged 13):

Michael:	'Four.'
Interviewer:	'...How did you work that out?'
Michael:	'Three times four.'
Interviewer:	'Three times four is what?'
Michael:	'12'
Interviewer:	'...how did you realise it was four?'
Michael:	'Just worked it out in tables...'
Interviewer:	'I see, OK. Which of these' (expressions)?'
Michael:	'Three fours (3×4)

Michael finally, when pushed, also accepts $12 \div 3$ but is then asked...

Interviewer:	'Which of those would you think the better one?'
Michael:	'Three times four.'

(Brown, 1981a)

Michael's choice of symbolic expression, into which to transform this problem, was influenced by his method of solution which had seemingly come about almost on an intuitive, automatic level. This was probably because the problem was immediately within his grasp in terms of his knowledge and understanding of the necessary number bonds.

If children rely on these less structured approaches to problem solving they inevitably experience great difficulty with situations which are less concrete, particularly where the numbers involved are large or relate to fractional quantities, or are replaced by letters, as in algebra. Therefore it is essential that they should learn how to transform problems into their symbolic representation, if only so as to be able to use a simple calculator efficiently.

Another study on children's strategies in solving word problems was carried out by Nesher and Teubal. They demonstrated the importance of the vocabulary used in the problem, examining in particular the use of the cue words 'more' and 'less' in addition and subtraction problems. They designed four similar problems of the following type:

The milkman brought 11 bottles of milk on Sunday. That was 4 *more* than he brought on Monday. How many bottles did he bring on Monday? (Nesher and Teubal, 1975)

One addition problem and one subtraction problem (the one above) contained the word 'more', while one problem of each type contained the word 'less'.

About 85 Israeli 7 year olds were tested, and the proportion succeeding in each case was as shown in the following table:

		Verbal cues	
		'more'	'less'
Correct arithmetic	+	87%	64%
operation	–	43%	81%

This appears to demonstrate the tendency for children to link addition with the cue word 'more' and subtraction with the cue word 'less', although some of the extra difficulty in the two hardest items might be accounted for by the slightly more complex problem structure.

4.3.3.2   *A Suggestion for Helping Children Solve Word Problems*

In Rothery is given the following example of an exercise children can undertake to help their problem solving. It originates from an American publication by Earle (1976). The first parts are to aid comprehension and part D is specially aimed at the transformational stage of the problem-solving process.

> Mary bought 2 yards of ribbon at 20 cents a yard and some cloth for 50 cents. How much change did she get from a dollar bill?
>
> A.   True or False? If an answer is false, cross out the wrong word or words and correct them.
>   .... Mary bought 2 yards of ribbon.
>   .... The ribbon cost 15 cents a yard.
>   .... The cost of two yards of ribbon was 50 cents.
>   .... She bought 2 yards of cloth at 50 cents a yard.
>   .... The price of the cloth was 50 cents.
>   .... She paid with a five dollar bill.
>
> B.   Underline the question in the problem.
>
> C.   Check ($\checkmark$) the correct answer below.
>   In solving this problem you are asked to find
>   .... How much change she received.
>   .... The price of ribbon and cloth.
>
> D.   Number the steps below in order.
>   .... Label the answer
>   .... Subtract the total amount from $1.00
>   .... Find how much the ribbon cost.
>   .... Find the total amount Mary spent.
>
> E.   Use the space below to solve the problem, using the steps from D above.
>
> (Rothery, Ed., 1980)

This system may indeed assist children to be systematic in extracting information from word problems, but it does not address the problem of selecting whichever operation is appropriate. As already noted, the calculator can certainly provide motivation for transforming the problem into symbols in order to press the correct buttons, but again it does not provide the strategy for doing this. The only real remedy would seem to lie in a deeper understanding of the operations themselves. The first step is to use apparatus to model a variety of different word problems, then, when children are ready, to progress to structural pictures (e.g.   ....   to represent 3×4),
                                                                    ....
                                                                    ....

and only finally to more abstract symbols. In the past, teachers may have been too hasty in rushing onto 'sums' given out of context, leaving many children with few strategies for linking a problem situation and the relevant symbolic expression. In Bruner's sequence of stages (Bruner, 1967)

enactive (action)	→	iconic (image)	→	symbolic (symbol)

the tendency has been to spend too little time for some children on the first two stages.

4.3.4 Attaching Meaning to Symbols

Sometimes children are more able to cope with a situation in a verbal form than when it is represented symbolically and in other situations the reverse occurs. The former appears to be particularly the case within the context of fractions. For example the APU Primary Report states:

> the use of verbal forms to stand for mathematical symbols substantially improved performance. This happened in the multiplication of fractions (using 'of' for the multiplication sign) and division of fractions (using 'how many halves in $2\frac{1}{2}$ for $2\frac{1}{2} \div \frac{1}{2}$') where the words make the situation more concrete.
>
> (APU, 1980a)

Clements (1980) presented 12 year olds with the following questions in a test:

Q.5   Write in the answer $1 - \frac{1}{4} =$

Q.18   A cake is cut into four equal parts and Bill takes one of the parts. What fraction of the cake is left?

About half the children were successful with the symbolic form whereas over three-quarters of them succeeded with the verbal form. When interviewing some of those who were correct on the written problem but not with the symbolic version, he found that the imagery evoked by the cake was helpful.

As noted in section 3 (page 304) Hart (1981) also demonstrated in the CSMS Fractions study that problem items were generally found to be easier than the parallel symbolically expressed computation. This suggests that children were unable to attach a concrete meaning to an expression like $\frac{3}{4} - \frac{1}{8}$.

On the other hand the APU Primary Report comments:

> However, in some items in computation with whole numbers a slight worsening of performance was the result of using verbal forms like 'what is the sum of' for the addition sign and 'what is 5 times as big as 2' in place of 5×2. This was probably because familiar and well understood symbols were replaced by less well known words and phrases. (APU, 1980a)

It may also have been because such familiar symbolic expressions are the subject of rote learning and little attention had been paid to the associated word forms.

The different linguistic translations of multiplication 'sums' can easily give rise to confusion. Duffin (1976) points to two possible ways of reading multiplication tables and their associated symbolism. If the child is confronted, say, with 3×4, how does he interpret this? Does he consider it to be part of the 4 times table i.e. '3 fours are 12', or, as part of the 3 times table i.e. '4 threes are 12'? The two forms suggest different concrete models. For instance the following linguistic forms relate to this model:

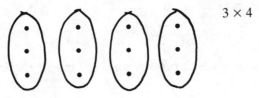

3×4

'3 multiplied by 4'
'3, times 4:   in the sense of 3 occurring on 4 occasions'
'4 threes'

whereas the next model relates to these other linguistic forms:

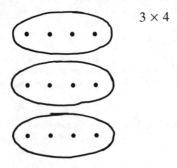

3×4

'3 fours'
'3 lots of 4'
'3 of 4'
'3 times 4:   in the sense of 4 occurring 3 times'

For division there are a variety of both linguistic and symbolic forms which relate to the same underlying mathematical structure.

e.g., '12 ÷ 3', '$\frac{12}{3}$', '12/3', '3 ⌊12', '3⟌12'

'12 shared by 3'
'12 divided by 3'
'3 divided into 12'
'3's into 12'
'12 over 3'
'12 thirds'

And at the same time these may refer to either a 'grouping' or 'sharing' model as described in Section 3 on page 236.

Children often regard '12 ÷ 3' and '3 ÷ 12' as being the same. This also occurs with subtraction where they do not distinguish between say '8 – 2' and '2 – 8'. The table that follows is adapted from Brown and it illustrates the frequency of secondary pupils not recognising the non-commutativity of subtraction and division. These results arose from the CSMS study where children were required to choose the symbolic expression they considered appropriate for solving a given word problem.

Expression (The first in each pair was the correct one for each problem)	Percentage which chose it in each age group	
	11 yrs	12 yrs
8 − 2	69	68
2 − 8	14	17
261 − 87	60	60
87 − 261	14	19
12 ÷ 3	32	32
3 ÷ 12	14	13
286 ÷ 26	34	34
26 ÷ 286	32	36
391 ÷ 23	44	47
23 ÷ 391	26	33

(Adapted from Brown, 1981c)

From her interviews Brown found evidence which suggested that linguistic factors played an important role in determining the child's choice of expression. These related to such alternative forms as:

'Take 87 from 261'

and

'261 take away 87',

where it appeared that sometimes the order of the verbal format decided the choice of symbolism. If 87 is said first then children are likely to choose the expression with 87 at the beginning. Also children used such phraseology as 'divided into', 'shared between' quite interchangeably. For instance for one of the division items, Ian, aged 12 years said:

391 divided into 17 (indicates 391 ÷ 17) and this one, 17 divided by 391 (indicates 17 ÷ 391).

(Brown, 1981a)

Brown also cites the case of one girl who, apparently for linguistic reasons, would only choose the reversed form of a division expression. This was in reply to the following item which is based on a 'grouping' model of division.

286 people are coming to see the school play. The chairs are arranged in rows and there are 13 chairs in each row. How do you work out how many rows are needed?

Pat: 'You just put 13... division, it would be division.'
Interviewer: 'Division. So what would you divide into what?'
Pat: '13 divided into 286.'
Interviewer: 'OK. Let's have a look at the back now. Which of those, if any, could do that for you?'
              (Showing various symbolic expressions)

Pat: 'That one.' (13 ÷ 286)
Interviewer: 'How does that read?'
Pat: '13 divided into 286.'
Interviewer: 'OK? any others?'
Pat: 'That one.' (286 ÷ 13)
Interviewer: 'How do you read that one?'
Pat: '286 divided into 13.'

(Interviewer questions further about whether both forms are acceptable)

Pat: 'Two ways, you can put 13 into 286; 286 can't go into 13.'
Interviewer: It *can't* go into 13?
Pat: 'No, that won't go into that (286 ÷ 13), but that will go into there (13 ÷ 286).'
Interviewer: 'I see, so how do you read that one then?' (286 ÷ 13)
Pat: '286 divided into 13.'
Interviewer: 'I see, and so ... you say that one won't go into that?'
Pat: 'No, that one's no good, that one.' (i.e., 286 ÷ 13)

(Brown, 1981c)

Pat was able to transform the problem into the appropriate operation but she could not translate this into the correct symbolism. The linguistic form she used for describing the operation and the manner in which she read and interpreted the symbolic expression, although consistent with each other were not in keeping with convention. Her misunderstanding of the symbolism prevented her from choosing the correct symbolic expression. This example illustrates well the possible inter-relatedness of the different categories of error in the Newman classification referred to in Section 4.2 on page 335.

Although it was not required during the study, it would have been interesting to see how Pat would have continued in trying to reach a solution to this problem. She might have chosen to adopt an informal strategy for finding the answer to '13 divided into 286' or to use the standard algorithmic approach in the form $13 \overline{)286}$; both of these methods would still be consistent with her verbal description and with her interpretation of 13 ÷ 286. (It may well have been that her past learning experiences of division in the form of $13 \overline{)286}$ played a major role in structuring her present linguistic interpretation of the symbolisation of the operation.) She would probably have encountered difficulties if she had wanted to use a calculator for the computation, as the answer would be obtained to 13 ÷ 286, assuming the buttons would be pressed in this order.

The frequency of reversals increased when decimal quantities were involved. For example Brown gave the item:

A table is 92.3 cm. long.
About how many inches is this?
(1 inch is about 2.54 cm.)

Ring the calculation you would need to find the answer.
2.54 + 92.3   92.3 ÷ 2.54
2.54 ÷ 92.3   29.3 − 2.54
2.54 − 92.3   92.3 × 2.54

During the interview Peter for example said:

'See how many twos go into 92',

but then ringed 2.54 ÷ 92.3 judging 92.3 ÷ 2.54 as inappropriate (Brown, 1981c)

Overall Brown concluded that hardly any 12 year olds and under 10 per cent of all 15 year olds were consistently correct in their choice of the order of symbols in expressions involving division. (It should be noted that Brown's sample was representive of the whole ability range, not just low attainers.) She points out that this would seem to have rather serious implications when viewed within the context of the increased use of calculators in the classroom. On the other hand this increased use of calculators may help children to distinguish between a division (or subtraction) expression and its reversed form more readily. They might thus gain a greater insight into the meaning of the symbols and be able to relate them more effectively to the mathematical operations required for solving a problem.

4.3.5 Some Difficulties Experienced by Children with Encoding their Answers

Sometimes children may read, understand and be able to solve a given problem but be unable to express their answer in a satisfactory form.

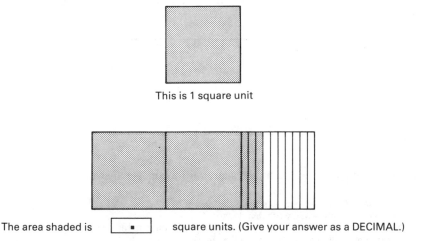

This is 1 square unit

The area shaded is ⬚ . square units. (Give your answer as a DECIMAL.)

Fig. 4.12 (reproduced from Brown 1981b)

For instance, in the answer to the item shown in Fig. 4.12, one secondary pupil gave the answer 2³/₁₀ which he then changed to 2.310. Others said, 'two and a third which one child recorded as 2^{3ard}. (See Brown, 1981c.)

For the item:

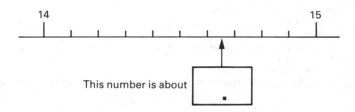

This number is about ⬚

Fig. 4.13 (reproduced from Brown 1981b)

A common response was 14.6½ which was given by the following percentages of each year group:

11	12	13	14	years
18%	14%	13%	9%	

Other answers were recorded as 14.6.5. All these children obviously had some understanding of the problems but were hampered by their inability to encode their answers into the conventional symbolic form. (The legitimate use of, for instance, £14.26½, may be partly responsible for their confusion.)

This kind of difficulty may have particularly adverse consequences within the topic of algebra. This is because, although children may understand what they are doing, the manner in which they record their answers may give rise to an expression which is not appropriate for the problem at hand.

In a study previously quoted in Section 4.3.2.1 on page 353, Booth interviewed 3rd, 4th and 5th year secondary pupils from 'middle ability' mathematics groups on finding the area of:

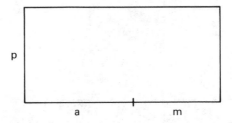

Fig. 4.14

One of the problems seemed to be the search for a concise answer. For example,

BR: 'Well I would add a onto m, and times the answer by p.'
LB: 'Do you know how you'd write that down?'
BR: 'No, I wouldn't... It'll be... I'm not sure really, how you'd do the plus m, to get the length.'
LB: 'Right, now what did you tell me you'd do first?'
BR: 'I'd add the a and m.'
LB: 'Right, well write that down first.'
BR: 'I'm not sure how you'd... (writes a + m doubtfully).'
LB: 'And then what would you do?'
....
BR: 'Yes, but I can't remember how you write the a and m down together, if you see what I mean. After a plussing...'
LB: 'Oh, I see...'
BR: 'It's a basic problem!' (Laughs)
LB: 'Oh I see, but you've put down a + m...'
BR: 'But I can't remember... It's presumably... I mean, times p is the question, but I don't know how you add up the a and the m, to get one thing' (writes a + mxp).
LB: 'So you're wondering how you can add the a and m together to get...'
BR: 'Yes, one thing, to times by p.'
LB: 'Well, can you get one thing? Can you?'
BR: 'Well, you should be able to! Presumably you can (laughs), but I can't think of any way...'

(Booth, 1982)

It seems then, that as mentioned at the beginning of this section, the need for negotiation of meanings of words and symbols is necessary throughout the teaching of mathematics. Not only does this allow for a deeper insight into the child's grasp of mathematics by the teacher but also facilitates a greater understanding of the subject for the child. Communication of quantitative and spatial ideas, especially the ability to listen and to talk as well as to read and to write, must be a central aim of mathematical education at all ages and ability levels.

References for Section 4

Assessment of Performance Unit (APU) – see Department of Education and Science.

Austin, J.L. and Howson, A.G. (1979) Language and Mathematical Education. *Educational Studies in Mathematics,* **10,** 161–197.

Behr, M; Erlwanger, S; Nichols, E. (1980) How Children View the Equals Sign. *Mathematics Teaching,* **92,** 13–15.

Booth, L. (1982) Getting the Answer Wrong. *Mathematics in School,* **11 (2),** 4–6.

Brown, M. (1978) *CSMS Number Operations Test.* National Foundation for Educational Research.

Brown, M. (1981a) Number Operations. In *Children's Understanding of Mathematics: 11–16* (Ed) Hart, K. John Murray.

Brown, M. (1981b) Place Value and Decimals. In *Children's Understanding of Mathematics: 11–16* (Ed) Hart, K. John Murray.

Brown, M. (1981c) *Levels of Understanding of Number Operations, Place Value and Decimals in Secondary School Children* – Ph.D. Thesis. University of London, Chelsea College.

Bruner, J.S. (1967) *Towards a Theory of Instruction.* Cambridge, Mass.: Belknap Press.

Bryant, P.E. (1974) *Perception and Understanding in Young Children.* Methuen.

Carpenter, T.P. and Moser, J.M. (1979) *The Development of Addition and Subtraction Concepts in Young Children* – presented at University of Warwick, Third International Group for the Psychology of Mathematics Education (IGPME).

Carpenter, T.P. et al. (1981) Table reprinted from *Results from the Second Mathematics Assessment of the National Assessment of Educational Progress.* © 1981 by the National Council of Teachers of Mathematics used by permission.

Choat, E. (1974) Johnnie is Disadvantaged, Johnnie is Backward. What Hope For Johnnie? *Mathematics Teaching,* **69,** 9–13.

Clements, M.A. (1980) Analyzing Children's Errors on Written Mathematical Tasks. *Educational Studies in Mathematics,* **11 (1),** 1–21.

Cribb, P. Learning Mathematical Language. *Vinculum, The Maths Association of Victoria,* **13,** 4.

CSMS (Concepts in Secondary Maths and Science). See Hart, K.M. (Ed, 1981).
Department of Education and Science, APU – Assessment of Performance Unit (1980a) *Mathematical Development, Primary Survey Report No. 1.* HMSO.

Department of Education and Science, APU – Assessment of Performance Unit (1980b) *Mathematical Development, Secondary Survey Report No. 1.* HMSO.

Donaldson, M. and Balfour, G. (1968) Less is More: A Study of Language Comprehension in Children. *British Journal of Psychology,* **59,** 461–471.

Duffin, J. (1976) Left-Right-Left: A Source of Confusion. *Mathematics Education for Teaching,* **2 (3),** 14–18.

Earle, R.A. (1976) *Teaching Reading and Mathematics* U.S.A: International Reading Association, Newark, Delaware.

Ferreiro, E. (1971) *Les Relations Temporelles dans le Language de l'Enfant.* Geneva: Librarie Droz.

Ginsburg, H. (1977) *Children's Arithmetic: How They Learn It and How You Teach It.* Austin, TX: PRO-ED, 1982.

Haigh, G. (1979) Reflections on Teaching Maths to Seven-Year Old Immigrant Children. *Remedial Education,* **14 (1)**, 14–16.

Hart, K.M. (1981) Fractions. In *Children's Understanding of Mathematics: 11–16.* (Ed) Hart, K.M. John Murray.

ILEA (Inner London Education Authority, 1979) *Checkpoints Assessment Cards.* London: ILEA Learning Material Service.

Kieran, C. (1979) *Children's Operational Thinking Within the Context of Bracketing and the Order of Operations.* University of Warwick: Proceedings of the Third International Conference for the Psychology of Mathematics Education.

Krulik, S. (1980) To Read or Not to Read that is the Question! *The Mathematics Teacher,* **73 (4)**, 248. Material reprinted from *The Mathematics Teacher,* copyright © 1980 by the National Council of Teachers of Mathematics, Virginia, USA. Used by permission.

Lawson, G; Baron, J; Siegel, L.S. (1974) The Role of Number and Length Cues in Children's Quantitive Judgements. *Child Development,* **45,** 731–736. *Mathematics Teaching, Journal of the Association of Teachers of Mathematics* (ATM) Kings Chambers, Queen Street, Derby.

Matthews, J. (1980) 'Five More Pages Fewer...', *Mathematics in School,* **9 (3)**, 24–25.

Matthews, J. (1981) An Investigation into Subtraction, *Educational Studies in Mathematics,* **12,** 327–338.

Moore, T.E. and Harris, A.E. (1978) Language and Thought in Piagetian Theory. In *Alternatives to Piaget: Critical Essays on the Theory* (Ed) Siegal, L.S. & Brainerd, C.J. New York: Academic Press.

Nesher, P. and Teubal, E. (1975) Verbal Cues as an Interfering Factor in Verbal Problem Solving, *Educational Studies in Mathematics,* **6,** 41–51.

Newman, M.A. (1977) An Analysis of Sixth-Grade Pupils' Errors in Written Mathematical Tasks. In *Research in Mathematics Education in Australia 1* (Ed) Clements, M. & Foyster, J.

Nicholson, A.R. (1977) Mathematics and Language, *Mathematics in School,* **6 (5)**, 32–34.

Nicholson, A.R. (1980) Mathematical Literacy, *Mathematics in School,* **9 (2)**, 33–34.

Otterburn, M.K. and Nicholson, A.R. (1976) The Language of (CSE) Mathematics, *Mathematics in School,* **5 (5)**, 18–20.

Piaget, J. (1972) *Problems of Equilibration.* In *Piaget and Inhelder on Equilibration* (Ed) Nodine, C; Gallagher, J; Humphrey, R. Philadelphia: Jean Piaget Society.

Rothery, A. (Ed, 1980) *Children Reading Maths.* Worcester: College of Higher Education.

SESM – Strategies and Errors in Secondary Mathematics Project. London: Chelsea College.

Siegel, L.S. (1978) The Relation of Language and Thoughts in the Pre-operational Child. In *Alternatives to Piaget: Critical Essays on the Theory* (Ed) Siegel, L.S. and Brainerd, C.J. New York: Academic Press.

Speer, W.R. (1979) Do you See What I Hear? A Look at Individual Learning Styles, *Arithmetic Teacher,* **27 (3)**, 22–26.

Sweet, A. (1972) Children Need Talk, *Mathematics Teaching,* **58,** 14–19.

Trivett, J. (1978) The End of the Three R's, *Mathematics Teaching,* **84,** 14–19.

Vygotsky, L.S. (1962) *Thought and Language,* M.I.T. Press.

Ward, M. (1979) *Mathematics and the 10-year old – Schools Council Working Paper 61.* Evans/Methuen for the Schools Council.

Watson, I. (1980) Investigating Errors of Beginning Mathematicians, *Educational Studies in Mathematics*, **11** (**3**), 319–329.

Wheatley, G.H. (1977) The Right Hemisphere's Role in Problem Solving *Arithmetic Teacher*, **25** (**2**), 37. Material reprinted from *Arithmetic Teacher*, copyright © 1977 by the National Council of Teachers of Mathematics, Virginia, USA. Used by permission.

Index